AF686128

Cellular Redox Mechanisms in Inflammation and Programmed Cell Death

Cellular Redox Mechanisms in Inflammation and Programmed Cell Death

Guest Editor

Irina I. Vlasova

MDPI

Basel • Beijing • Wuhan • Barcelona • Belgrade • Novi Sad • Cluj • Manchester

Guest Editor
Irina I. Vlasova
Institute for Regenerative
Medicine
Sechenov First Moscow State
Medical University
Moscow
Russia

Editorial Office
MDPI AG
Grosspeteranlage 5
4052 Basel, Switzerland

This is a reprint of the Special Issue, published open access by the journal *International Journal of Molecular Sciences* (ISSN 1422-0067), freely accessible at: https://www.mdpi.com/journal/ijms/special_issues/QNT7P5P1RO.

For citation purposes, cite each article independently as indicated on the article page online and as indicated below:

Lastname, A.A.; Lastname, B.B. Article Title. *Journal Name* **Year**, *Volume Number*, Page Range.

ISBN 978-3-7258-6530-7 (Hbk)
ISBN 978-3-7258-6531-4 (PDF)
https://doi.org/10.3390/books978-3-7258-6531-4

© 2026 by the authors. Articles in this book are Open Access and distributed under the Creative Commons Attribution (CC BY) license. The book as a whole is distributed by MDPI under the terms and conditions of the Creative Commons Attribution-NonCommercial-NoDerivs (CC BY-NC-ND) license (https://creativecommons.org/licenses/by-nc-nd/4.0/).

Contents

About the Editor

Irina I. Vlasova

Irina I. Vlasova received her PhD in biophysics from M.V. Lomonosov Moscow State University; after that, she worked as a leading researcher and project leader at the Federal Research and Clinical Center of Physical–Chemical Medicine (Moscow, Russia). From 2004 to 2008, Dr. Vlasova worked as a research scholar at the Department of Environmental and Occupational Health, University of Pittsburgh (PA, USA). Her research focused on the oxidation of proteins and lipids by reactive oxygen species, the peroxidase activity of heme-containing proteins, and activated neutrophils capable of secreting myeloperoxidase and its product, hypochlorous acid, which is a powerful oxidizing agent that degrades not only collagen-based biomaterials, but also rigid nanoparticles such as carbon nanotubes and graphene oxide. Currently, Dr. Vlasova works as a leading researcher and project leader at the Institute for Regenerative Medicine, I. M. Sechenov First Moscow State Medical University (Sechenov University). Her research focuses on redox and phagocytic activities of different phenotypes of human macrophages and their ferroptosis.

International Journal of *Molecular Sciences*

Editorial

Special Issue "Cellular Redox Mechanisms in Inflammation and Programmed Cell Death"

Irina I. Vlasova

Institute for Regenerative Medicine, Sechenov First Moscow State Medical University (Sechenov University), 119991 Moscow, Russia; iivlasova08@gmail.com or vlasova_i_i@staff.sechenov.ru

1. Introduction

Redox reactions play a central role in the regulation of physiological processes in cells and tissues. Reactive oxygen species (ROS) and reactive nitrogen species (RNS) are finely integrated in cellular signaling pathways [1]. They participate in gene regulation, cellular communication and host defense, orchestrating the balance between cellular homeostasis and stress response [2,3]. The main source of ROS is plasma membrane NADPH oxidase [4]. The enzyme catalyzes the formation of superoxide anion ($O_2\bullet-$), which is the primary ROS formed from molecular oxygen. It is unstable and dismutates spontaneously, or is converted by superoxide dismutase (SOD) into hydrogen peroxide [5]. H_2O_2 is a stable compound capable of migrating relatively long distances until it encounters iron in low-molecular iron complexes or in the active centers of some enzymes initiating the formation of other ROS (Figure 1) [6]. The primary RNS is nitric oxide (NO•, NO), which is synthetized in cells by nitric oxide synthases (NOSs) [7]. NO• is a relatively stable compound with an array of health benefits: control of cardiovascular system through modulation of vasodilation, regulation of immune response, neurotransmission, and inhibition of ferroptosis [8,9]. When interacting with superoxide anion, NO• is converted into highly reactive peroxynitrite, switching the biochemistry of NO• toward nitro-oxidative processes [10].

The innate immune response is initiated by "alarm signals" from the damaged tissues and pathogens. The main participants in the immediate and non-specific innate immunity are neutrophils, monocytes, and macrophages [11]. They are professional phagocytes, and their main functions are ROS production and phagocytosis of pathogens and dead cells. Macrophages participate in and orchestrate all stages of inflammation. The traditional M1/M2 paradigm distinguishes two polar macrophage subtypes based on their phenotypic characteristics, cytokine secretion, and transcriptional profiles: classically activated M1 phenotype and alternatively activated M2 macrophages [12]. M1 macrophages produce ROS and pro-inflammatory cytokines to promote inflammation [13]. M2 macrophages exhibit anti-inflammatory and immunoregulatory activity mediating resolution of inflammation and tissue regeneration [14].

Along with low molecular antioxidants, antioxidant enzymes play a crucial role in maintaining cellular redox homeostasis. SOD, catalase, glutathione peroxidases (GPX), thioredoxin, and heme oxygenase control the redox status of the cells. The search for bioactive molecules capable of buffering redox imbalance is a critical task [15]. Flavonoids are an important family of polyphenolic compounds classified as secondary plant metabolites and widely present in fruits and vegetables. Numerous studies have demonstrated their protective role in pathologies caused by oxidative stress [16]. Exogenous antioxidants with therapeutic potential, which introduces the concept of hormesis applied to flavonoids, also called "hormetic nutrients", induce health benefits by activating the Nrf2 pathway and related antioxidant enzymes for protection during oxidative damage, apoptosis/ferroptosis, and chronic inflammatory disorders [17,18]. The optimal dose, which is the minimum dose capable of producing beneficial effects, is crucial to achieve protection and prevent harmful effects and must be carefully identifying.

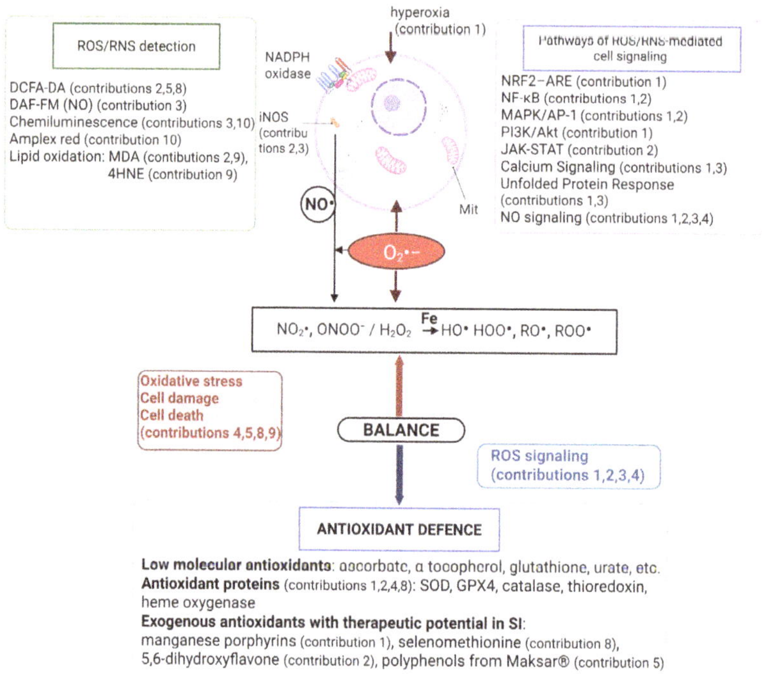

Figure 1. The balance between ROS/RNS and antioxidant system determines the involvement of reactive species participation in cell signaling or cell damage. The top part of the scheme presented ROS and RNS, the sources of their formation in cells, methods of RONS detection, and signaling pathways. The lower part of the scheme presented antioxidant systems—exogenous low molecular compounds, antioxidant enzymes, and compounds proposed in SI papers capable of buffering redox imbalance. References from the SI are provided for each topic. Abbreviations are listed at the end of text. Mit—mitochondria.

When ROS/RNS production in cell and tissues exceeds antioxidant capacity, reactive species initiate oxidative damage through modification of lipids, proteins, and nucleic acids, triggering pathology of tissues and programs of cell death [19]. Apoptosis is the most common form of cell death for damaged or aged cells. The primary mechanism for apoptosis is caspase activation, and its hallmark is cytochrome *c* (cyt *c*)-induced oxidation of mitochondrial cardiolipin [20]. Cyt *c* and other mitochondrial pro-apoptotic factors released into cytosol, where cyt *c* can bind to and oxidize an anionic phosphatidylserine (PS) changing the cell membrane properties. PS and oxidized PS exposed to cell surface serve as eat me signals in efferocytosis [21]. The term "ferroptosis" comes from the dependence of this type of cell death on intracellular iron. The program's key regulatory mechanisms involve the Fe-dependent generation of phospholipid hydroperoxides and their reduction to alcohols by GPX4 via the oxidation of glutathione [21–23].

The Special Issue (SI) contains two reviews and eight original research. The goal of the Editorial is to summarize and classify the contents of all the studies in order to guide the readers through the SI book (Figure 1, Table 1).

https://doi.org/10.3390/ijms27010274

Table 1. Summary of topics highlighted in the articles of the Special Issue. "Sections" are the Sections in the text where the articles are discussed.

Contribution/ Sections in Chapter 2.	Cells/Model	ROS, RNS Activation	Pathology/ Problem	ROS/RNS Detection	A Form of Cell Death	Antioxidants/ Effectors
(contr. 1) Review Silva et al. Section 2.1, Section 2.2.1, Section 2.4, and Section 2.5	neurons and glial cells (including microglia)	-	neuronal pain	review of all RONS, their chemistry and origins, mechanisms	-	manganese porphyrins
(contr. 2) Cao et al. Sections 2.1, 2.4 and 2.5	RAW 264.7 macrophages	LPS	inflammation	DCFH-DA, mtROS—MitoSOX red; MDA	-	5,6-dihydroxy flavone
(contr. 3) Zavadskis et al. Sections 2.1 and 2.2.3	precisely cut liver slices HUVECs	NO release: tunicamycin acetylcholine	tunicamycin effects on liver	chemiluminescence DAF-FM		NO-dependent vasodilation
(contr. 4) Review Vaglio-Garro et al. Sections 2.2.1 and 2.3	neurons, astrocytes	mitochondria disfunctions	brain pathologies		ferroptosis	
(contr. 5) Tarbeeva et al. Section 2.2.1, Section 2.2.3, Section 2.3, and Section 2.4	neuro-2a cells Vero cells	6-OHDA, paraquat; HSV-1 virus	neurotoxicity	DCFH-DA	apoptosis	polyphenols from amurensis heartwood (Maksar®)
(contr. 6) Beloborodova et al. Section 2.2.2	COX Lysates of THP-1, monocytes		sepsis	COX activity, TMPD		microbiota metabolites
(contr. 7) Bestavashvili et al. Section 2.2.2	a prospective randomized study, TMAO is a risk factor of CVDs	hypoxic–hyperoxic exposures	cardiovascular diseases and metabolic syndrome			hyperoxia activates the antioxidant defense systems of cells
(cont. 8) Assabayev et al. Sections 2.2.3, 2.3 and 2.4	mammary epithelial cells (bMECs)	Nocardia cyriacigeorgica	bovine mastitis	DCFH-DA	apoptosis	seleno methionine
(contr. 9) Gartzke Section 2.3	Human Embryonic Kidney (HEK) 293 cells	cell cooling-rewarming	cell preservation	Western blot against 4HNE and MDA measurements	ferroptosis	ferrostatin-1, 6-chromanol SUL150—mitochondrial functions
(contr. 10) Suleimanov Section 2.5	myeloid-derived macrophages	PMA, opsonized zymosan	Inflammation	chemiluminescence Amplex red		

Abbreviations: Contr.—contribution; CVDs—Cardiovascular Diseases; 4HNE—4-Hydroxy-Nonenal; DAF-FM—4-Amino-5-Methylamino-2′,7′-Difluorofluorescein Diacetate; DCFH-DA—2,7-Dichlorodihydrofluorescein Diacetate; PMA—Phorbol 12-Myristate 13-Acetate; TMPD—N,N,N′,N′-Tetramethyl-p-Phenylenediamine; TMAO—Trimethylamine-N-Oxide.

https://doi.org/10.3390/ijms27010274

2. An Overview of Published Articles

The articles presented in the SI highlight the pathways of cellular stress response to bacterial infections, toxic compounds, metabolic imbalance, or cell cooling. The articles do not merely highlight the mechanisms of redox-dependent damages but show how targeted antioxidant strategies can restore cellular homeostasis, considering the redox distress as both a driver of pathology and a target for therapeutic intervention.

The review by Silva et al. evaluates the role of reactive species in cell signaling and pathologies, focusing on neuropathic pain (contribution 1). The first part of the review can serve as an Introduction to the SI. This article is suitable for introducing redox reactive species, the redox interactions and regulations in cells. For each ROS and RNS type (RONS), the chemical processes and reactions are described. The authors considered the sources of ROS in cells: plasma membrane NADPH oxidase and the mitochondrial transport chain, as well as, to a lesser extent, the endoplasmic reticulum, peroxisomes, and cytosol. The review summarizes mechanisms of cell signaling, and discusses the mechanisms of antioxidant enzyme activity.

2.1. Pathways of ROS/RNS-Mediated Cell Signaling

RONS act as mediators of cellular signaling pathways that are involved in the modulation of normal physiological functions and the initiation of the response to cellular injury.

The mechanisms of activation of the main ROS/RNS-mediated cellular signaling pathways and their role in neuropathic pain are discussed in the review (contribution 1) (Section 2.2.1, Section 2.4, and Section 2.5). The authors provide a detailed description and diagrams of nuclear factor erythroid 2-related factor 2/antioxidant response elements (NRF2–ARE), nuclear factor kappa-B (NF-κB), mitogen-activated protein (MAPK/AP-1) and phosphoinositide 3-Kinase (PI3K)/Akt) signaling pathways, calcium signaling, and unfolded protein response. The mediating factors are translocated to the nucleus where they activate gene transcription and signal transduction.

Activation of the cell signaling pathways was studied in RAW 264.7 macrophages treated with lipopolysaccharides (LPS) (contribution 2). The interaction of LPS with the cell surface toll-like receptor 4 (TLR4) triggers the activation of signaling cascades, including the NF-κB, MAPK pathways, and the Janus kinase (JAK)-signal transducer and activator of the transcription (STAT) pathway. Protein phosphorylation is an important mechanism underlying the activation of all of these pathways and the transmission of cellular signals. Blocking TLR4 activation with 5,6-dihydroxyflavone suppresses cell activation by inhibiting the phosphorylation of c-JNK and p38 in the MAPK pathway, JAK2 and STAT3 in the JAK-STAT pathway, and p65 from the NF-κB pathway (Sections 2.4 and 2.5). Importantly, MAPK, JAK-STAT, and NF-κB are non-overlapping pathways, but they act in concert to regulate the macrophage response to LPS.

Endoplasmic reticulum (ER) stress accompanied by the accumulation of unfolded/misfolded proteins in the ER lumen induces the unfolded protein response (UPR), which aims to resolve ER stress or initiates cell death if ER stress cannot be resolved [24,25]. However, the biological effects of ER stress are not always mediated by UPR. Induction of ER stress in liver tissue by the antibiotic tunicamycin resulted in vasodilation of blood vessels in hepatic parenchyma in an unconventional way independent of UPR (contribution 3). Vasodilation can be caused by ER stress itself due to disruption of Ca^{2+} homeostasis and triggering of NO-generation mechanisms, which are key regulators of vascular tone. Interestingly, the elevated NO level is provided by its release from intracellular NO store in smooth muscle cells, rather than from endothelial NOS activity (Section 2.2.3).

https://doi.org/10.3390/ijms27010274

2.2. ROS/RNS-Related Pathological Conditions

Oxidative stress and inflammation are often inseparable partners in response to infections or injury. If left uncontrolled, these processes lead to tissue disfunction and the development of pathologies.

2.2.1. Neuronal Disorders

Nitroxidative stress in neuronal cells involves the complex interactions between RONS, antioxidant systems, and cell signaling that underlie the pathophysiology of neuropathic pain (contribution 1) (Section 2.1, Section 2.4, and Section 2.5). Immune cells of the central nervous system play a crucial role in the development and maintenance of neuropathic pain as nerve injury results in a neuroinflammatory response and activation of glial cells. Microglia, which are resident macrophages in the brain, initiate the development of neuropathic pain by releasing inflammatory mediators that not only transmit signals to other cells but also cause cell damage and neuronal death. Activated glial cells release chemokines, ROS, and neurotransmitters that modulate synaptic transmission, causing pain amplification and promoting the death of dopaminergic neurons.

Glutamate, a vital mitochondrial substrate fueling oxidative phosphorylation, also plays a central role in neuronal signaling [26], but high levels of extracellular glutamate are toxic, causing dysregulation of intracellular redox homeostasis, mitochondrial dysfunction, and cell death (contribution 4) (Section 2.3). Under physiological conditions, glutamate is stored inside cells in what's known as an easily releasable pool. In pathological conditions such as neuroinflammation, brain trauma, and neurodegenerative diseases, glutamate concentrations in extracellular space increase sharply. Disruption of glutamate metabolism in mitochondria leads to a vicious cycle: disruption of the tricarboxylic acid cycle due to neuroinflammation or elevated cortical NO level increases glutamate concentration, which leads to mitochondria damage due to the impaired oxidative phosphorylation [27].

Treatment of Neuro-2a cells by 6-hydroxydopamine (6-OHDA) and paraquat (PQ) caused an increase in intracellular ROS, a decrease in mitochondrial membrane potential, and apoptotic cell death (contribution 5) (Sections 2.2.3, 2.3 and 2.4). Herpes simplex virus type 1 (HSV-1) infection is a risk factor for neurodegenerative diseases, primarily due to the virus-induced mitochondrial disfunction, accumulation of ROS in neurons, and lipid peroxidation.

2.2.2. Microbiota-Related Metabolites in Inflammation and Pathologies

The human microbiota produces metabolites that can enter the bloodstream and participate in the regulation of immune system, central nervous system, and metabolism.

Cyclooxygenase (COX) is one of the key enzymes in the inflammatory process. It catalyzes the synthesis of prostaglandin H2, a precursor of all prostaglandins and thromboxanes [28,29]. The influence of some selected compounds from microbiota-related metabolites on the activity of COX was tested using pure enzyme and in monocyte lysates (contribution 6). Microbial metabolites lowered the activity of COX mainly due to their effect on the peroxidase activity of the enzyme. The study demonstrates how systemic inflammation may be fine-tuned by small molecules of microbial origin. Under septic conditions, the metabolites may shift COX peroxidase activity, modulating prostaglandin synthesis and the downstream inflammatory cascade.

Risk factors for cardiovascular diseases are associated not only with metabolic syndrome but also with some gut microbiota metabolites, including trimethylamine-N-oxide (TMAO), which stimulates an active search for TMAO inhibitors (contribution 7). Hypoxic conditions significantly and multifacetedly alter the ratio of intestinal microbiome strains and their metabolites in feces and blood. Intermittent normobaric hypoxic–hyperoxic exposures (IHHEs) are successfully used in the treatment of various pathologies. Moderate hypoxic conditions can activate various transcriptional pathways that initiate the synthesis of proteins with antioxidant activity, while episodes of hyperoxia activate ROS-dependent mechanisms stimulating the antioxidant defense systems of cells (Figure 1). The prospective randomized pilot study demonstrated that IHHEs tend to reduce blood TMAO and cause a more significant reduction in cardiometabolic and hepatic markers of metabolic syndrome compared with the placebo group.

2.2.3. Cellular and Tissue Models of Pathologies

Models of 6-OHDA and PQ-induced neurotoxicity in Neuro-2a cells and HSV-1 infection in Vero cells were used (contribution 5) (Section 2.2.1, Section 2.3, and Section 2.4).

Mammary epithelial cells (bMECs) were infected with Nocardia cyriacigeorgica to create a bovine mastitis model (contribution 8) (Sections 2.3 and 2.4). Mastitis in cattle reduces milk quantity and quality, and it is accompanied by mammary dysfunction. The bacteria-induced oxidative stress and inflammatory response in bMECs accompanied by elevating mRNA production and the expression of the key pro-inflammatory cytokines, such as tumor necrosis factor-α (TNF-α), interleukin (IL)-1, IL-6, and IL-8, decreasing the activity of antioxidants enzymes such as SOD and GPX and generating destructive levels of malondialdehyde (MDA) and ROS. This resulted in mitochondrial damage and apoptosis in cells.

Precision cut tissue slices have emerged as a physiologically relevant ex vivo model for studying human diseases. Precisely cut liver tissue slices preserving both liver structure and cellular functions were used to study the functionality of hepatic vessels and their response to various treatments (contribution 3) (Section 2.1). The effect of antibiotic tunicamycin was similar to that of specific vasodilator acetylcholine: both compounds induced ER stress in all types of liver cells, disrupting Ca^{2+} transport and increasing NO level, and caused vasodilation of liver vessels increasing their volume. Unlike acetylcholine, tunicamycin did not induce NO release in HUVECs cells, where endothelial NOS is the sole source of NO. Possibly, tunicamycin induces NO production in the smooth muscle layer of the vessels.

2.3. *ROS-Mediated Programmed Forms of Cell Death: Apoptosis and Ferroptosis*

Although reactive oxygen species are necessary for physiological processes, their excess can cause cell damage and trigger a program of cell death.

After infection of bMECs with N. cyriacigeorgica, MDA and ROS levels were significantly upregulated, and cell damage and apoptosis increased in a time-dependent manner (contribution 8) (Sections 2.2.3 and 2.4). Transmission electron microscopy revealed mitochondrial and endoplasmic reticulum swelling, loss of organelles, cell membrane rupture, cristae degeneration, all markers of apoptosis. Bacteria were detected in the nucleus and cytoplasm of the cells. Annexin V-FITC/propidium iodide assay confirmed apoptotic cell death in infected bMECs. Selenomethionine inhibits bacteria-induced apoptosis in bMECs (Section 2.4).

In Neuro-2a cells, a significant increase in the percentages of Annexin V-detected early and late apoptotic cells was observed after treatment with 6-OHDA (contribution 5). Polyphenols extracted from Maackia amurensis heartwood are registered in the Russian Federation as a drug named Maksar®. Some of these polyphenols reduced the neurotoxic effect of 6-OHDA treatment (Section 2.2.1, Section 2.2.3, and Section 2.4).

Human Embryonic Kidney (HEK) 293 cells were studied during a cooling–rewarming procedure (contribution 9). University of Wisconsin Cold Storage Solution substantially increased cell survival during hypothermia (4°C) compared to DMEM, but upon rewarming, lipid peroxidation and ATP depletion rapidly occurred, leading to complete cell death.

https://doi.org/10.3390/ijms27010274

Cell death was not prevented by an apoptosis inhibitor. In contrast, blocking of ferroptosis by the specific ferroptosis inhibitor ferrostatin-1 or maintaining mitochondrial functions by the 6-chromanol SUL150 effectively prevented ferroptosis. The study provides new approaches for organ preservation in clinical transplantation surgery.

Primary pathways of neuron death induced by glutamate include excitotoxicity and ferroptosis (contribution 4) (Section 2.2.1). High levels of extracellular glutamate block the cysteine–glutamate antiporter, leading to the depletion of intracellular cysteine and subsequently glutathione. The depletion of glutathione inactivates GPX-inducing oxidation of lipids and ferroptosis.

2.4. Exogenous Antioxidants with Therapeutic Potential

An important theme of the issue is the search for bioactive molecules capable of buffering redox imbalance and control inflammation.

Manganese porphyrins (MnPs) are a class of low molecular weight synthetic metalloproteins containing manganese in the porphyrin ring at different oxidative states (contribution 1) (Section 2.1, Section 2.2.1, and Section 2.5). Initially, MnPs were considered simply as $O_2\bullet-$ scavengers (as SOD mimics), but new knowledge about their redox chemistry has revealed that MnPs influence signaling pathways and cellular functions such as proliferation, differentiation, and cell death. Unlike iron porphyrins, MnPs have a very limited ability to react with H_2O_2 and produce the dangerous hydroxyl radical (HO•) in a Fenton-like reaction.

Selenomethionine (SeMet) is a form of selenium with antioxidant properties. The therapeutic effect of SeMet at different concentrations was studied in a mastitis model using bMECs infected with N. cyriacigeorgica, which is resistant to most antibiotics (contribution 8) (Sections 2.2.3 and 2.3). Pretreatment of cells with SeMet mitigated bacterial-induced inflammatory response and oxidative damage in bMECs by reducing MDA and ROS concentrations, decreasing the release of pro-inflammatory cytokines, upregulating SOD and GPX activities, and reducing apoptosis.

5,6-dihydroxyflavone (5,6-DHF) is a flavonoid with anti-inflammatory and antioxidant activity. In RAW 264.7 macrophages, 5,6-DHF significantly blocked LPS-induced activation of cell surface receptor TLR4, preventing stimulation of cell activation pathways (contribution 2) (Sections 2.1 and 2.5). 5,6-DHF reduced cytoplasmic ROS production in a dose-dependent manner, blocked mtROS formation, hindered the expression of iNOS, thereby reducing NO release, and suppressed the expression of pro-inflammatory cytokines IL-1β, IL-6, and TNF-α. Furthermore, 5,6-DHF scavenged ROS and induced heme oxygenase-1 expression. 5,6-DHF had the highest anti-inflammatory and antioxidant properties in RAW 264.7 cells compared to its natural analogues, baicalein and chrysin.

Maksar®, a polyphenol-based drug, possesses hepatoprotective, antithrombogenic, antiplatelet, antitumor, and antioxidant properties (contribution 5). Fourteen compounds were isolated from the drug and tested for neuroprotective potential and anti-HSV-1 activity (Section 2.2.1, Section 2.2.3, and Section 2.3). The most active compounds were found. These compounds regulated mitochondrial membrane potential and significantly reduced intracellular ROS level increasing the viability of 6-OHDA-treated Neuro-2a cells. Maksar® is therefore a promising drug for the treatment of herpetic infections and neuronal pathologies.

2.5. Macrophages

Macrophages are the main participants and conductors of all stages of inflammation. Under the influence of tissue factors, they can change their phenotype over a wide range between the M1 and M2 states, which represent the polar phenotypes in the continuum of

https://doi.org/10.3390/ijms27010274

phenotypic variability of macrophages. When characterizing the macrophage phenotype, the main focus is usually on secreted cytokines and expressed membrane proteins: THF-, IL-6, and CD86 are markers of M1 phenotype, while IL-10, IL-1Ra, and CD206, CD163 are markers of M2 phenotypes (contribution 10). Studies of four different macrophage phenotypes revealed their differences not only in surface markers and cytokine profiles but also in their functional activity, namely radical-generating and phagocytic activities, and even in mechanical properties. This extends our appreciation of the M1/M2 dichotomy toward a continuum of functional states with varying combinations of both markers and functional activities.

Morphologically, microglial cells are described as having three phenotypes similar to those in peripheral macrophages: M0 phenotype for the resting state, M1 and M2 phenotypes for activated cells (contribution 1) (Section 2.1, Section 2.2.1, and Section 2.4). The M0 state is characterized by the expression of genes associated with neuronal function and development, whereas the M1 and M2 macrophages are induced by CNS injury and chronic pain, appearing as mixed, time-varying phenotypes that are not in equilibrium on an activation continuum.

Bacterial LPS interacts with TLR4 receptor on the surface of RAW 264.7 cells and triggers inflammatory cascade in murine macrophages through the key signaling pathways (contribution 2) (Sections 2.1 and 2.4).

3. Conclusions

Exploring the articles of the SI provides a panoramic view of redox biology as a unifying concept encompassing seemingly disparate fields: infection biology, immunology, vascular physiology, metabolism, transplantation, and neurology. Each article of the SI addresses a distinct topic of redox biology, but together they reinforce the central principle that redox signals are integral to every stage of inflammation and cell death, from initiation and progression to resolution, which makes them key targets for controlling inflammation, preventing tissue damage, and promoting repair.

Funding: Work was supported by the state assignment of the Ministry of Health of the Russian Federation, Theme No. NZAF-2024-0016.

Conflicts of Interest: The authors declare no conflicts of interest.

Abbreviations

5,6-DHF—5,6-Dihydroxyflavone; 6-OHDA—6-Hydroxydopamine; AP-1—Activator Protein 1; bMECs—Bovine Mammary Epithelial Cells; CNS—Central Nervous System; Contr—Contribution; COX—Cyclooxygenase; ER—Endoplasmic reticulum; GPX—Glutathione Peroxidases;HSV-1—Herpes Simplex Virus Type 1; IHHEs—Hypoxic–Hyperoxic Exposures; IL—interleukin;JAK-STAT—Janus Kinase-Signal Transducer and Activator of Transcription; LPS—Lipopolysaccharides; MAPK—Mitogen-Activated Protein Kinase; MDA—Malondialdehyde; Mit—mitochondrion, MnPs—Manganese Porphyrins; NO—Nitric Oxide; NO_2•—Nitrogen Dioxide; NOSs—Nitric Oxide Synthases; NF-κB—Nuclear Factor Kappa-B; O_2•−—Superoxide Anion; ONOO−—Peroxynitrite; 6-OHDA—6-hydroxydopamine; PI3K—Phosphoinositide 3-Kinase; PQ—Paraquat; RNS—Reactive Nitrogen Species; ROS—Reactive Oxygen Species; RONS—Reactive Oxygen and Nitrogen Species; mtROS—mitochondrial Reactive Oxygen Species; SeMet—Selenomethionine; SI—Special Issue; SOD—Superoxide Dismutase; TLR4—toll-like receptor 4; TNF-α—tumor necrosis factor-α; UPR—Unfolded Protein Response.

https://doi.org/10.3390/ijms27010274

List of Contributions:

1. Silva, Á.J.C.; De Lavor, M.S.L. Nitroxidative Stress, Cell—Signaling Pathways, and Manganese Porphyrins: Therapeutic Potential in Neuropathic Pain. *Int. J. Mol. Sci.* **2025**, *26*, 2050. https://doi.org/10.3390/ijms26052050.
2. Cao, Y.; Tan, Y.-J.; Huang, D. Molecular Mechanism of 5,6-Dihydroxyflavone in Suppressing LPS-Induced Inflammation and Oxidative Stress. *Int. J. Mol. Sci.* **2024**, *25*, 10694. https://doi.org/10.3390/ijms251910694.
3. Zavadskis, S.; Shiganyan, A.; Müllebner, A.; Oesterreicher, J.; Holnthoner, W.; Duvigneau, J.C.; Kozlov, A.V. Endoplasmic Reticulum Stress Induces Vasodilation in Liver Vessels That Is Not Mediated by Unfolded Protein Response. *Int. J. Mol. Sci.* **2024**, *25*, 3865. https://doi.org/10.3390/ijms25073865.
4. Vaglio-Garro, A.; Kozlov, A.V.; Smirnova, Y.D.; Weidinger, A. Pathological Interplay between Inflammation and Mitochondria Aggravates Glutamate Toxicity. *Int. J. Mol. Sci.* **2024**, *25*, 2276. https://doi.org/10.3390/ijms25042276.
5. Tarbeeva, D.V.; Pislyagin, E.A.; Menchinskaya, E.S.; Berdyshev, D.V.; Krylova, N.V.; Iunikhina, O.V.; Kalinovskiy, A.I.; Shchelkanov, M.Y.; Mishchenko, N.P.; Aminin, D.L.; et al. Polyphenols from Maackia Amurensis Heartwood Protect Neuronal Cells from Oxidative Stress and Prevent Herpetic Infection. *Int. J. Mol. Sci.* **2024**, *25*, 4142. https://doi.org/10.3390/ijms25084142.
6. Beloborodova, N.; Fadeev, R.; Fedotcheva, N. Influence of Microbiota-Related Metabolites Associated with Inflammation and Sepsis on the Peroxidase Activity of Cyclooxygenase in Healthy Human Monocytes and Acute Monocytic Leukemia Cells. *Int. J. Mol. Sci.* **2023**, *24*, 16244. https://doi.org/10.3390/ijms242216244.
7. Bestavashvili, A.; Glazachev, O.; Ibragimova, S.; Suvorov, A.; Bestavasvili, A.; Ibraimov, S.; Zhang, X.; Zhang, Y.; Pavlov, C.; Syrkina, E.; et al. Impact of Hypoxia–Hyperoxia Exposures on Cardiometabolic Risk Factors and TMAO Levels in Patients with Metabolic Syndrome. *Int. J. Mol. Sci.* **2023**, *24*, 14498. https://doi.org/10.3390/ijms241914498.
8. Assabayev, T.; Han, J.; Bahetijiang, H.; Abdrassilova, V.; Khan, M.A.; Barkema, H.W.; Liu, G.; Kastelic, J.P.; Zhou, X.; Han, B. Selenomethionine Mitigates Effects of Nocardia Cyriacigeorgica-Induced Inflammation, Oxidative Stress, and Apoptosis in Bovine Mammary Epithelial Cells. *Int. J. Mol. Sci.* **2024**, *25*, 10976. https://doi.org/10.3390/ijms252010976.
9. Gartzke, L.P.; Hendriks, K.D.W.; Hoogstra-Berends, F.; Joschko, C.P.; Strandmoe, A.-L.; Vogelaar, P.C.; Krenning, G.; Henning, R.H. Inhibition of Ferroptosis Enables Safe Rewarming of HEK293 Cells Following Cooling in University of Wisconsin Cold Storage Solution. *Int. J. Mol. Sci.* **2023**, *24*, 10939. https://doi.org/10.3390/ijms241310939.
10. Suleimanov, S.K.; Efremov, Y.M.; Klyucherev, T.O.; Salimov, E.L.; Ragimov, A.A.; Timashev, P.S.; Vlasova, I.I. Radical-Generating Activity, Phagocytosis, and Mechanical Properties of Four Phenotypes of Human Macrophages. *Int. J. Mol. Sci.* **2024**, *25*, 1860. https://doi.org/10.3390/ijms25031860.

References

1. Schieber, M.; Chandel, N.S. ROS Function in Redox Signaling and Oxidative Stress. *Curr. Biol.* **2014**, *24*, R453–R462. [CrossRef]
2. Hong, Y.; Boiti, A.; Vallone, D.; Foulkes, N.S. Reactive Oxygen Species Signaling and Oxidative Stress: Transcriptional Regulation and Evolution. *Antioxidants* **2024**, *13*, 312. [CrossRef] [PubMed]
3. Mackova, V.; Raudenska, M.; Polanska, H.H.; Jakubek, M.; Masarik, M. Navigating the Redox Landscape: Reactive Oxygen Species in Regulation of Cell Cycle. *Redox Rep.* **2024**, *29*, 2371173. [CrossRef] [PubMed]
4. Nauseef, W.M. The Phagocyte NOX2 NADPH Oxidase in Microbial Killing and Cell Signaling. *Curr. Opin. Immunol.* **2019**, *60*, 130–140. [CrossRef] [PubMed]
5. Kettle, A.J.; Ashby, L.V.; Winterbourn, C.C.; Dickerhof, N. Superoxide: The Enigmatic Chemical Chameleon in Neutrophil Biology. *Immunol. Rev.* **2023**, *314*, 181–196. [CrossRef]
6. Winterbourn, C.C. The Biological Chemistry of Hydrogen Peroxide. In *Methods in Enzymology*; Elsevier: Amsterdam, The Netherlands, 2013; Volume 528, pp. 3–25. ISBN 978-0-12-405881-1.
7. Alderton, W.K.; Cooper, C.E.; Knowles, R.G. Nitric Oxide Synthases: Structure, Function and Inhibition. *Biochem. J.* **2001**, *357*, 593–615. [CrossRef]

https://doi.org/10.3390/ijms27010274

8. Kapralov, A.A.; Yang, Q.; Dar, H.H.; Tyurina, Y.Y.; Anthonymuthu, T.S.; Kim, R.; St. Croix, C.M.; Mikulska-Ruminska, K.; Liu, B.; Shrivastava, I.H.; et al. Redox Lipid Reprogramming Commands Susceptibility of Macrophages and Microglia to Ferroptotic Death. *Nat. Chem. Biol.* **2020**, *16*, 278–290. [CrossRef]
9. Andrabi, S.M.; Sharma, N.S.; Karan, A.; Shahriar, S.M.S.; Cordon, B.; Ma, B.; Xie, J. Nitric Oxide: Physiological Functions, Delivery, and Biomedical Applications. *Adv. Sci.* **2023**, *10*, 2303259. [CrossRef]
10. Piacenza, L.; Zeida, A.; Trujillo, M.; Radi, R. The Superoxide Radical Switch in the Biology of Nitric Oxide and Peroxynitrite. *Physiol. Rev.* **2022**, *102*, 1881–1906. [CrossRef]
11. Hossain, M.; Kubes, P. Innate Immune Cells Orchestrate the Repair of Sterile Injury in the Liver and Beyond. *Eur. J. Immunol.* **2019**, *49*, 831–841. [CrossRef]
12. Murray, P.J. Macrophage Polarization. *Annu. Rev. Physiol.* **2017**, *79*, 541–566. [CrossRef] [PubMed]
13. Yurkanova, M.D.; Kosheleva, N.V.; Teplova, A.A.; Timashev, P.S.; Vlasova, I.I. Pro-Inflammatory Properties of M1 Phenotypes of Human Macrophages: Prolongation of Myeloperoxidase-Mediated Oxidative Stress. *Free Radic. Res.* **2025**, *59*, 452–461. [CrossRef] [PubMed]
14. Pérez, S.; Rius-Pérez, S. Macrophage Polarization and Reprogramming in Acute Inflammation: A Redox Perspective. *Antioxidants* **2022**, *11*, 1394. [CrossRef] [PubMed]
15. Flieger, J.; Flieger, W.; Baj, J.; Maciejewski, R. Antioxidants: Classification, Natural Sources, Activity/Capacity Measurements, and Usefulness for the Synthesis of Nanoparticles. *Materials* **2021**, *14*, 4135. [CrossRef]
16. Lang, Y.; Gao, N.; Zang, Z.; Meng, X.; Lin, Y.; Yang, S.; Yang, Y.; Jin, Z.; Li, B. Classification and Antioxidant Assays of Polyphenols: A Review. *J. Future Foods* **2024**, *4*, 193–204. [CrossRef]
17. Scuto, M.C.; Anfuso, C.D.; Lombardo, C.; Di Fatta, E.; Ferri, R.; Musso, N.; Zerbo, G.; Terrana, M.; Majzúnová, M.; Lupo, G.; et al. Neuronutrition and Nrf2 Brain Resilience Signaling: Epigenomics and Metabolomics for Personalized Medicine in Nervous System Disorders from Bench to Clinic. *Int. J. Mol. Sci.* **2025**, *26*, 9391. [CrossRef]
18. Scuto, M.; Majzúnová, M.; Torcitto, G.; Antonuzzo, S.; Rampulla, F.; Di Fatta, E.; Trovato Salinaro, A. Functional Food Nutrients, Redox Resilience Signaling and Neurosteroids for Brain Health. *Int. J. Mol. Sci.* **2024**, *25*, 12155. [CrossRef]
19. Newton, K.; Strasser, A.; Kayagaki, N.; Dixit, V.M. Cell death. *Cell.* **2024**, *187*, 235–256. [CrossRef]
20. Vitale, I.; Pietrocola, F.; Guilbaud, E.; Aaronson, S.A.; Abrams, J.M.; Adam, D.; Agostini, M.; Agostinis, P.; Alnemri, E.S.; Altucci, L.; et al. Apoptotic Cell Death in Disease—Current Understanding of the NCCD 2023. *Cell Death Differ* **2023**, *30*, 1097–1154. [CrossRef]
21. Kagan, V.E.; Tyurina, Y.Y.; Vlasova, I.I.; Kapralov, A.A.; Amoscato, A.A.; Anthonymuthu, T.S.; Tyurin, V.A.; Shrivastava, I.H.; Cinemre, F.B.; Lamade, A.; et al. Redox Epiphospholipidome in Programmed Cell Death Signaling: Catalytic Mechanisms and Regulation. *Front. Endocrinol.* **2021**, *11*, 628079. [CrossRef]
22. Dixon, S.J.; Olzmann, J.A. The Cell Biology of Ferroptosis. *Nat. Rev. Mol. Cell Biol.* **2024**, *25*, 424–442. [CrossRef] [PubMed]
23. Ashraf, A.; Jeandriens, J.; Parkes, H.G.; So, P.-W. Iron Dyshomeostasis, Lipid Peroxidation and Perturbed Expression of Cystine/Glutamate Antiporter in Alzheimer's Disease: Evidence of Ferroptosis. *Redox Biol.* **2020**, *32*, 101494. [CrossRef]
24. Duvigneau, J.C.; Luís, A.; Gorman, A.M.; Samali, A.; Kaltenecker, D.; Moriggl, R.; Kozlov, A.V. Crosstalk between Inflammatory Mediators and Endoplasmic Reticulum Stress in Liver Diseases. *Cytokine* **2019**, *124*, 154577. [CrossRef]
25. Almanza, A.; Carlesso, A.; Chintha, C.; Creedican, S.; Doultsinos, D.; Leuzzi, B.; Luís, A.; McCarthy, N.; Montibeller, L.; More, S.; et al. Endoplasmic Reticulum Stress Signalling—From Basic Mechanisms to Clinical Applications. *FEBS J.* **2019**, *286*, 241–278. [CrossRef]
26. Gundersen, V.; Storm-Mathisen, J.; Bergersen, L.H. Neuroglial Transmission. *Physiol. Rev.* **2015**, *95*, 695–726. [CrossRef]
27. Isaev, N.K.; Andreeva, N.A.; Stel'mashuk, E.V.; Zorov, D.B. Role of Mitochondria in the Mechanisms of Glutamate Toxicity. *Biochemistry* **2005**, *70*, 611–618. [CrossRef]
28. Fitzpatrick, F. Cyclooxygenase Enzymes: Regulation and Function. *Curr. Pharm. Des.* **2004**, *10*, 577–588. [CrossRef]
29. Faki, Y.; Er, A. Different Chemical Structures and Physiological/Pathological Roles of Cyclooxygenases. *Rambam Maimonides Med. J.* **2021**, *12*, e0003. [CrossRef]

Disclaimer/Publisher's Note: The statements, opinions and data contained in all publications are solely those of the individual author(s) and contributor(s) and not of MDPI and/or the editor(s). MDPI and/or the editor(s) disclaim responsibility for any injury to people or property resulting from any ideas, methods, instructions or products referred to in the content.

https://doi.org/10.3390/ijms27010274

International Journal of
Molecular Sciences

MDPI

Article

Selenomethionine Mitigates Effects of *Nocardia cyriacigeorgica*-Induced Inflammation, Oxidative Stress, and Apoptosis in Bovine Mammary Epithelial Cells

Talgat Assabayev [1], Jinge Han [2], Halihaxi Bahetijiang [1], Venera Abdrassilova [3], Muhammad Asfandyar Khan [1], Herman W. Barkema [4], Gang Liu [1], John P. Kastelic [4], Xueying Zhou [1,*] and Bo Han [1,*]

1 College of Veterinary Medicine, China Agricultural University, Beijing 100193, China; talgata@cau.edu.cn (T.A.); halihaxi@cau.edu.cn (H.B.); asfand@cau.edu.cn (M.A.K.); gangliu@cau.edu.cn (G.L.)

2 College of Animal Science and Veterinary Medicine, Tianjin Agricultural University, Tianjin 300384, China; 2203020138@stu.tjau.edu.cn

3 Department of Normal Physiology with Biophysics Course, Asfendiyarov Kazakh National Medical University, Almaty 050012, Kazakhstan; abdrasilova.v@kaznmu.kz

4 Faculty of Veterinary Medicine, University of Calgary, Calgary, AB T2N 4N1, Canada; barkema@ucalgary.ca (H.W.B.); jpkastel@ucalgary.ca (J.P.K.)

* Correspondence: xueying@cau.edu.cn (X.Z.); hanbo@cau.edu.cn (B.H.); Tel.: +86-10-6273-3801 (B.H.)

Abstract: *Nocardia cyriacigeorgica* causes bovine mastitis, reduces milk quantity and quality, and is often resistant to antimicrobials. Selenomethionine (SeMet) is a form of selenium, which reduces reactive oxygen species (ROS)-mediated apoptosis and intramammary infections. However, the protective effects of SeMet on *N. cyriacigeorgica*-infected bovine mammary epithelial cells (bMECs) are unclear. The objective of this study was to evaluate whether SeMet mitigated *N. cyriacigeorgica*-induced inflammatory injury, oxidative damage and apoptosis in bMECs. Cells were cultured with or without being pretreated with 40 μM of SeMet for 12 h, then challenged with *N. cyriacigeorgica* (multiplicity of infection = 5:1) for 6 h. Although *N. cyriacigeorgica* was resistant to lincomycin, erythromycin, enrofloxacin, penicillin, amoxicillin, cephalonium, cephalexin, and ceftriaxone, 40 μM SeMet increased cell viability and inhibited lactate dehydrogenase release in infected bMECs. Furthermore, *N. cyriacigeorgica* significantly induced mRNA production and protein expression of TNF-α, IL-1β, IL-6, and IL-8 at 6 h. Cell membrane rupture, cristae degeneration and mitochondria swelling were evident with transmission electron microscopy. Superoxide dismutase (SOD) and glutathione peroxidase (GSH-px) activities were down-regulated after 3, 6, or 12 h, whereas malondialdehyde (MDA) and ROS contents were significantly upregulated, with cell damage and apoptosis rapidly evident (the latter increased significantly in a time-dependent manner). In contrast, bMECs pretreated with 40 μM SeMet before infection, SOD, and GSH-px activities were upregulated ($p < 0.05$); MDA and ROS concentrations were downregulated ($p < 0.05$), and apoptosis was reduced ($p < 0.05$). In conclusion, 40 μM SeMet alleviated inflammation, oxidative stress and apoptosis induced by *N. cyriacigeorgica* in bMECs *cultured* in vitro.

Keywords: bovine mastitis; *Nocardia cyriacigeorgica*; selenomethionine; oxidative stress; apoptosis

1. Introduction

Bovine mastitis affects dairy cows worldwide, with Nocardia spp. often causing mastitis in dairy cows, and enormous economic losses [1]. Furthermore, Nocardia can cause various local and systemic infections in humans and animals, and is transmitted in a variety of ways, including inhalation, ingestion, or traumatic implantation, with dissemination via lymphatics and blood [2]. *Nocardia cyriacigeorgica* causes chronic inflammation that does not respond to therapy [3] and is the species of Nocardia most commonly linked to bovine mastitis [4–6]. Nocardia mastitis causes granulomatous inflammation in cattle, which is initially acute and eventually becomes chronic [3,7].

As Nocardia spp. are resistant to many antimicrobials commonly used in veterinary medicine, mastitis caused by these bacteria results in chronic intramammary infection, mammary dysfunction, and cow culling [8]. Longstanding antibiotic use has increased antimicrobial resistance in *N. cyriacigeorgica*, creating global concerns for human and animal health [9]. However, antimicrobial resistance data on isolates associated with mastitis are limited.

Bacterial mastitis causes oxidative stress by promoting production of reactive oxygen species (ROS). Malondialdehyde (MDA) is frequently used to assess lipid peroxidation [10], which is counteracted by specific enzymes [11]. Oxidative stress can damage udder tissues and decrease milk production [12]. With cells cultured in vitro, damaged membranes release intracellular lactate dehydrogenase (LDH), which is an indicator of cytopathic effects [13]. In addition, ROS generated during oxidative stress could result in cytotoxic effects through protein and amino acid oxidation, DNA damage, altered intracellular signaling, inhibition of antioxidant defense systems, and tissue damage.

Cell death and inflammation are closely linked during mastitis due to *Staphylococcus aureus* [14]. Pyroptosis, a programmed necrosis triggered by the gasdermin protein family, often occurs after inflammatory caspase activation. Many pathogens invade host cells and activate cell-intrinsic death mechanisms, including pyroptosis, apoptosis, and necroptosis. Apoptosis occurs in response to various stimuli, including infection and injury [15]. Apoptosis has characteristic ultrastructural changes, including cell shrinkage, chromatin condensation, nuclear fragmentation, and apoptotic bodies [16]. The role of apoptosis in the context of infection or injury, particularly in bovine mastitis caused by *N. cyriacigeorgica*, is an area of ongoing research [3,16]. Although apoptosis has a role in host defense mechanisms against microbial invasion, specific interactions between *N. cyriacigeorgica* and bMECs regarding apoptosis are not well studied. Investigations in bovine mastitis caused by *N. cyriacigeorgica* have primarily focused on prevalence, molecular characterization, immunogenic proteins, and virulence determinants [3,15,16]. Further research on the apoptotic response of bMECs to *N. cyriacigeorgica* infection may provide insights into host–pathogen interactions and pathogenesis.

Selenomethionine (SeMet) is an important micronutrient that can prevent mastitis in dairy cows through various biological pathways, including suppressing apoptosis [17]. Furthermore, selenium deficiency increases susceptibility to various diseases and decreases productivity globally [18]. Nevertheless, the effects of SeMet on apoptosis in *N. cyriacigeorgica*-infected bMECs are unknown. The objective of this study was to evaluate whether SeMet mitigated *N. cyriacigeorgica*-induced inflammatory injury, oxidative damage, and apoptosis in bMECs.

2. Results

2.1. Antimicrobial Susceptibility

The *N. cyriacigeorgica* was resistant to most antibiotics tested, although it was susceptible to gentamycin (66.7%) and ceftiofur (93.3%) (Table 1).

2.2. Cytotoxicity Effects of SeMet on bMECs

Based on a CCK-8 assay, bMECs treated with 40 μM of SeMet had higher viability ($p < 0.01$; Figure 1A) compared to the control group. However, 60 or 80 μM of SeMet significantly damaged bMECs at 12 h. In bMECs infected with *N. cyriacigeorgica*, 40 μM SeMet increased cell viability ($p < 0.05$; Figure 1B). Although 40 μM SeMet inhibited LDH release, > 60 μM SeMet increased LDH release ($p < 0.01$; Figure 1C). Furthermore, 40 μM SeMet reduced LDH release ($p < 0.01$) in bMECs infected with *N. cyriacigeorgica* (Figure 1D) without cytotoxicity.

Table 1. Antimicrobials used for MIC assay.

Antimicrobial	*Nocardia cyriacigeorgica*—MIC (μg/mL)											Resistance Rate (%)	MIC_{90} (μg/mL)
	0.5	1	2	4	8	16	32	64	128	256	512		
Gentamycin				5	2			1	2		5	10/15 (66.7)	512
Enrofloxacin				1		1				13		15/15 (100)	256
Penicillin											15	15/15 (100)	512
Amoxicillin							2			13		15/15 (100)	256
cephalonium						1	2			12		15/15 (100)	256
Cefalexin											15	15/15 (100)	512
Ceftriaxone				1					2	12		15/15 (100)	256
Ceftiofur			1						2	12		14/15 (93.3)	256
Lincomycin											15	15/15 (100)	-
Erythromycin											14	15/15 (100)	-

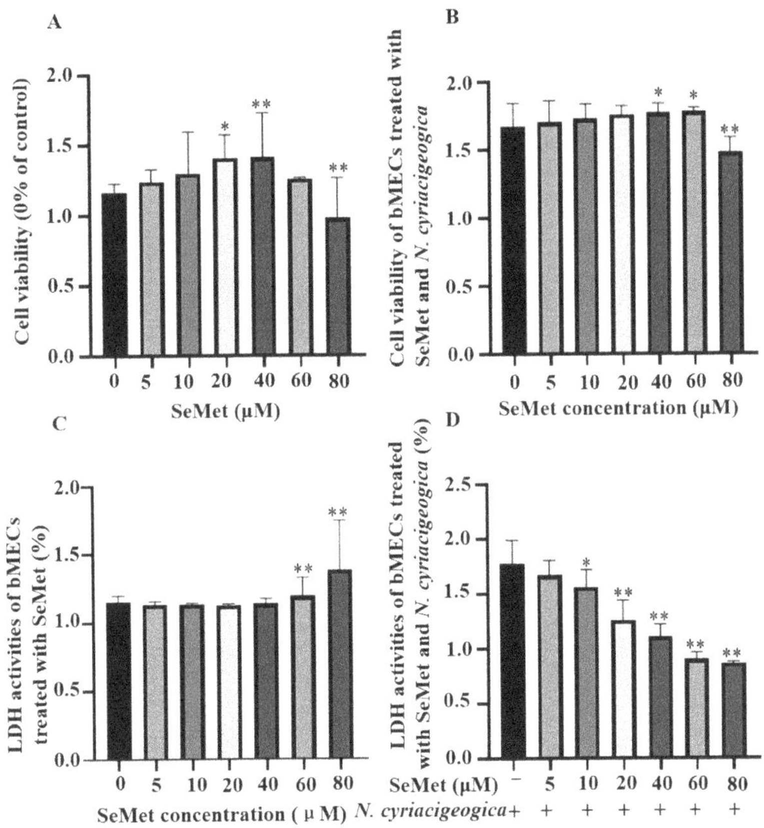

Figure 1. Effects of SeMet with or without *N. cyriacigeorgica* on viability of bMECs and cytotoxicity of *N. cyriacigeorgica*. Cell viability of bMECs exposed to various concentrations of SeMet for 12 h, followed by exposure to *N. cyriacigeorgica* for 6 h. (**A**) Cell viability was determined with a CCK-8 assay in SeMet-treated bMECs. (**B**) Cell viability was determined with a CCK-8 assay in SeMet-treated bMECs subsequently infected with *N. cyriacigeorgica*. (**C**) LDH release detected by an LDH assay kit in SeMet-treated bMECs. (**D**) LDH activities determined by an LDH assay kit in SeMet-treated bMECs that were subsequently infected with *N. cyriacigeorgica*. Data are means ± SD of 3 independent experiments. * $p < 0.05$; ** $p < 0.01$.

2.3. Transmission Electron Microscopy (TEM) of bMECs

In bMECs, at 6 h post-infection with *N. cyriacigeorgica*, ultrastructure changes included mitochondria and endoplasmic reticulum swelling, organelle loss, cell membrane rupture, perinuclear space, and cellular dysfunction (Figure 2). Moreover, Nocardia were in the nucleus and cytoplasm, with indications that this bacterium caused apoptosis. In contrast, the SeMet-treated group had fewer intracellular bacteria, and cell morphology was similar to the untreated group.

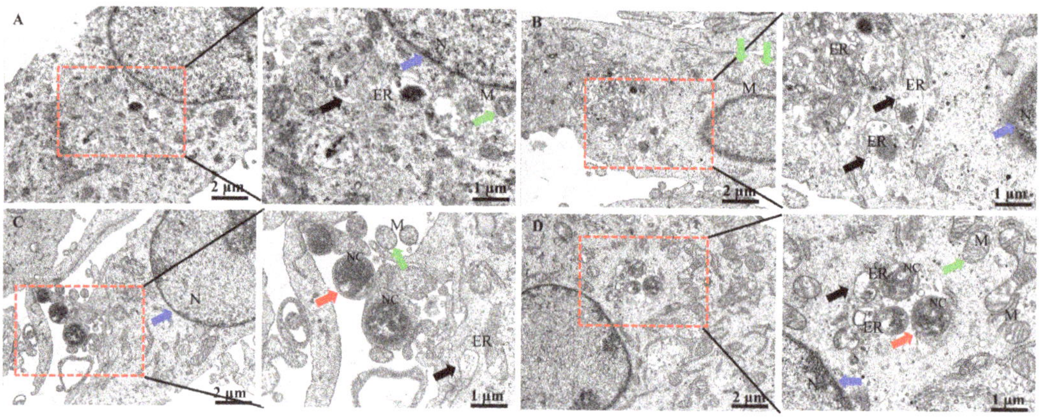

Figure 2. Ultrastructural changes of bMECs infected with *N. cyriacigeorgica* and examined with transmission electron microscopy. (**A**) Control group bMECs had abundant mitochondria (green arrow) in cytoplasm, and normal nucleus (blue arrow), and normal endoplasmic reticulum (black arrow); (**B**) in the SeMet-treated group, all bMECs components appeared unaltered; (**C**) *N. cyriacigeorgica* infection group, the red arrow indicates *N. cyriacigeorgica* (NC); note that the endoplasmic reticulum (black arrow) was swollen, indicating the cells were under stress. The green arrow indicates mitochondria, which appeared enlarged, with disruption of mitochondrial cristae (compared to the control group), whereas the blue arrow indicates the nucleus; (**D**) pretreatment with SeMet before *N. cyriacigeorgica* infection. The red arrow indicates *N. cyriacigeorgica* (NC), the green arrow indicates enlarged mitochondria, and black arrows indicate swollen endoplasmic reticulum which is under stress. Note the protective effects of SeMet on bMECs, including mitochondria (green arrow), endoplasmic reticulum (black arrow), and nucleus (blue arrow), with fewer changes compared to the infection group.

2.4. Gene Expression and Production of Pro-Inflammatory Cytokines

Compared to the control group, both mRNA and protein expression levels of TNF-α, IL-1β, IL-6, and IL-8 in bMECs were downregulated and decreased in cells pretreated with 40 μM of SeMet and then infected with *N. cyriacigeorgica* (Figure 3, $p < 0.01$; Figure 4, $p < 0.05$).

2.5. Superoxide Dismutase (SOD) and Glutathione Peroxidase (GSH-px) Activities and MDA Concentration

The activities of antioxidant enzymes SOD and GSH-px were decreased ($p < 0.01$), whereas MDA concentration was increased ($p < 0.01$) in infected cells versus controls. However, in cells pretreated with 40 μM SeMet and later challenged with *N. cyriacigeorgica*, activities of both SOD and GSH-px increased ($p < 0.05$), whereas MDA content decreased ($p < 0.01$) compared to the infection group (Figure 5).

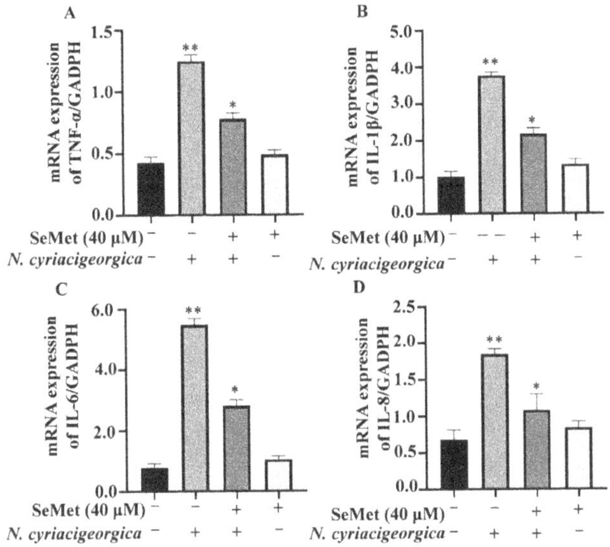

Figure 3. Effects of SeMet on gene expression of proinflammatory cytokines. Expression of: (**A**) TNF-α; (**B**) IL-1β; (**C**) IL-6; and (**D**) IL-8 in bMECs pretreated with 40 μM SeMet or nothing for 6 h before infection with *N. cyriacigeorgica* (multiplicity of infection (MOI) = 5:1) for 4 h. The mRNA expression of TNF-α, IL-1β, IL-6, and IL-8 were downregulated compared to the control group. Data are means ± SD of three independent experiments. * $p < 0.05$; ** $p < 0.01$.

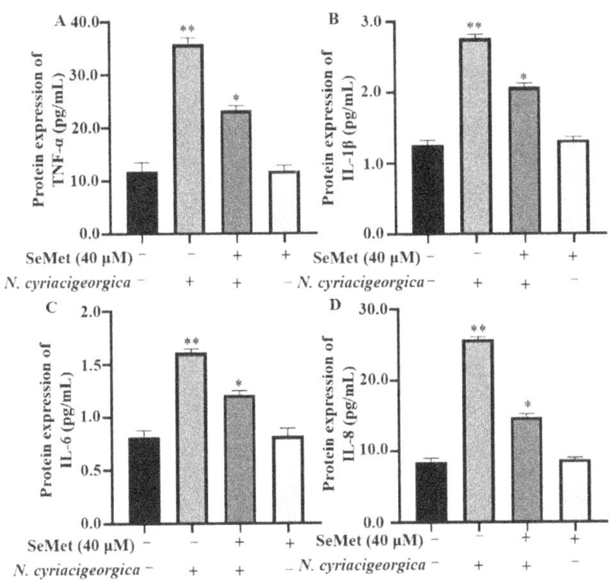

Figure 4. SeMet alleviated *N. cyriacigeorgica*-induced inflammatory responses in bMECs. The bMECs were treated with 40 μM of SeMet for 6 h and then infected with *N. cyriacigeorgica* for 4 h. SeMet pretreatment alleviated *N. cyriacigeorgica*-induced release of inflammatory cytokines in bMECs. (**A**) TNF-α; (**B**) IL-1β; (**C**) IL-6; and (**D**) IL-8. Data are means ± SD of three independent experiments. * $p < 0.05$; ** $p < 0.01$.

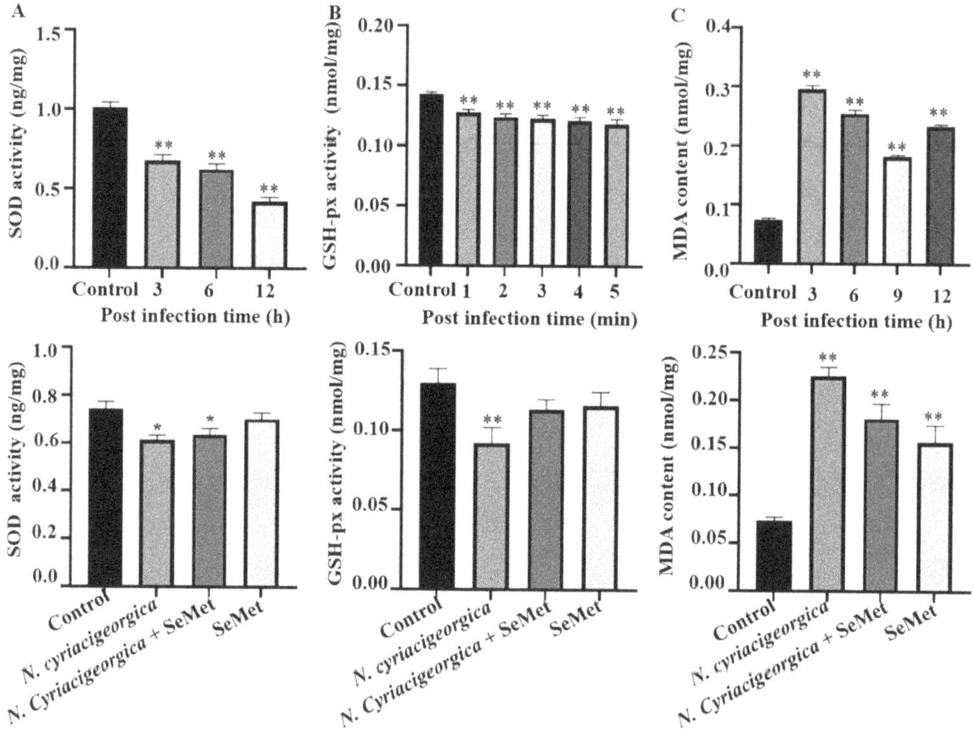

Figure 5. Effects of SeMet on *N. cyriacigeorgica* induced oxidative stress indicators in bMECs. (**A**) Superoxide dismutase (SOD) activities of bMECs infected *N. cyriacigeorgica* for 3, 6, or 12 h were decreased gradually ($p < 0.01$) but increased ($p < 0.05$) with 40 μM SeMet pretreament; (**B**) glutathione peroxidase GSH-px activity for 1, 2, 3, 4 or 5 min were decreased ($p < 0.01$) but increased ($p < 0.05$) with 40 μM SeMet pretreatment; (**C**) malondialdehyde (MDA) content in bMECs infected with *N. cyriacigeorgica* for 3, 6, 9, or 12 h were increased ($p < 0.01$), but were decreased ($p < 0.05$) with 40 μM SeMet pretreatment. Means ± SD of three independent experiments. Data are means ± SD of three independent experiments. * $p < 0.05$, ** $p < 0.01$.

2.6. N. Cyriacigeorgica Stimulated Reactive Oxygen Species (ROS)

At 6 h after *N. cyriacigeorgica* infection, ROS concentrations in bMECs increased ($p < 0.01$) compared to the control group. However, in cells pretreated with 40 μM SeMet, ROS concentrations were decreased ($p < 0.01$) compared to the infection group. Furthermore, ROS concentration was not different between the group pretreated with SeMet and uninfected bMECs ($p > 0.05$, Figure 6).

2.7. N. cyriacigeorgica Induced Apoptosis in bMECs

The percentage of apoptosis in bMECs infected with *N. cyriacigeorgica* was increased in a time-dependent manner. At 3 h, apoptosis of the infection group was higher than the control group ($p < 0.05$, Figure 7). In addition, apoptosis in the infection group was higher ($p < 0.01$) than in the SeMet-treated group.

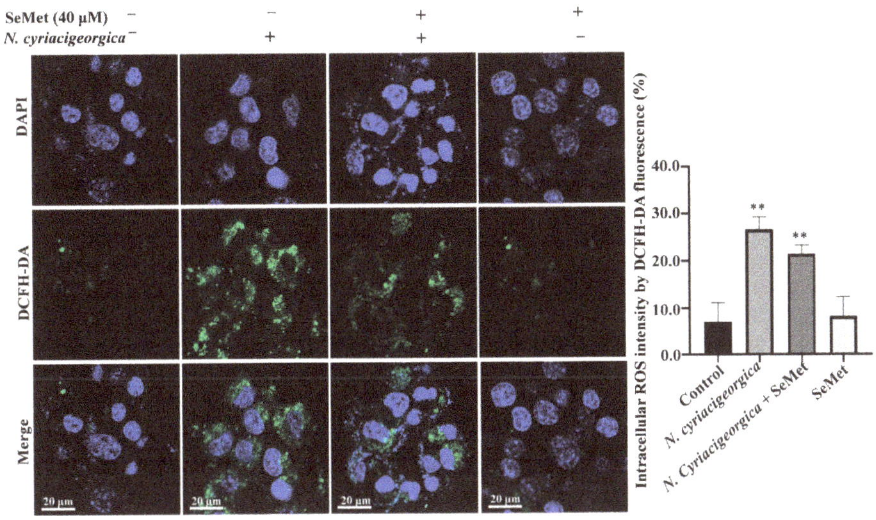

Figure 6. Effects of SeMet on *N. cyriacigeorgica*-induced increased in ROS concentrations in bMECs. In bMECs infected with *N. cyriacigeorgica*, ROS was increased ($p < 0.01$). However, in bMECs pretreated with 40 μM SeMet for 6 h before infection with *N. cyriacigeorgica* (MOI = 5:1), the ROS concentration was decreased ($p < 0.01$). In addition, ROS concentration of bMECs only pretreated with 40 μM SeMet for 6 h was not significantly different from the control. Data are means ± SD of three independent experiments. ** $p < 0.01$.

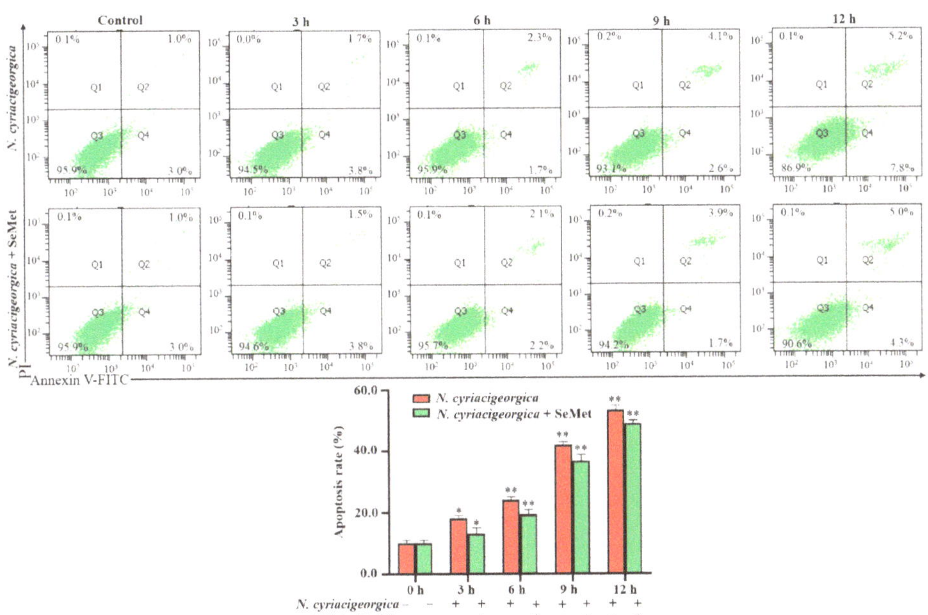

Figure 7. Apoptosis of bMECs treated with 40 μM SeMet for 12 h and subsequently infected with *N. cyriacigeorgica*. The bMECs were treated with 40 μM SeMet for 12 h and then infected with *N. cyriacigeorgica*

at MOI of 0 and 5 for 3, 6, 9, or 12 h. The percentage of apoptosis in bMECs infected with *N. cyriacigeorgica* was increased in a time-dependent manner. Apoptosis of the infection group was higher (p < 0.01) at various time points compared to the group given 40 µM SeMet. Therefore, SeMet mitigated apoptosis induced by *N. cyriacigeorgica*. Data are means ± SD of three independent experiments. * p < 0.05; ** p < 0.01.

3. Discussion

Outbreaks of bovine mastitis caused by *Nocardia* spp. are common worldwide, including in China [3]. Key factors contributing to *Nocardia*-induced bovine mastitis include poor management and hygiene, such as inadequate disinfection of teats before and after milking, lack of proper sanitation, and use of contaminated drugs or containers for treatments [19]. Additionally, factors such as decreased immunity due to high milk production, calving, and other diseases can increase the susceptibility of cows to *Nocardia* infection.

As *N. cyriacigeorgica* adheres to and penetrate bMECs, causing cell damage and apoptosis [20], it is necessary to seek effective treatments. Although *Nocardia* spp. have diverse sensitivity to antibiotics, which can be used for phenotypic identification [21], strains affecting cattle are usually resistant to common veterinary antimicrobials. Antimicrobial susceptibility (MIC method) of *N. cyriacigeorgica* to 10 antimicrobials was determined. The *N. cyriacigeorgica* were susceptible only to gentamycin and ceftiofur, but were resistant to eight other microbials. *N. puris, N. veterana, N. cyriacigeorgica, N. arthritidis,* and *N. africana* were resistant to cloxacillin (75%), ampicillin (55%), and cefoperazone (47%), but were less resistant to amikacin, cefuroxime, or gentamicin [7]. However, there are limited reports characterizing antimicrobial sensitivity/resistance of *Nocardia* spp. from bovine mastitis [22]. Moreover, based on the intracellular location of the pathogen and virulence characteristics that cause a pyogranulomatous reaction, treatment of *Nocardia* mastitis is usually not curative. Therefore, early identification of clinical and subclinical cases of bovine mastitis by microbiological identification and drug sensitivity tests in vitro is crucial to improve response to treatment or promote timely culling [23]. Although ceftriaxone and imipenem were used for the treatment of moderate and severe nocardiosis in Australia, only 58.8% of isolates were sensitive to ceftriaxone and 41.0% to imipenem [2]. A similar susceptibility pattern was reported in other Australian studies for ceftriaxone (44–60%) and imipenem (33.7–48%), based on a bone mineral density test [8,24]. Therefore, SeMet was assessed as a potential alternative treatment method for *Nocardia* mastitis in the current research.

In this study, the effects of various concentrations of SeMet on cell viability, LDH assay, antioxidant status and oxidative stress in bMECs were determined. Although 40 µM SeMet had protective effects on bMECs, higher concentrations (i.e., 60 or 80 µM) were toxic. It was reported that SeMet alleviated ESBL *E. coli*-induced inflammation by activating the selenoprotein S-mediated TLR4/NF-κB signaling pathway in bMECs [25], which is consistent with the present results. Furthermore, SeMet enhanced expression of glutathione peroxidase 1 and 3 both in vivo and in vitro, and improved the growth and viability of bMECs [26]; whereas selenium concentrations from 50 to 100 nmol/L improved antioxidant function but did not affect milk protein synthesis in healthy bMECs. In addition, selenium mitigated lipopolysaccharide-induced damage by increasing antioxidant markers and stimulating milk protein synthesis and expression of milk protein-related genes in bMECs [27].

Oxidative stress impacts various physiological cell functions, including the immune response and the reproductive system [28,29]. Selenomethionine is regarded as an important element of the antioxidant system, as selenium enhances bMECs' antioxidant function and protects cells from diethylenetriamine/NO-induced oxidative damage, mainly by increasing TrxR activity and reducing NO concentration via modulation of Nrf2 and MAPK signalling pathways [30]. Dietary intake of selenomethionine is directly correlated with the activities of various antioxidant enzymes (GSH-px, SOD, MDA, and ROS). In this study, SOD and GSH-px activities were downregulated by *N. cyriacigeorgica* infection. However, pretreatment with 40 μm SeMet upregulated SOD and GSH-px activities. Furthermore, in bMECs, MDA is considered a major product of lipid peroxidation and a marker of oxidative stress and antioxidant status [29]. In this study, whereas MDA and ROS contents were upregulated by *N. cyriacigeorgica* infection, SeMet pretreatment downregulated MDA and ROS, implying SeMet had a protective effect. Similarly, dairy cows given selenium yeast supplements (SeMet is the predominant form of selenium) had increased serum GSH-px activity compared to cows given sodium selenite supplements [31]. Furthermore, 0.3 mg/kg DM of hydroxy-selenomethionine enhanced GSH-px activity in dairy cow serum compared to the same dose of sodium selenite [32]. In studies using primary rainbow trout hepatocytes to determine the effects of SeMet and sodium selenite on arsenite cytotoxicity, SeMet attenuated oxidative stress by increasing SOD [32,33]. Furthermore, SeMet increased the capacity of bMECs to withstand oxidative stress, as indicated by decreased ROS and MDA [32], which is consistent with current results. Additionally, selenium suppressed *S. aureus*-induced inflammation by inhibiting the NLRP3 inflammasome and ROS activation in bMECs [34].

The current study aimed to determine whether SeMet mitigated inflammatory and oxidative damage in bMECs caused by *N. cyriacigeorgica* strains isolated from mastitic milk. Expression of TNF-α, IL-1β, IL-6, and IL-8 were significantly upregulated in bMECs infected by *N. cyriacigeorgica*. In contrast, pretreatment with 40 μM SeMet significantly downregulated gene expression of TNF-α, IL-1β, IL-6, and IL-8. Therefore, SeMet reduced production of inflammatory cytokines and decreased inflammatory damage to bMECs induced by *N. cyriacigeorgica*. Similarly, SeMet increased the autophagic flux and inhibited the NF-κB-mediated inflammatory response of bMECs caused by *Klebsiella pneumoniae* [35]. Furthermore, gene expression of TNF-α, IL-1β, IL-6, and IL-8 were also significantly upregulated 1 to 5 h after infection by *K. pneumoniae* [36], although SeMet reduced inflammation and protein expression of TLR4, Ikappa-B, NF-κB, and TNF-α in bMECs and macrophages infected with ESBL *E. coli* [17]. Increased selenium concentrations decreased expression of IL-1β and IL-6 [37]. In addition, selenium reduced gene expression of inflammatory mediators TNF-α, IL-1β, and IL-6 through the TLR2, NF-κB, and MAPK signaling pathway in *S. aureus*-infected bMECs, implying this was the basis of the anti-inflammatory effect of selenium [38]. Pre-treatment with various concentrations of SeMet significantly downregulated protein expression in ESBL-producing *E. coli*-infected bMECs [17], which is consistent with current findings. IL-1β triggers neutrophil migration and its expression level is related to severity of inflammation. Moreover, IL-6 promotes cell proliferation and differentiation and induces acute gene expression in hepatocytes, increasing inflammation [39].

In the current study, SeMet clearly had a protective effect against apoptosis induced by *N. cyriacigeorgica* infection of bMECs. Although apoptosis in the *N. cyriacigeorgica* infection group was significantly upregulated in a time-dependent manner, this was prevented by pretreatment with 40 μM SeMet, indicating that SeMet alleviated *N. cyriacigeorgica*-induced cell apoptosis, as this bacterium induces both apoptotic and necrotic changes in bMECs after 6 h [20]. Based on transcription levels of apoptosis-associated genes, overexpression of heme-binding protein could induce cell apoptosis [40]. MTSP3141, a secretory protein from *N. seriolae*, can target mitochondria and induce apoptosis in host cells, and function as a virulence factor [41]. However, selenium-related treatments protected against apoptosis, whereas selenium-enriched plant supplements significantly downregulated apoptosis of ovarian cells [42]. SeMet supplementation protected bMECs from H_2O_2-induced apoptosis and enhanced cell proliferation and viability under oxidative stress [43]. Furthermore, 500 nM of $Na_2 SeO_3$ or 30 μM of L-selenium-methylselenocysteine significantly suppressed apoptosis and reduced mitochondrial membrane potential in murine liver cells, whereas 0.45 mg/kg dietary selenium also remarkably ameliorated liver damage in mice fed high-fat diets [44]. Extracellular vesicles derived from selenium-deficient bMECs induced oxidative stress, provoking apoptosis and inflammation via endoplasmic reticulum stress and PI3K/Akt/mTOR signaling pathway, contributing to explain susceptibility to mastitis due to selenium deficiency [45]. SeMet decreased apoptosis by increasing the overall antioxidant capacity and reducing cellular ROS [46], which is consistent with current results.

4. Materials and Methods

4.1. Cell Culture

A bovine mammary epithelial cell line (MAC-T) (Shanghai Jingma Biological Technology Co., Ltd., Minhang district, Shanghai, China) was cultured in Dulbecco's modified Eagle medium (DMEM) containing 10% fetal bovine serum and 1% antibiotic mix (100 μg/mL streptomycin and 100 U/mL penicillin), in an incubator with 5% CO_2 at 37 °C. At 70–80% confluence, cells were detached and seeded into 6- or 96-well plates for assays (Figure 8).

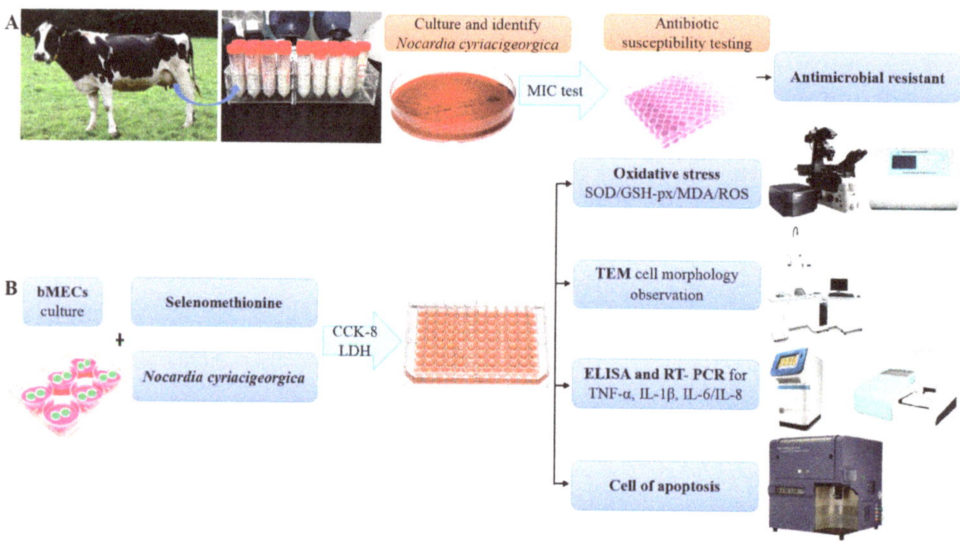

Figure 8. Schematic diagram of the experimental protocol. (**A**) Identification of bovine mastitis milk samples and antimicrobial resistance. (**B**) Selenomethionine alleviates effects of *Nocardia cyriacigeorgica*-induced inflammation, oxidative stress, and apoptosis in bovine mammary epithelial cells.

4.2. Bacterial Culture and Basic Tests

N. cyriacigeorgica, previously isolated in our laboratory from milk samples with bovine mastitis, was activated from frozen stocks by culturing on tryptose soya agar (DifcoTM, Becton Dickison, Sparks, MD, USA) supplemented with 5% defibrinated sheep blood. It was maintained in aerobic conditions at 37 °C for 72 h, then sub-cultured in Brain Heart Infusion (BHI) broth for the following experiments. The detailed procedure was the same as described by Chen (2017) [3].

4.3. Minimum Inhibitory Concentration (MIC) Determination

The antimicrobial susceptibility of *N. cyriacigeorgica* was assessed with a broth microdilution assay, recommended by the Clinical and Laboratory Standards Institute (CLSI, 2017) [47]. *Staphylococcus aureus* ATCC 29213 was used for quality control [48]. The MIC was the lowest concentration of antimicrobial that totally inhibited bacterial growth. For MIC analyses, 96-well microplates were used, with 200 μL of BHI and antimicrobial added. After incubation of *N. cyriacigeorgica* in BHI broth for 48 h, MIC values were the lowest concentrations of the antimicrobial without visible bacterial growth. We tested the following antimicrobial agents: gentamycin, lincomycin, erythromycin, enrofloxacin, penicillin, amoxicillin, cephalonium, cephalexin, ceftriaxone, and ceftiofur; with MIC values based on visual observation. The concentrations needed to inhibit visible growth of 50% and 90% of isolates were defined as MIC_{50} and MIC_{90}, respectively.

4.4. Assessment of Cell Viability by CCK-8

Cell Counting Kit-8 (CCK-8 kit; Beyotime Institute of Biotechnology, Beijing, China) is a rapid and highly sensitive method for measuring cell proliferation and cytotoxicity. MAC-T cells (~5000 cells/well) were plated in 96-well plates with 0, 5, 10, 20, 40, 60, or 80 μmol/L SeMet for 12 h. After treatment, cells were infected with *N. cyriacigeorgica* (MOI = 5:1) and incubated at 37 °C for 6 h and a CCK-8 kit used to assess cell viability, in accordance with the manufacturer's directions. After 10 μL of CCK-8 solution was added in each well, cells were incubated at 37 °C for 2 h. Absorbance was recorded at 450 nm in a DNM 9602 microplate reader. Cell viability was evaluated as a percentage (%) relative to the control.

4.5. Assessment of Lactate Dehydrogenase (LDH) Release

Cytotoxic effects of *N. cyriacigeorgica* on bMECs were evaluated using an LDH assay kit (Beyotime Institute of Biotechnology). The bMECs cells were cultured at 37 °C with 5% CO_2 in 96-well plates and incubated with 0, 5, 10, 20, 40, 60, or 80 μmol/L SeMet for 12 h. Afterwards, treated cells were infected with *N. cyriacigeorgica* at a MOI of 5:1 for 4 or 6 h. After incubation, the cell culture plate was centrifuged at 400× *g* for 5 min, and then 120 μL of supernatant was transferred to a new 96-well plate, with 60 μL of reaction mixture added to each well, followed by incubation at room temperature in the dark for 30 min on a horizontal shaker (150 rpm). Optical density (OD) was measured by a spectrophotometer at 490 nm.

4.6. Transmission Electron Microscopy (TEM) of BMECs

For TEM, bMECs were seeded on 6-well plates and infected with *N. cyriacigeorgica* (MOI = 5:1) for 6 h. After infection, cells were collected with a cell scraper and centrifuged at 1000× *g* for 5 min. They were washed twice with PBS, fixed in 2.5% glutaraldehyde for 2 h and 1% osmium tetroxide for 2 h at 4 °C, dehydrated in a graded ethanol series, embedded in epoxy resin acetone mixtures (2:1) for 2 h, then immersed in pure resin for 12 h at 37 °C. After polymerization, ultrathin sections (50~70 nm) were cut, stained with 1% uranyl acetate and lead citrate, and examined with transmission electron microscopy (JEM-1400, Tokyo, Japan) at 200 KeV.

4.7. Determination of Pro-Inflammatory Cytokines

BMECs were infected with *N. cyriacigeorgica* (MOI = 5:1) for 4 h and remained in culture medium with antibiotics and lysozyme. The supernatant was collected by centrifugation at 12,000× *g* for 10 min at 4 °C. An ELISA kit (Jiangsu Meimian Industrial Co., Ltd., Yancheng, China) with a detection range of 0.1 to 20.0 ng/mL was used (according to manufacturer's instructions) to assess concentrations of pro-inflammatory cytokines TNF-α, IL-1β, IL-6, and IL-8 in cell supernatant. Absorbance at 450 nm was measured on a DNM 9602 microplate reader. All measurements were done 3 times, and the mean of 3 independent readings used for statistical analyses.

4.8. RNA Extraction and Reverse Transcriptase Quantitative Polymerase Chain Reaction (RT-qPCR)

Briefly, bMECs were washed once with PBS and harvested with 1 mL of TransZol Up lysis solution to extract total RNA, according to label instructions (TransGen Biotech Co., Ltd., Beijing, China). The cDNA was synthesized by TransScript® II All-in-One First-Strand cDNA Synthesis SuperMix for qPCR (One-Step gDNA Removal; TransGen Biotech). Concentrations of total RNA and cDNA were determined by absorbance at 260 and 280 nm on a NanoDrop-2000 spectrophotometer (Thermo Fisher Scientific, Waltham, MA, USA). A

20-μL RT-qPCR mixture was used, comprised of 10 μL of 2 × RealStar Fast SYBR qPCR mix (GenStar, Beijing, China), 10 μM (0.5 μL) of each primer, 1 μL cDNA, and 8 μL of nuclease-free water. Expression levels of mRNA were tested by real-time PCR as follows: predenaturation at 94 °C for 2 min; 40 cycles of denaturation at 95 °C for 5 s; and annealing at 60 °C for 30 s using an Applied Biosystems StepOnePlus Real Time PCR system (Thermo Fisher Scientific). Primers for a housekeeping gene (GAPDH) and genes encoding TNF-α, IL-1β, IL-6, and IL-8 were used (Table 2). Cycle threshold (Ct) values and $2^{-\Delta\Delta Ct}$ values were calculated to determine expression.

Table 2. Primer sequences of bovine genes used for quantitative PCR.

Gene	Primer Sequence		Reference
TNF-α	Upstream	5′-ACGGGCTTTACCTCATCTACTC	Liu et al., 2020 [49]
	Downstream	3′-GCTCTTGATGGCAGACAGG	
IL-1β	Upstream	5′-AGGTGGTGTCGGTCATCGT	Liu et al., 2020 [49]
	Downstream	3′-GCTCTCTGTCCTGGAGTTTGC	
IL-6	Upstream	5′-ATCAGAACACTGATCCAGATCC	Liu et al., 2020 [49]
	Downstream	3′-CAAGGTTTCTCAGGATGAGG	
IL-8	Upstream	5′-ACACATTCCACACCTTTCCA	Liu et al., 2020 [49]
	Downstream	3′-GGTTTAGGCAGACCTCGTTT	
GAPDH	Upstream	5′-CATTGACCTTCACTACATGGT	Liu et al., 2020 [49]
	Downstream	3′-ACCCTTCAAGTGAGCCCCAG	

4.9. Determination of MDA Concentration, SOD and GSH-px Activities

The bMECs were cultured in 6-well plates and incubated with *N. cyriacigeorgica* (MOI 0 and 5) for 6 h. After infection, bMECs were washed 3 times with ice-cold PBS, lysed in SOD sample preparation solution, then centrifuged at 12,000× *g* at 4 °C for 3–5 min and the supernatant was collected for a subsequent assay. Then, MDA concentrations and activities of GSH-px and SOD were detected using Cell MDA, GSH-px and SOD assay kits (Cat. no. S0131, S0056 and S0101, respectively; Beyotime Biotechnology, Shanghai, China), according to the manufacturer's instructions. Readings were recorded with a DNM 9602 microplate reader at 450 and 340 nm. Each group was assayed 30 times (10 replicates per group and each sample assayed in triplicate).

4.10. Measurement of Reactive Oxygen Species (ROS) Concentrations

Intracellular ROS were measured by the reactive oxygen species assay kit (Beyotime Institute of Biotechnology, Shanghai, China) with an oxidation-sensitive fluorescent probe (DCFH-DA). Cells were seeded into 6-well plates with glass coverslips, treated, then cells were washed once using PBS and stained in darkness with 100 μM of diluted DCFH-DA for 20 min at 37 °C and then washed 3 times with DMEM. The DCFH-DA was diluted 1000 times with serum-free DMEM/F12 culture medium, according to the manufacturer's instructions. Finally, cells were stained with DAPI for 10 min. After washing 3 times in ice-cold PBS, samples were viewed with a laser confocal microscope (NIKON A1 HD25, Tokyo, Japan).

4.11. Determination of Apoptosis

Apoptosis induced by *N. cyriacigeorgica* in bMECs was determined with an Annexin V-FITC/propidium iodide (PI) analysis kit (Yeasen Biotechnology (Shanghai) Co., Ltd., Shanghai, China) according to the manufacturer's instructions. Cells were plated into 6-well plates and infected with *N. cyriacigeorgica* at MOI of 0 or 5 for 3, 6, 9, or 12 h. After infection, cell culture medium was discarded and cells were harvested with 0.25% trypsin without EDTA and collected by pipetting into 1.5 mL PCR tubes. Cells were centrifuged at 1000× *g* 4 °C for 5 min, washed twice with ice-cold PBS, re-suspended in 100 μL of 1 × binding buffer, and stained with 5 μL of Annexin V (FITC) and 10 μL of PI. After 15 min in the dark at room temperature, cells were assessed (FAC Scan Flow Cytometer; Becton,

Dickinson and Co., East Rutherford, NJ, USA), as follows: Q1 = represents necrotic cells; Q2 = late apoptotic cells; Q3 = early apoptotic cells; and Q4 = late live cells or normal cells.

4.12. Statistical Analyses

Results were expressed as the mean ± standard deviation (SD) from at least 3 independent experiments. GraphPad Prism 8.0 (GraphPad Software, Inc., San Diego, CA, USA) was used and significance was assessed using SPSS 26.0 (IBM Corp., Armonk, NY, USA). Statistical differences were assessed with one-way analysis of variance (ANOVA), followed by the Tukey's multiple comparison post hoc test. Significance was accepted at * $p < 0.05$; ** $p < 0.01$.

5. Conclusions

In conclusion, *Nocardia cyriacigeorgica* infection of bMECs induced structural disruption and mitochondrial degeneration. However, pretreatment with 40 μM SeMet had the following protective actions: upregulated antioxidant enzyme activities, improved mRNA expressions and protein synthesis rates, reduced oxidative stress and apoptosis, and protected against inflammatory damage. An optimal dose of selenomethionine had cytoprotective roles in bMECs, although higher doses were cytotoxic. These findings highlighted the potential therapeutic benefits of selenomethionine supplementation in mitigating oxidative stress, inflammation, apoptosis, and structural damage induced by bacterial infection in bMECs.

Author Contributions: Conceptualization, B.H.; methodology, T.A. and V.A.; software, T.A. and V.A.; validation, T.A., J.H., M.A.K., H.B. and G.L.; formal analysis, T.A., J.H., V.A. and M.A.K.; investigation, T.A.; resources, G.L.; data curation, T.A., J.H. and M.A.K.; writing—original draft preparation, T.A., J.H., V.A., M.A.K., G.L. and X.Z.; writing—review and editing, J.P.K., X.Z. and H.W.B.; visualization, T.A.; supervision, X.Z. and B.H.; project administration, B.H.; funding acquisition, B.H. All authors have read and agreed to the published version of the manuscript.

Funding: This research was funded by: National Natural Science Foundation of China grant number [32172928] and the High-End Foreign Experts Recruitment Program grant number [G2022108009L].

Institutional Review Board Statement: The research was conducted in accordance with the Ethical Committee at the College of Veterinary Medicine, China Agricultural University (CAU), Beijing (Protocol AW01703202-2-1), and adhered to CAU's established ethical standards.

Informed Consent Statement: Informed consent was obtained from all subjects involved in the study.

Data Availability Statement: The raw data supporting the conclusions of this manuscript will be made available by the authors, without undue reservation, to any qualified researcher.

Conflicts of Interest: The authors declare no conflicts of interest.

Abbreviations

bMECs: bovine mammary epithelial cells; MOI: multiplicity of infection; PBS: phosphate-buffered saline; *N. cyriacigeorgica*: *Nocardia cyriacigeorgica*; mTOR: Mammalian target of rapamycin; ESBL: extended-spectrum β-lactamase; TLR4/NF-κB: Toll-like receptor 4/nuclear factor kappa B; TLR2: Toll-like receptor 2; Nrf2: nuclear factor-E2-related factor 2; MAPK: mitogen-activated protein kinase; NLRP3: NOD-like receptor protein 3; TNF-α: tumor necrosis factor-α; IL-1β: interleukin-1β; IL-6: interleukin-6; IL-8: interleukin-8; PI3K/Akt/mTOR: phosphatidylinositol 3-kinase/protein kinase B/mammalian target of rapamycin; MDA: malondialdehyde; GSH-px: glutathione peroxidase; SOD: superoxide dismutase; ROS: reactive oxygen species; LDH: Lactate Dehydrogenase; SeMet: Selenomethionine; TEM: transmission electron microscope; MIC: minimum inhibitory concentration.

References

1. Mushtaq, S.; Shah, A.M.; Shah, A.; Lone, S.A.; Hussain, A.; Hassan, Q.P.; Ali, M.N. Bovine mastitis: An appraisal of its alternative herbal cure. *Microb. Pathog.* **2018**, *114*, 357–361. [CrossRef] [PubMed]
2. Yeoh, K.; Globan, M.; Naimo, P.; Williamson, D.A.; Lea, K.; Bond, K. Identification and antimicrobial susceptibility of referred *Nocardia* isolates in Victoria, Australia 2009–2019. *J. Med. Microbiol.* **2022**, *71*, 1–7. [CrossRef] [PubMed]
3. Chen, W.; Liu, Y.X.; Barkema, H.W.; Gao, J.; De Buck, J.; Kastelic, J.P.; Liu, G.; Ali, T.; Shahid, M.; Han, B. Short communication: Molecular characteristics, antimicrobial susceptibility, and pathogenicity of clinical *Nocardia cyriacigeorgica* isolates from an outbreak of bovine mastitis. *J. Dairy Sci.* **2017**, *100*, 8414–8842. [CrossRef] [PubMed]
4. Lu, S.H.; Qian, Z.W.; Mou, P.P.; Xie, L. Clinical Nocardia species: Identification, clinical characteristics, and antimicrobial susceptibility in Shandong, China. *Bosn. J. Basic Med. Sci.* **2020**, *20*, 531–538. [CrossRef]
5. Conville, P.S.; Brown-Elliott, B.A.; Smith, T.; Zelazny, A.M. The complexities of Nocardia taxonomy and identification. *J. Clin. Microbiol.* **2017**, *6*, e01419-17. [CrossRef]
6. Stevanović, O.; Milanov, D.; Prošić, I.; Gajdov, V.; Nedić, D.; Sladojević, Ž.; Radalj, A. Multilocus sequence analysis (MLSA) of a *Nocardia cyriacigeorgica* strain causing severe bovine mastitis in Bosnia and Herzegovina. *Acta Vet. Hung.* **2023**, *71*, 65–70. [CrossRef]
7. Condas, L.A.Z.; de Farias, M.R.; Siqueira, A.K.; Salerno, T.; Chi, K.D.; Werner, J.; de Vargas, A.C.; Bond, G.B.; Gonoi, T.; Matsuzawa, T.; et al. Molecular identification and antimicrobial resistance pattern of Nocardia isolated from 14 diseased dogs and cats. *Braz. J. Microbiol.* **2023**, *54*, 1287–1294. [CrossRef]
8. Tan, Y.E.; Chen, S.C.A.; Halliday, C.L. Antimicrobial susceptibility profiles and species distribution of medically relevant Nocardia species: Results from a large tertiary laboratory in Australia. *J. Glob. Antimicrob. Resist.* **2020**, *20*, 110–117. [CrossRef]
9. Lebeaux, D.; Bergeron, E.; Berthet, J.; Djadi-Prat, J.; Mouniée, D.; Boiron, P.; Lortholary, O.; Rodriguez-Nava, V. Antibiotic susceptibility testing and species identification of Nocardia isolates: A retrospective analysis of data from a French expert laboratory, 2010–2015. *Clin. Microbiol. Infect.* **2019**, *25*, 489–495. [CrossRef]
10. Li, H.; Song, F.; Duan, L.R.; Sheng, J.J.; Xie, Y.H.; Yang, Q.; Chen, Y.; Dong, Q.Q.; Zhang, B.L.; Wang, S.W. Paeonol and danshensu combination attenuates apoptosis in myocardial infarcted rats by inhibiting oxidative stress: Roles of Nrf2/HO-1 and PI3K/Akt pathway. *Sci. Rep.* **2016**, *6*, 23693. [CrossRef]
11. Rojo de la Vega, M.; Dodson, M.; Gross, C.; Mansour, H.M.; Lantz, R.C.; Chapman, E.; Wang, T.; Black, S.M.; Garcia, J.G.; Zhang, D.D. Role of Nrf2 and autophagy in acute lung injury. *Curr. Pharmacol. Rep.* **2016**, *2*, 91–101. [CrossRef] [PubMed]
12. Ayemele, A.G.; Tilahun, M.; Lingling, S.; Elsaadawy, S.A.; Guo, Z.; Zhao, G.; Xu, J.; Bu, D. Oxidative stress in dairy cows: Insights into the mechanistic mode of actions and mitigating strategies. *Antioxidants* **2021**, *10*, 1918. [CrossRef]
13. Kaja, S.; Payne, A.J.; Naumchuk, Y.; Koulen, P. Quantification of lactate dehydrogenase for cell viability testing using cell lines and primary cultured astrocytes. *Curr. Protoc. Toxicol.* **2017**, *72*, 1–15. [CrossRef] [PubMed]
14. Wang, X.; Liu, M.; Geng, N.; Du, Y.; Li, Z.; Gao, X.; Han, B.; Liu, J.; Liu, Y. *Staphylococcus aureus* mediates pyroptosis in bovine mammary epithelial cell via activation of NLRP3 inflammasome. *Vet. Res.* **2022**, *53*, 10. [CrossRef] [PubMed]
15. Damien, B.; Eike, L.; Bernardo, S.F. Necroptosis, pyroptosis and apoptosis: An intricate game of cell death. *Cell. Mol. Immunol.* **2021**, *18*, 1106–1121.
16. Xue, B.X.; Yue, Y.L.; Zi, C.H. Apoptosis and apoptotic body: Disease message and therapeutic target potentials. *Biosci. Rep.* **2019**, *39*, BSR20180992. [CrossRef] [PubMed]
17. Zhuang, C.; Gao, J.; Liu, G.; Zhou, M.; Yang, J.; Wang, D.; Kastelic, J.; Han, B. Selenomethionine activates selenoprotein S, suppresses Fas/FasL and the mitochondrial pathway, and reduces *Escherichia coli*-induced apoptosis of bovine mammary epithelial cells. *J. Dairy Sci.* **2021**, *104*, 10171–10182. [CrossRef]
18. Zou, Y.; Shao, J.; Li, Y.; Zhao, F.; Liu, J.; Liu, H. Protective effects of inorganic and organic selenium on heat stress in bovine mammary epithelial cells. *Oxid. Med. Cell. Long.* **2019**, *2019*, 1503478. [CrossRef]
19. Sharun, K.; Dhama, K.; Tiwari, R.; Gugjoo, M.B.; Iqbal Yatoo, M.; Patel, S.K.; Pathak, M.; Karthik, K.; Khurana, S.K.; Singh, R.; et al. Advances in therapeutic and managemental approaches of bovine mastitis: A comprehensive review. *VetlQ* **2021**, *41*, 107–136. [CrossRef]
20. Chen, W.; Liu, Y.; Zhang, L.; Gu, X.; Liu, G.; Shahid, M.; Gao, J.; Ali, T.; Han, B. *Nocardia cyriacigeogica* from bovine mastitis induced in vitro apoptosis of bovine mammary epithelial cells via activation of mitochondrial-caspase pathway. *Front. Cell. Infect. Microbiol.* **2017**, *7*, 194. [CrossRef]
21. Toyokawa, M.; Ohana, N.; Ueda, A.; Imai, M.; Tanno, D.; Honda, M.; Takano, Y.; Ohashi, K.; Saito, K.; Shimura, H. Identification and antimicrobial susceptibility profiles of Nocardia species clinically isolated in Japan. *Sci. Rep.* **2021**, *11*, 16742. [CrossRef] [PubMed]
22. Huang, L.; Chen, X.; Xu, H.; Sun, L.; Li, C.; Guo, W.; Xiang, L.; Luo, G.; Cui, Y.; Lu, B. Clinical features, identification, antimicrobial resistance patterns of Nocardia species in China: 2009–2017. *Diagn. Microbiol. Infect. Dis.* **2019**, *94*, 165–172. [CrossRef] [PubMed]
23. Amano, T.; Nishikawa, T.; Oka, K.; Ota, K.; Shimizu, T. How an antimicrobial stewardship team treated a *Nocardia farcinica*-associated brain abscess: A case report. *Cureus* **2024**, *16*, e54605. [CrossRef] [PubMed]
24. Paige, E.K.; Spelman, D. Nocardiosis: 7-year experience at an Australian tertiary hospital. *Intern. Med. J.* **2019**, *49*, 373–379. [CrossRef]

25. Zhuang, C.; Liu, G.; Barkema, H.W.; Zhou, M.; Xu, S.; ur Rahman, S.; Liu, Y.; Kastelic, J.P.; Gao, J.; Han, B. Selenomethionine suppressed TLR4/NF-κB pathway by activating selenoprotein S to alleviate ESBL *Escherichia coli*-induced inflammation in bovine mammary epithelial cells and macrophages. *Front. Microbiol.* **2020**, *11*, 1461. [CrossRef]
26. Miranda, S.G.; Wang, Y.J.; Purdie, N.G.; Osborne, V.R.; Coomber, B.L.; Cant, J.P. Selenomethionine stimulates expression of glutathione peroxidase 1 and 3 and growth of bovine mammary epithelial cells in primary culture. *J. Dairy Sci.* **2009**, *92*, 2670–2683. [CrossRef]
27. Zhang, B.; Guo, Y.; Yan, S.; Guo, X.; Zhao, Y.; Shi, B. The protective effect of selenium on the lipopolysaccharide-induced oxidative stress and depressed gene expression related to milk protein synthesis in bovine mammary epithelial cells. *Biol. Trace Elem. Res.* **2020**, *197*, 141–148. [CrossRef]
28. Mavangira, V.; Sordillo, L.M. Role of lipid mediators in the regulation of oxidative stress and inflammatory responses in dairy cattle. *Res. Vet. Sci.* **2018**, *116*, 4–14. [CrossRef]
29. Sordillo, L.M. Nutritional strategies to optimize dairy cattle immunity. *J. Dairy Sci.* **2016**, *99*, 4967–4982. [CrossRef]
30. Guo, Y.; Guo, X.; Yan, S.; Zhang, B.; Shi, B. Mechanism underlying the protective effect of selenium on NO-induced oxidative damage in bovine mammary epithelial cells. *Biol. Trace Elem. Res.* **2019**, *191*, 104–114. [CrossRef]
31. Gong, J.; Ni, L.; Wang, D.; Shi, B.; Yan, S. Effect of dietary organic selenium on milk selenium concentration and antioxidant and immune status in midlactation dairy cows. *Livest. Sci.* **2014**, *70*, 84–90. [CrossRef]
32. Sun, P.; Wang, J.; Liu, W.; Bu, D.P.; Liu, S.J.; Zhang, K.Z. Hydroxy-selenomethionine: A novel organic selenium source that improves antioxidant status and selenium concentrations in milk and plasma of mid-lactation dairy cows. *J. Dairy Sci.* **2017**, *100*, 9602–9610. [CrossRef] [PubMed]
33. Jamwal, A.; Niyogi, S. Dose and chemical species-specific effects of selenium against arsenite toxicity in cultured hepatocytes of rainbow trout (*Oncorhynchus mykiss*). *Metallomics* **2017**, *9*, 744–756. [CrossRef] [PubMed]
34. Yang, Y.; Lv, S.; Wang, Z.; Liu, J. Selenium ameliorates *S. aureus*-induced inflammation in bovine mammary epithelial cells by regulating ROS-induced NLRP3 inflammasome. *Biol. Trace Elem. Res.* **2022**, *200*, 3171–3175. [CrossRef] [PubMed]
35. Xu, S.; Miao, Y.; Dong, J.; Cui, L.; Liu, K.; Li, J.; Meng, X.; Zhu, G.; Wang, H. Selenomethionine inhibits NF-κB-mediated inflammatory responses of bovine mammary epithelial cells caused by *Klebsiella pneumoniae* by increasing autophagic flux. *Biol. Trace Elem. Res.* **2024**, *202*, 1568–1581. [CrossRef]
36. Ma, X.M.; Xu, S.Y.; Li, J.J.; Cui, L.Y.; Dong, J.S.; Meng, X.; Zhu, G.Q.; Wang, H. Selenomethionine protected bMECs from inflammatory injury and oxidative damage induced by *Klebsiella pneumoniae* by inhibiting the NF-κB and activating the Nrf2 signaling pathway. *Inter. Immunopharmacol.* **2022**, *110*, 109027. [CrossRef]
37. Mahmoodpoor, A.; Hamishehkar, H.; Shadvar, K.; Ostadi, Z.; Sanaie, S.; Saghaleini, S.H.; Nader, N.D. The effect of intravenous selenium on oxidative stress in critically ill patients with acute respiratory distress syndrome. *Immunol. Investig.* **2019**, *48*, 147–159. [CrossRef]
38. Wang, H.; Bi, C.; Wang, Y.; Sun, J.; Meng, X.; Li, J. Selenium ameliorates *Staphylococcus aureus*-induced inflammation in bovine mammary epithelial cells by inhibiting activation of TLR2, NF-κB and MAPK signaling pathways. *BMC Vet. Res.* **2018**, *14*, 197. [CrossRef]
39. Ulasov, A.V.; Rosenkranz, A.A.; Georgiev, G.P.; Sobolev, A.S. Nrf2/Keap1/are signaling: Towards specific regulation. *Life Sci.* **2022**, *291*, 120111. [CrossRef]
40. Wen, Y.; Kang, X.; Li, Z.; Xia, L.; Lu, Y. Identification of a secretory heme-binding protein from *Nocardia seriolae* involved in cell apoptosis. *J. Fish Dis.* **2022**, *45*, 1189–1199. [CrossRef]
41. Chen, J.; Xia, L.; Wang, W.; Wang, Z.; Hou, S.; Xie, C.; Cai, J.; Lu, Y. Identification of a mitochondrial-targeting secretory protein from Nocardia seriolae which induces apoptosis in fathead minnow cells. *J. Fish Dis.* **2019**, *42*, 1493–1507. [CrossRef] [PubMed]
42. Wang, Y.; Liu, B.B.; Wu, P.X.; Chu, Y.; Gui, S.S.; Zheng, Y.Z.; Chen, X.D. Dietary selenium alleviated mouse liver oxidative stress and NAFLD induced by obesity by regulating the KEAP1/NRF2 pathway. *Antioxidants* **2022**, *11*, 349. [CrossRef] [PubMed]
43. Miranda, S.G.; Purdie, N.G.; Osborne, V.R.; Coomber, B.L.; Cant, J.P. Selenomethionine increases proliferation and reduces apoptosis in bovine mammary epithelial cells under oxidative stress. *J. Dairy Sci.* **2011**, *94*, 165–173. [CrossRef] [PubMed]
44. Wang, H.; Cong, X.; Qin, K.; Yan, M.K.; Xu, X.F.; Liu, M.K.; Xu, X.; Zhang, Y.; Gao, Q.Y.; Cheng, S.Y.; et al. Se-enriched cardamine violifolia improves laying performance and regulates ovarian antioxidative function in aging laying hens. *Antioxidants* **2023**, *12*, 450. [CrossRef]
45. Chen, Y.; Zhang, X.; Yang, J.; Feng, W.; Deng, G.; Xu, S.; Guo, M. Extracellular vesicles derived from selenium-deficient MAC-T cells aggravated inflammation and apoptosis by triggering the endoplasmic reticulum (ER) stress/PI3K-AKT-mTOR pathway in bovine mammary epithelial cells. *Antioxidants* **2023**, *12*, 2077. [CrossRef]
46. Xie, L.; Xu, Y.; Ding, X.; Li, K.; Liang, S.; Li, D.; Wang, Y.; Fu, A.; Yu, W.; Zhan, X. Selenomethionine attenuated H2O2-induced oxidative stress and apoptosis by Nrf2 in chicken liver cells. *Antioxidants* **2023**, *12*, 1685. [CrossRef]
47. *CLSI M07*; Methods for Dilution Antimicrobial Susceptibility Tests for Bacteria That Grow Aerobically, 12th edtion. CLSI: Berwyn, PA, USA, 2024.

48. Ross, J.E.; Scangarella-Oman, N.E.; Flamm, R.K.; Jones, R.N. Determination of disk diffusion and MIC quality control guidelines for GSK2140944, a novel bacterial type II topoisomerase inhibitor antimicrobial agent. *J. Clin. Microbiol.* **2014**, *52*, 2629–2632. [CrossRef]
49. Liu, Y.; Zhou, M.; Xu, S.; Khan, M.A.; Shi, Y.; Qu, W.; Gao, J.; Liu, G.; Kastelic, J.P.; Han, B. *Mycoplasma bovis*-generated reactive oxygen species and induced apoptosis in bovine mammary epithelial cell cultures. *J. Dairy Sci.* **2020**, *103*, 10429–10445. [CrossRef]

Disclaimer/Publisher's Note: The statements, opinions and data contained in all publications are solely those of the individual author(s) and contributor(s) and not of MDPI and/or the editor(s). MDPI and/or the editor(s) disclaim responsibility for any injury to people or property resulting from any ideas, methods, instructions or products referred to in the content.

International Journal of
Molecular Sciences

MDPI

Article

Molecular Mechanism of 5,6-Dihydroxyflavone in Suppressing LPS-Induced Inflammation and Oxidative Stress

Yujia Cao [1], Yee-Joo Tan [2] and Dejian Huang [1,3,*]

1 Department of Food Science and Technology, National University of Singapore, Singapore 117542, Singapore; yujia.cao@u.nus.edu
2 Infectious Diseases Translational Research Program, Department of Microbiology and Immunology, Yong Loo Lin School of Medicine, National University of Singapore, Singapore 117545, Singapore; mictyj@nus.edu.sg
3 National University of Singapore (Suzhou) Research Institute, 377 Linquan Street, Suzhou 215123, China
* Correspondence: dejian@nus.edu.sg; Tel.: +65-82836618

Abstract: 5,6-dihydroxyflavone (5,6-DHF), a flavonoid that possesses potential anti-inflammatory and antioxidant activities owing to its special catechol motif on the A ring. However, its function and mechanism of action against inflammation and cellular oxidative stress have not been elucidated. In the current study, 5,6-DHF was observed inhibiting lipopolysaccharide (LPS)-induced nitric oxide (NO) and cytoplasmic reactive oxygen species (ROS) production with the IC_{50} of 11.55 ± 0.64 μM and 0.8310 ± 0.633 μM in murine macrophages, respectively. Meanwhile, 5,6-DHF suppressed the overexpression of pro-inflammatory mediators such as proteins and cytokines and eradicated the accumulation of mitochondrial ROS (mtROS). The blockage of the activation of cell surface toll-like receptor 4 (TLR4), impediment of the phosphorylation of c-Jun N-terminal kinase (JNK) and p38 from the mitogen-activated protein kinases (MAPK) pathway, Janus kinase 2 (JAK2) and signal transducer and activator of transcription 3 (STAT3) from the JAK-STAT pathway, and p65 from nuclear factor-κB (NF-κB) pathways were involved in the process of 5,6-DHF suppressing inflammation. Furthermore, 5,6-DHF acted as a cellular ROS scavenger and heme-oxygenase 1 (HO-1) inducer in relieving cellular oxidative stress. Importantly, 5,6-DHF exerted more potent anti-inflammatory activity than its close structural relatives, such as baicalein and chrysin. Overall, our findings pave the road for further research on 5,6-DHF in animal models.

Keywords: 5,6-dihydroxyflavone; flavonoids; anti-inflammation; oxidative stress; signaling pathway

1. Introduction

Inflammation is a complicated immune response to defense noxious foreign agents and internal abnormal conditions [1]. Inflammation-derived oxidative stress that basically attributes to the reactive oxygen species (ROS) leads to dangerous damages on cellular essential components such as lipids, proteins, and DNA and further aggravates the pathological inflammation [2,3]. Also, the positive feedback loop between inflammation and oxidative stress has been established [4].

Flavonoids are an important family of polyphenolic compounds belonging to plant secondary metabolites and widely distributed in fruits, vegetables, and medicinal plants [5]. Flavonoids have been reported to have a variety of health-promoting effects, including antioxidant, anti-inflammatory, and anti-tumor properties [6]. In particular, flavonoids have strong influence on the inflammatory processes with diverse identified mechanisms, such as the nuclear factor kappa-B (NF-κB) pathway, the mitogen-activated protein kinase (MAPK) pathway, and the Janus kinase (JAK)-signal transducer and activator of the transcription (STAT) pathway, that are attributable to their structural and functional diversity [6–8]. For example, catecholic moieties played an indispensable role as a functional group for the anti-inflammatory activity of flavonoids [9]. As a class of polyphenols, flavonoids widely receive attention for their antioxidant functions as radical scavengers [10]. Apart from numerous flavonoids isolated from natural resources, chemically synthesized or structurally modified flavonoids have also promoted the development of research on the biological and pharmacological activity of flavonoids to a certain extent [11].

5,6-dihydroxyflavone (5,6-DHF), a rare flavonoid aglycone found in nature, is obtained commercially through chemical synthesis as there are few reports on its natural occurrence [12]. It has been found in rat urine after oral administration of the traditional Chinese medicine formula Xiao-Ai-Jie-Du, decoction which is prescribed to reduce fever, detoxify, and resolve phlegm and congestion [13]. Possessing the catechol motif at the A ring (Figure 1A), 5,6-DHF showed radical scavenging activity with a Trolox equivalent antioxidant capacity (TEAC) of 0.98 ± 0.05 μM [14]. Results from intracellular fluorescent probes and electron paramagnetic resonance spectroscopy revealed the intercellular antioxidant effect of 5,6-DHF by eliminating ROS in L-6 and THP-1 cells at concentrations as low as 100 pM [15]. Additionally, its potential utilization in treating Alzheimer's disease, a neuroinflammatory disease, was demonstrated with a half-maximal inhibitory concentration (IC_{50}) of 4.51 ± 0.03 μM on acetylcholinesterase (AChE), which is a cholinergic enzyme in the nervous system [14]. Collectively, 5,6-DHF holds high potency in modulating cellular inflammatory responses and oxidative stress due to its anti-inflammatory and antioxidant function, yet its performance on immune cells in inflammatory conditions and its underlying mechanisms of action have nevertheless not been elucidated. Herein, we investigated the effects of 5,6-DHF on lipopolysaccharide (LPS)-induced inflammation and oxidative stress in RAW 264.7 cells and elucidated its mechanisms of action.

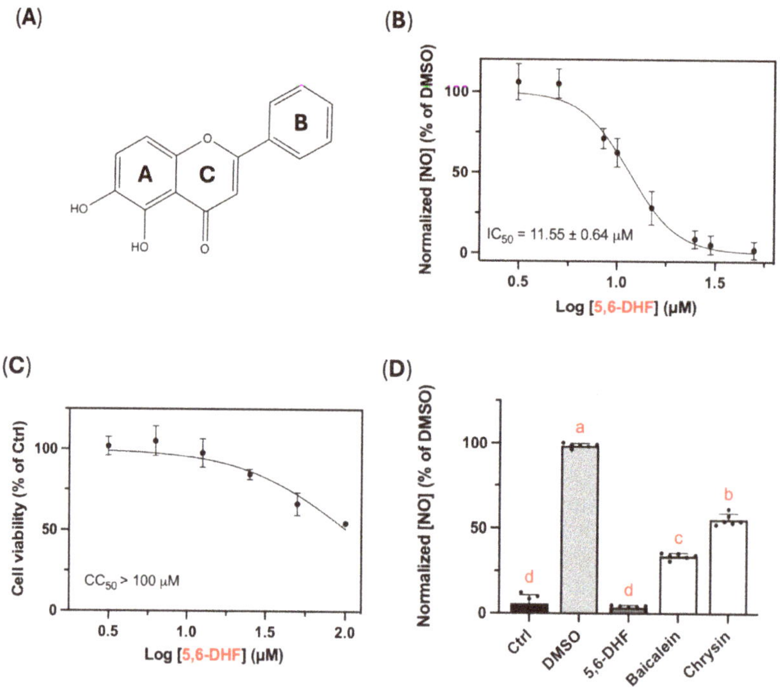

Figure 1. Chemical structure of 5,6-DHF (**A**). Dose response curves of 5,6-DHF on LPS-induced NO production (**B**) and cell cytotoxicity of 5,6-DHF on RAW 264.7 cells (**C**). Comparison of the anti-NO activity of 5,6-DHF and its structural analogues at 25 μM (**D**). The values shown are the mean ± SD of three independent experiments in duplicate. Different letters indicate statistical significance ($p < 0.05$).

2. Results

2.1. Effect of 5,6-DHF on LPS-Activated Nitric Oxide Production

To preliminarily screen the inhibitory effects of 5,6-DHF on LPS-activated inflammation in RAW 264.7 cells, the nitric oxide (NO) concentration in the cell culture medium was measured. Figure 1B showed that 5,6-DHF significantly attenuated the secreted NO from LPS-induced RAW 264.7 cells in a dose-dependent manner. The preliminary results demonstrated 5,6-DHF was a potential anti-inflammatory candidate that inhibited LPS-induced NO production with an IC_{50} of 11.55 ± 0.64 μM (Figure 1B). In addition, 5,6-DHF slightly impacted cell growth in a dose-dependent manner, but the cytotoxic concentration (CC_{50}) was greater than 100 μM (Figure 1C). Interestingly, compared with its structural analogues, baicalein and chrysin, 5,6-DHF exerted significantly high potency on the suppression of NO at the concentration of 25 μM, which is about twice the IC_{50} of 5,6-DHF.

2.2. Suppressive Effects of 5,6-DHF on Key Inflammatory-Related Proteins and Pro-Inflammatory Cytokines

To further confirm the anti-inflammatory effect of 5,6-DHF, we assessed its impact on inflammatory-related proteins including cyclooxygenase-2 (COX-2), inducible nitric oxide synthase (iNOS), and toll-like receptor 4 (TLR4) at its IC_{50} against NO production and twice the amount of IC_{50} using Western blotting (Figure 2A). We found 5,6-DHF could not alleviate the LPS-induced expression of COX-2 in RAW 264.7 cells (Figure 2B). In contrast, 5,6-DHF significantly hindered the expression of iNOS (Figure 2C). We also evaluated the role that 5,6-DHF played in the interaction between LPS and cell surface recognitive receptors. Figure 2D indicates that 5,6-DHF could significantly block the LPS-induced activation of TLR4 expression on the cell surface, therefore protecting RAW 264.7 cells from stimulation by foreign agents.

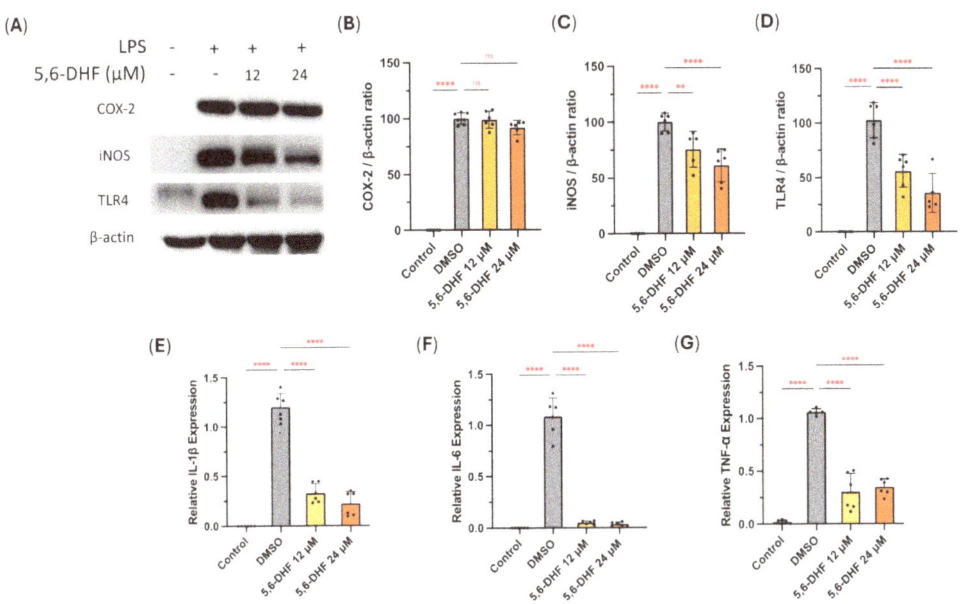

Figure 2. The suppressive effects of 5,6-DHF on pro-inflammatory protein expression in LPS-stimulated RAW 264.7 models (**A**). The expression levels of COX-2 (**B**), iNOS (**C**), and TLR4 (**D**) were determined by Western blot. The inhibitory effects of 5,6-DHF on mRNA expression levels of IL-1β (**E**), IL-6 (**F**), and TNF-α (**G**) tested by qRT-PCR. Data points and bar represent arithmetic means ± SD. Ns, not significant. ** $p < 0.01$ and **** $p < 0.0001$ as compared to DMSO group.

Furthermore, the alterations of three predominant pro-inflammatory cytokines, interleukin (IL)-1β, IL-6, and tumor necrosis factor-alpha (TNF-α), were analyzed at 5,6-DHF's IC_{50} against NO production and twice the amount of IC_{50}. The mRNA expression of these pro-inflammatory cytokines in RAW 264.7 macrophages dramatically increased after LPS stimulation. However, the treatment of 5,6-DHF with cells at both 12 and 24 μM could significantly resolve this phenomenon (Figure 2E–G).

2.3. Effects of 5,6-DHF on MAPK Pathway

The activation of MAPKs is recognized by the phosphorylation of three families of proteins, namely extracellular signal-regulated kinase 1/2 (ERK1/2), p38 MAPK, and c-Jun N-terminal kinase (JNK) [16]. As Figure 3A–D illustrates, LPS triggered a significant phosphorylation of ERK1/2, p38, and JNK as compared with the non-treated control group, yet 5,6-DHF downregulated the expression level of phosphorylated typed p38 and JNK. By calculating the ratio of phosphorylated MAPKs to their prototype ones, we noticed 5,6-DHF obstructed the phosphorylation process of p38 and JNK, thereby blocking the detrimental actions of MAPKs in the inflammatory response cascade (Figure 3E–G).

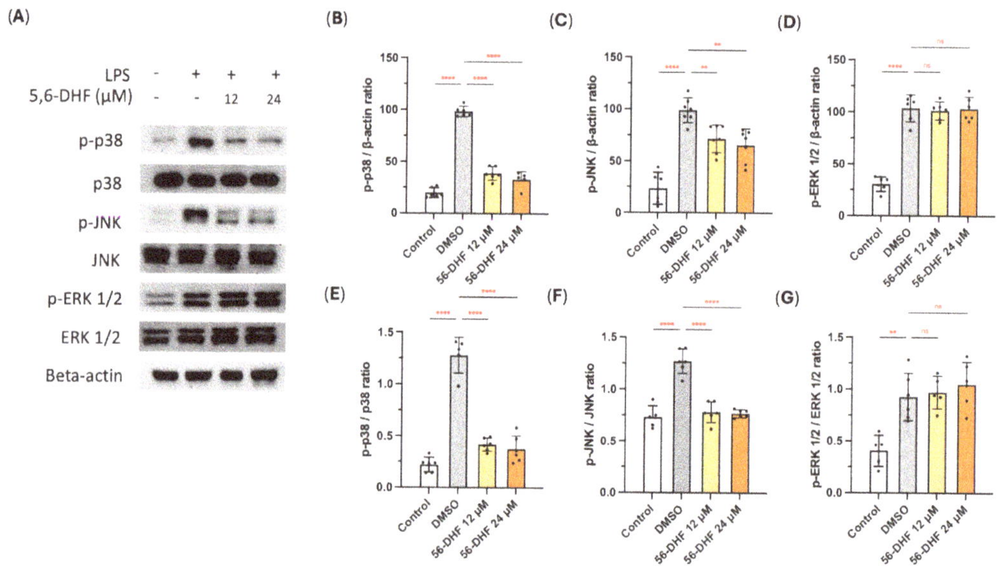

Figure 3. Effects of 5,6-DHF on MAPK pathway in LPS-induced RAW 264.7 cells (**A**). Suppressive effects of 5,6-DHF on the LPS-induced p-p38 (**B**), p-JNK, (**C**) and p-ERK1/2 (**D**) expressions and the phosphorylation level of p38 (**E**), JNK (**F**), and ERK1/2 (**G**). All expressions were normalized to that of DMSO treatment. Data points and bar represent arithmetic means ± SD. Ns, not significant. ** $p < 0.01$ and **** $p < 0.0001$ as compared to DMSO group.

2.4. Effects of 5,6-DHF on JAK-STAT Pathway

The influence of 5,6-DHF on the JAK-STAT pathway was analyzed by measuring the phosphorylation expression of key proteins (Figure 4A). Cells stimulated by LPS showed an obvious increment in the expression of phosphorylated JAK2 and STAT3 levels. 5,6-DHF reversed this trend in a concentration-dependent manner (Figure 4B,C). In comparison with the untreated group, 5,6-DHF significantly reduced the LPS-induced phosphorylation ratios of JAK2 and STAT3 (Figure 4D,E).

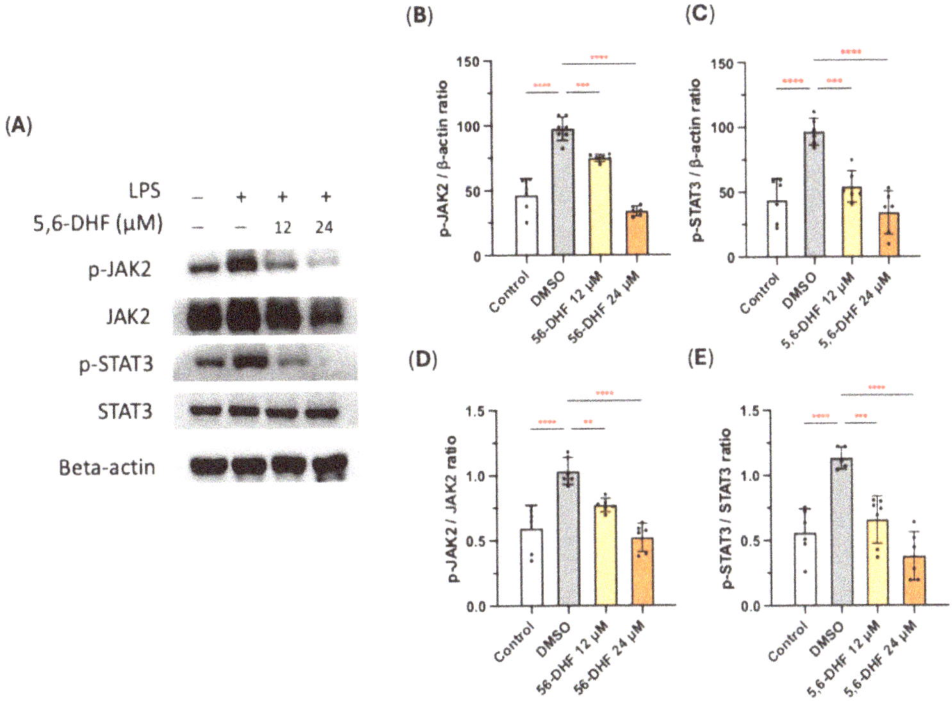

Figure 4. Effects of 5,6-DHF on JAK-STAT pathway in LPS-induced RAW 264.7 cells (**A**). Suppressive effects of 5,6-DHF on the LPS-induced p-JAK2 (**B**) and p-STAT3 (**C**) expression and the phosphorylation level of JAK2 (**D**) and STAT3 (**E**). All expressions were normalized to that of DMSO treatment. Data points and bar represent arithmetic means ± SD. Ns, not significant. ** $p < 0.01$, *** $p < 0.001$, and **** $p < 0.0001$ as compared to DMSO group.

2.5. Effects of 5,6-DHF on NF-κB Pathway

The phosphorylation of the p65 (p-p65) subunit belonging to the NF-κB family in particular marks NF-κB activation by translocating to the nucleus and triggering target gene expressions. To investigate the influence of 5,6-DHF on the NF-κB pathway, the expression and phosphorylation level of p65 were determined (Figure 5). There was no difference in activation levels between cells incubated with a low dose (12 μM) of 5,6-DHF and those incubated with DMSO, whereas a higher dose (24 μM) of 5,6-DHF successfully hindered the activation of NF-κB, showing significantly suppressive effects on the p-p65 expression and phosphorylation level of p65 (Figure 5B,C).

2.6. Effect of 5,6-DHF on LPS-Induced Cytoplasmic and Mitochondrial ROS

Since 5,6-DHF is a radical scavenger, its capability to inhibit LPS-induced oxidative stress in RAW 264.7 cells was first investigated by measuring intracellular ROS levels that predominantly originated from NADPH oxidases (Figure 6A). Our previous study showed 1 μM flavones could reduce intracellular ROS levels caused by tert-butyl hydroperoxide (tBHP) [17]. We observed that LPS led to intensive accumulation of ROS signal in the intercellular region of cells treated with DMSO, while cells incubated with 1 μM 5,6-DHF remarkedly attenuated cytoplasmic ROS signal (Figure 6A). Meanwhile, the 5 μM 5,6-DHF group exhibited a negligible ROS signal compared to that of the positive control group, demonstrating its ROS scavenging activity. Dose-dependent manner of 5,6-DHF on scavenging LPS-induced ROS was revealed, and its IC_{50} value was 0.8310 ± 0.633 μM (Figure 6B).

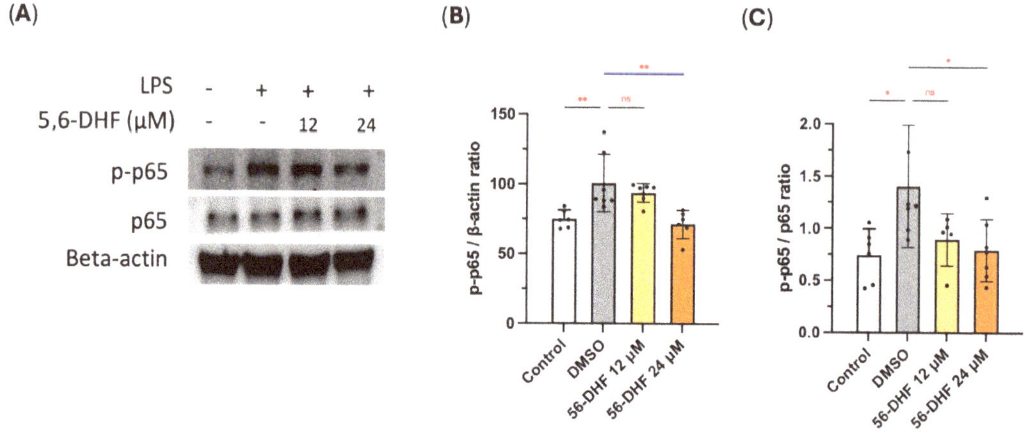

Figure 5. Effects of 5,6-DHF on NF-κB pathway in LPS-induced RAW 264.7 cells (**A**); suppressive effects of 5,6-DHF on the LPS-induced p-p65 expression (**B**); and the phosphorylation level of p65 (**C**). All expressions were normalized to that of DMSO treatment. Data points and bar represent arithmetic means ± SD. Ns, not significant. * $p < 0.05$ and ** $p < 0.01$ as compared to DMSO group.

In addition to cytoplasmic ROS, the alteration of mitochondrial ROS (mtROS) characterized by superoxide in live cells was also visualized by a redox-sensitive fluorescent probe, MitoSOX red. LPS produced a strong oxidative stress in the DMSO group in RAW 264.7 cells, while the mtROS content in the 12 μM 5,6-DHF group was significantly lower, indicating that 5,6-DHF attenuated the overproduction of mtROS (Figure 6C).

2.7. 5,6-DHF Exhibited Antioxidant Activity by Inducing HO-1 Expression

To understand the other mechanism of 5,6-DHF suppressing cellular oxidative stress, the impact of 5,6-DHF on heme oxygenase 1 (HO-1) expression was assessed in RAW 264.7 cells. Remarkably, 5,6-DHF significantly induced the generation of HO-1 at both tested concentrations, suggesting its function in counteracting LPS-caused cellular oxidative stress in macrophages (Figure 6D).

Taken together, the overall mechanisms of 5,6-DHF on relieving LPS-induced inflammatory response and oxidative stress are summarized in Figure 7. Specifically, 5,6-DHF inhibited LPS-induced activation of TLR4 and phosphorylation of p38 and JNK in the MAPK pathway, JAK2 and STAT3 in the JAK-STAT pathway, and p65 in the NF-κB pathways. Its role in suppressing cytokine expression, including IL-1β, IL-6, and TNF-α, and downregulating downstream pro-inflammatory mediators such as iNOS and NO in murine macrophages is also attributed to its anti-inflammatory activity. The reduction of LPS-induced cellular oxidative stress by 5,6-DHF was manifested by scavenging cytoplasmic ROS and mtROS. The induction of HO-1 by 5,6-DHF was also responsible for protecting cells from oxidative damage.

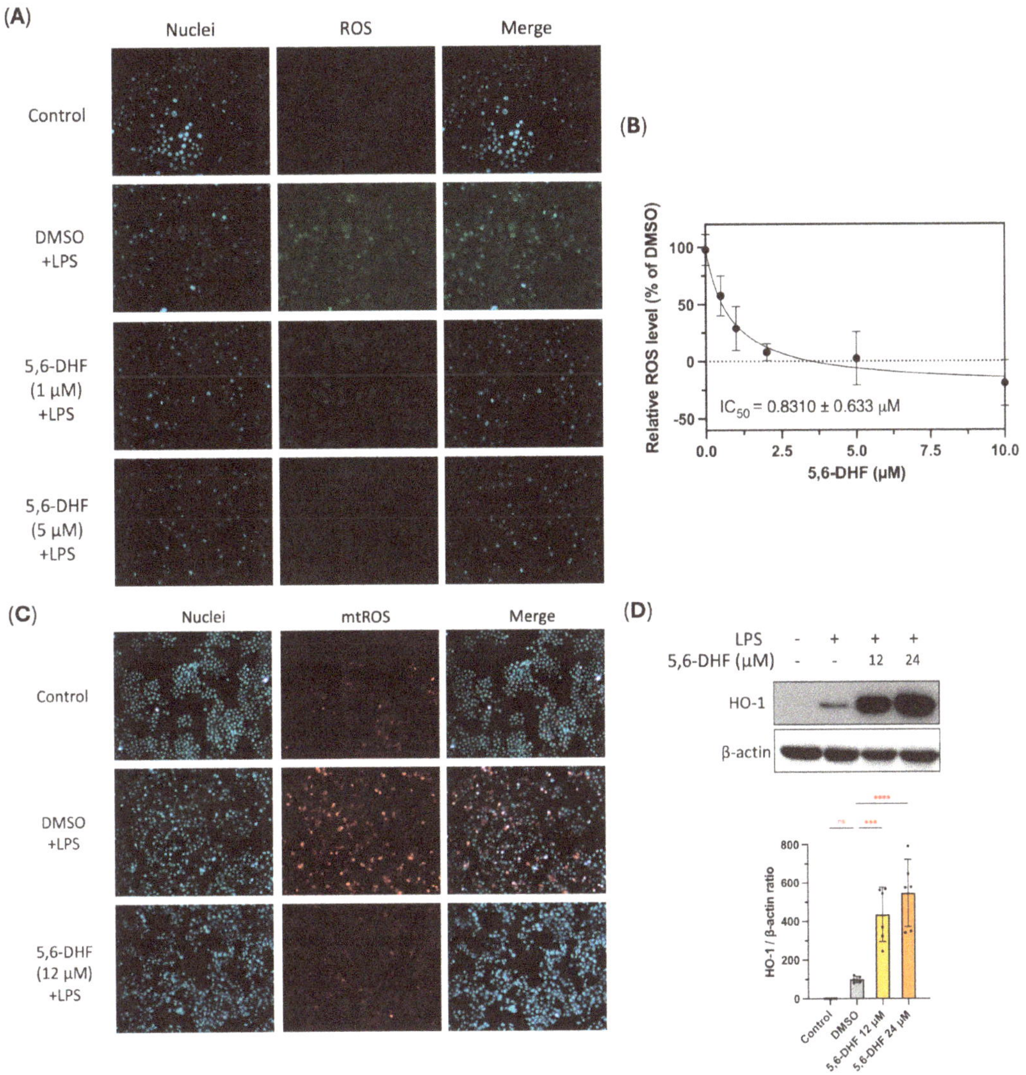

Figure 6. Cellular ROS-scavenging activity of 5,6-DHF on LPS-induced RAW264.7 cells, images were captured by fluorescent microscope (**A**); blue: cell nuclei stained by H33342; and green: cellular ROS stained by H_2DCFDA. Dose-response curve of ROS-scavenging activity of 5,6-DHF; scale bar 20 μm (**B**). Mitochondrial ROS-scavenging activity of 5,6-DHF on LPS-induced RAW264.7 cells (**C**). The fluorescence was visualized by fluorescent microscopy (blue: cell nuclei stained by H33342; and red: mitochondrial ROS stained by MitoSOX red). The influence of 5,6-DHF on the expression levels of HO-1; scale bar 20 μm (**D**) Data points and bar represent arithmetic means ± SD. Ns, not significant. *** $p < 0.001$ and **** $p < 0.0001$ as compared to DMSO group.

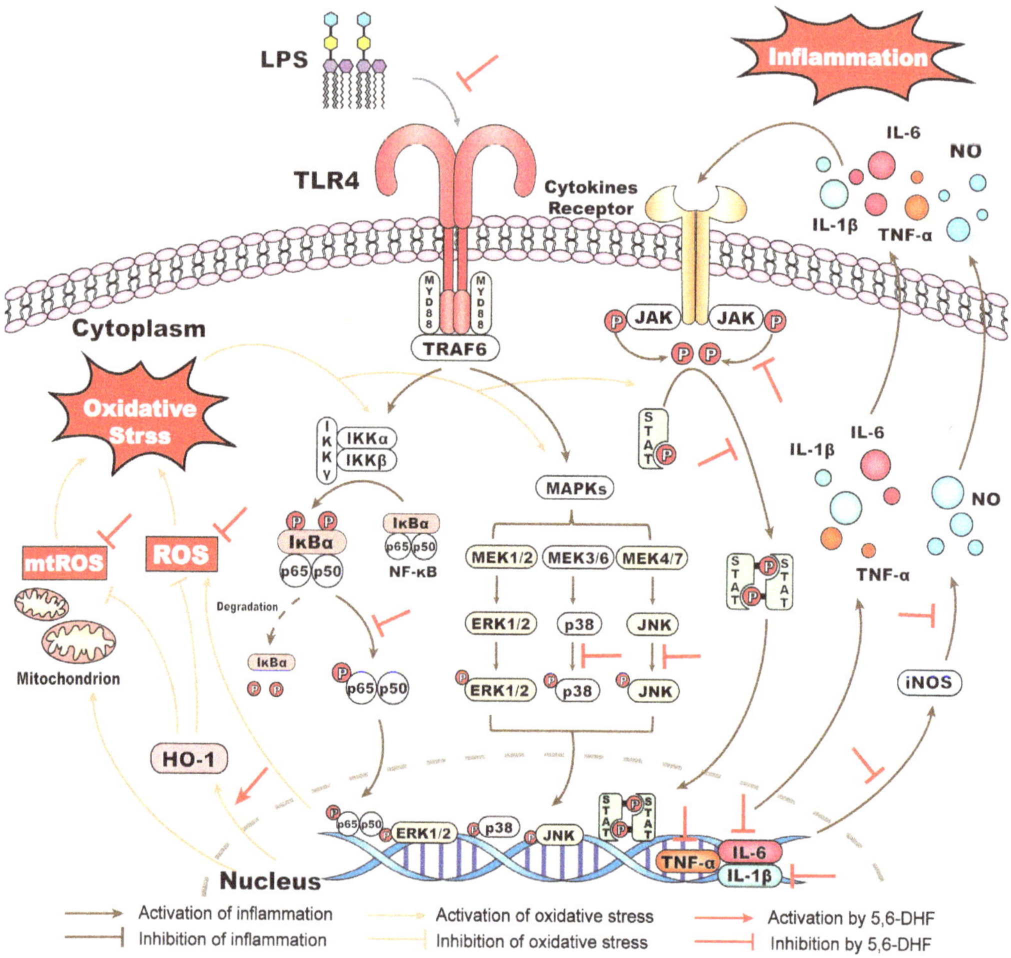

Figure 7. Schematic diagram of 5,6-DHF mechanisms in inhibiting LPS-induced inflammatory responses and oxidative stress in RAW 264.7 cells.

3. Discussion

Inflammation disturbs the intercellular redox equilibrium, resulting in uncontrollable oxidation of pivotal cellular components and severe inflammation [4]. Flavonoids, as a class of polyphenolic compounds, have earned a favorable reputation regarding their anti-inflammatory and antioxidant properties [18]. Particularly, the catechol moiety at the A ring renders 5,6-DHF a promising contribution to resolving inflammation and eradicating cellular oxidative radicals [19].

The inflammatory mediator NO is a clinical biomarker of several inflammatory diseases [20]. It is biosynthesized under the catalysis of iNOS under harmful stimulation. 5,6-DHF was discovered to attenuate the expression of iNOS, thus prohibiting the production of NO. However, unlike commonly prescribed conventional nonsteroidal anti-inflammatory drugs (NSAIDs) that target COX-2, 5,6-DHF could not diminish COX-2 expression level in RAW 264.7 cells. In addition, dysregulated pro-inflammatory cytokine production contributes to the pathogenesis of inflammation-related disorders like neurodegenerative diseases, cardiovascular disease, and cancer [21–23]. We observed macrophage-derived pro-inflammatory cytokines, such as IL-1β, IL-6, and TNF-α, tumbled down with the co-treatment of 5,6-DHF under LPS-activated circumstances, therefore relieving inflammation.

Surface receptor, TLR4, specifically recognizes LPS upon stimulation and triggers the activation of molecular determinants in the LPS-induced signaling cascade, including MAPK, JAK-STAT, and NF-κB pathways, resulting in the translocation of mediating factors to the nucleus, where they activate gene transcription and signal transduction [24–27]. Importantly, MAPK, JAK-STAT, and NF-κB pathways are different, but they interconnect in regulating cellular responses to stimulants such as LPS. Preventing the interaction of LPS and TLR4, 5,6-DHF diminished the activation of surface TLR4 and restrained the following cellular responses to LPS stimulation. The inflammatory signal transduction is marked principally by protein phosphorylation as post-translational modification (PTM) [28]. The phosphorylation levels of predominant markers in MAPK pathway, JAK-STAT pathway, and NF-κB pathway were downregulated by 5,6-DHF treatment, therefore stopping unwanted inflammatory responses.

It has been established that chronic oxidative stress is a causative factor for aging and age-related diseases such as type II diabetes and chronic kidney disease [29–31]. The production of ROS, including superoxide, hydrogen peroxide, and hydroxyl radicals commonly happens in the cytoplasm, mitochondria, and endoplasmic reticulum [28]. Mitochondrion, the energy-producing organelle, contributes the most to the cellular ROS level through the electron transport chain [32]. And ROS derived by electron transfer in the NADPH-oxidase (NOX) bound to cytoplasm membranes is the other prime source for oxidative stress [33]. By clearing both cytoplasmic ROS and mtROS, 5,6-DHF could eliminate oxidative stress induced by LPS and impede unfavorable cell damage.

The attenuated cytoplasmic ROS and mtROS levels could be credited to the radical scavenging property of 5,6-DHF and its regulatory roles on cellular expression of proteins related to inflammation. Under the condition of LPS stimulation, HO-1, an enzyme that is involved in diverse protective actions for redox balance, was considerably expressed in response to 5,6-DHF treatment in the fight against oxidative stress. Given that excessive ROS is an inducer of inflammation, the cellular oxidative stress orchestrates MAPK, JAK-STAT, and NF-κB pathways and then leads to the production of pro-inflammatory cytokines [34–36]. Thus, the impact of 5,6-DHF on these signaling pathways could also hinder the activation of inflammation caused by oxidative stress.

It is worth pointing out that antioxidant and anti-inflammatory studies of flavonoids primarily lie in flavonoids found in nature because they can be safely consumed as coexisting with the food containing them. Nevertheless, they are only generally safe in low concentrations in natural foods since isolated and concentrated flavonoid compounds have shown cytotoxicity and elevated cellular ROS levels in normal human cells, which hinders their long-term application and weakens their safety advantages [37]. Flavonoids are privileged structural motif for medicinal molecules. In particular, chemical modifications of the structures of flavonoids sheds light on their structure-activity relationship (SAR) [9]. Compared to its natural structural analogues, baicalein and chrysin, 5,6-DHF with modification on the hydroxyl groups in the A ring had the highest potency in blocking inflammatory responses in RAW 264.7 cells. This pinpointed the importance of the catechol motif (5,6-dihydroxyl group) in the A-ring of flavones for their bioactivity. The finding of the predominant role of catechol motif was consistent with our previous study [8]. It is worth mentioning that 5,6-DHF had a CC_{50} greater than 100 μM on RAW 264.7 cells and did not show significant cytotoxicity on myoblasts and monocytes at the concentration that 5,6-DHF exerted its antioxidant effect [15].

To take advantage of the biological and pharmaceutical functions of flavonoids, extensive explorations of structurally different flavonoids will contribute to the optimization of drug or dietary supplement candidates. Also, the combinational use of flavonoids or combining flavonoids with conventional therapeutic agents should be taken into consideration to improve therapeutic efficacy and drug resistance [38–40].

4. Materials and Methods

4.1. Materials

5,6-dihydroxyflavone (purity: 98%) was purchased from Indofine Chemical Co., Inc., Hilsborough, NJ, USA. Chrysin (5,7-dihydroxyflavone, purity: 98%) was supplied by Shanghai Yien Chemical Technology Co., Ltd. (Shanghai, China). Baicalein (5,6,7-trihydroxyflavone, purity: 98%), LPS (*Escherichia coli* serotype 055: B5), paraformaldehyde (PFA), 4′,6-diamidino-2-phenylindole (DAPI), dichlorofluorescin diacetate (DCFDA) (purity $\geq$ 97%), bisbenzimide H33342 (purity $\geq$ 97%), and dimethyl sulfoxide (DMSO) were obtained from Sigma-Aldrich Co., Ltd. (Singapore). Antibodies used for Western blot were purchased from Cell Signaling Technology Inc. (Danvers, MA, USA). Primers were purchased from Integrated DNA Technologies Pte. Ltd. (Singapore).

4.2. Cell Culture

The macrophage cell line RAW 264.7 was obtained from the American Type Culture Collection (ATCC). Cells were cultured with Dulbecco's Modified Eagle Medium (DMEM) supplemented with 10% fetal bovine serum (FBS) and 1% penicillin–streptomycin antibiotics and maintained in the 37 °C incubator with 5% CO_2.

4.3. Cytotoxicity Test

RAW 264.7 cells were seeded in a 96-well plate at a density of 3×10^4 cells/well and incubated overnight. Until wells filled with a monolayer, compounds were co-incubated with cells for 24 h. Afterwards, the cell viability was determined with a fresh mixture of cell counting kit 8 (Dojindo Molecular Technologies, Kumamoto, Japan) and DMEM (1:10, v/v) [9]. The absorbance at 450 nm was measured using the microplate reader (Tecan Infinite F200, Mannedorf, Switzerland). The cell viability was calculated as a percentage of the control group.

4.4. Quantitation of Nitric Oxide Concentration

RAW 264.7 cells were pretreated with 5,6-DHF at concentrations of 1, 5, 8.5, 10, 15, 25 and 30 μM for 1 h before the stimulation by LPS (100 ng/mL). After a 23 h incubation, the nitric oxide concentrations were quantified by Griess reagent (Promega, Singapore) according to the manufacturer's instructions [9]. The absorbance was measured in a microplate reader at 540 nm using Tecan Infinite F200 (Tecan, Mannedorf, Switzerland). The normalized NO concentrations were calculated as a ratio to that of the DMSO group.

4.5. Western Blotting Analysis

RAW 264.7 cells were seeded in a 6-well plate at the density of 1×10^6 cells/well and incubated overnight until cells formed a monolayer. After stimulation with LPS, cells were washed and lysed using the radio-immunoprecipitation assay buffer (Thermo Fisher Scientific, Singapore) at 4 °C for 15 min. Western blotting was conducted as reported before [9]. The blots were imaged by ChemiDoc Imaging System (Bio-Rad, Singapore) with Amersham ECL (Cytiva) or WesternBright™ Sirius (Advansta Inc., San Jose, CA, USA) and quantified by ImageJ 1.53a. All protein expressions were normalized to that of the DMSO group.

4.6. Real-Time Quantitative Polymerase Chain Reaction (qRT-PCR)

Total RNA of treated RAW 264.7 cells was extracted by Trizol reagent (Thermo Fisher Scientific) according to its protocols. The qRT-PCR process was conducted based on the reported method [17]. Primer sequences used were listed in Table 1. The RNA expression levels were analyzed by the $2^{-\Delta\Delta C_T}$ method with β-actin as the internal standard. All gene expressions were normalized to that of the DMSO group.

Table 1. Primer sequences for RT-qPCR.

Gene	Forward Primer	Reverse Primer
β-actin	5′-CCACAGCTGAGAGGGAAATC-3′	5′-AAGGAAGGCTGGAAAAGAGC-3′
IL-1β	5′-GGGCCTCAAAGGAAAGAATC-3′	5′-TACCAGTTGGGGAACTCTGC-3′
IL-6	5′-AGTTGC CTTCTTGGGACTGA-3′	5′-CAGAATGCCATTGCACAAC-3′
TNF-α	5′-AGCCCCCAGTCTGTATCCTT-3′	5′-CATTCGAGGCTCCAGTGAAT-3′

4.7. Determination of Cytoplasmic ROS and Mitochondrial ROS Levels

The cellular antioxidant activities were analyzed by reacting the cytoplasmic ROS with a redox-sensitive probe H_2DCFDA. Macrophages monolayer in a 96-well plate was pretreated with 5,6-DHF at indicated concentrations for 1 h before the 23 h stimulation of LPS (100 ng/mL). Afterwards, H_2DCFDA (10 μM) was incubated with cells at 37 °C for 15 min in the dark [41]. The endpoint fluorescence was measured at the wavelengths of 485 nm (excitation) and 535 nm (emission) using the BioTek Cytation 5 Cell Imaging Multimode Reader (Agilent, Santa Clara, CA, USA) after washing with PBS.

Cells cultured on coverslips in a 24-well plate were subjected to the same treatment as the 96-well plate. The cell nuclei were stained with 10 μg/mL Hochest 33342 for 15 min. After washing to remove excessive dye, the cytoplasmic and mitochondrial ROS levels were determined by incubating cells with 10 μM H_2DCFDA and 5 μM MitoSOX™ (Thermo Fisher Scientific) for 15 min in the incubator, respectively [42]. Glass coverslips were then transferred to a microscope slide and imaged under an Olympus CKX53 inverted microscope (EVIDENT, Tokyo, Japan) with an integrated 100 W mercury lamp (BH2-RFL-T2, Ushio, Tokyo, Japan) in the dark room. The images were captured and analyzed by EVIDENT cell Sens Entry 1.18.

4.8. Data Analysis

The multi-group data comparison was analyzed by Prism 10 using one-way analysis of variance (ANOVA). * $p < 0.05$ was considered statistically significant.

5. Conclusions

5,6-DHF blocked the activation cell surface stimuli receptor, hindered phosphorylation steps in MAPK, JAK-STAT, and NF-κB pathways, and downregulated pro-inflammatory mediators such as proteins and cytokines in the LPS-induced murine macrophage model. Its antioxidant function was revealed to act as a scavenger to eliminate cytoplasmic ROS and mitochondrial ROS and consequently alleviate oxidative stress. The overexpression of HO-1 and attenuation of inflammatory determinants by 5,6-DHF also contribute to the cellular redox balance. However, the specific cellular binding target of 5,6-DHF is still unclear and requires further investigation. Our findings demonstrated the potential application of 5,6-DHF as an anti-inflammatory and antioxidant agent and provide evidence for future investigation in animal models and humans.

Author Contributions: Conceptualization, Y.C. and D.H.; methodology, Y.C. and D.H.; software, Y.C.; validation, Y.C., Y.-J.T. and D.H.; formal analysis, Y.C.; investigation, Y.C.; resources, Y.C. and D.H.; data curation, Y.C.; writing—original draft preparation, Y.C.; writing—review and editing, Y.C. and D.H.; visualization, Y.C.; supervision, Y.-J.T. and D.H.; project administration, D.H.; funding acquisition, Y.C. and D.H. All authors have read and agreed to the published version of the manuscript.

Funding: This project is funded by the Singapore Ministry of Education with a scholarship to Yujia Cao and a tier-one grant (A-8000983-01-00) for financial support.

Institutional Review Board Statement: Not applicable.

Informed Consent Statement: Not applicable.

Data Availability Statement: All data presented this study are available from the corresponding author upon responsible request.

Conflicts of Interest: The authors declare no conflicts of interest.

References

1. Gusev, E.; Zhuravleva, Y. Inflammation: A new look at an old problem. *Int. J. Mol. Sci.* **2022**, *23*, 4596. [CrossRef] [PubMed]
2. Lugrin, J.; Rosenblatt-Velin, N.; Parapanov, R.; Liaudet, L. The role of oxidative stress during inflammatory processes. *Biol. Chem.* **2014**, *395*, 203–230. [CrossRef] [PubMed]
3. Leyane, T.S.; Jere, S.W.; Houreld, N.N. Oxidative stress in ageing and chronic degenerative pathologies: Molecular mechanisms involved in counteracting oxidative stress and chronic inflammation. *Int. J. Mol. Sci.* **2022**, *23*, 7273. [CrossRef] [PubMed]
4. Li, Y.; Chen, L.; Gao, Y.; Zou, X.; Wei, F. Oxidative stress and intervertebral disc degenersation: Pathophysiology, signaling pathway, and therapy. *Oxid. Med. Cell. Longev.* **2022**, *2022*, 1984742. [CrossRef]
5. Panche, A.N.; Diwan, A.D.; Chandra, S.R. Flavonoids: An overview. *J. Nutr. Sci.* **2016**, *5*, e47. [CrossRef]
6. Maleki, S.J.; Crespo, J.F.; Cabanillas, B. Anti-inflammatory effects of flavonoids. *Food Chem.* **2019**, *299*, 125124. [CrossRef]
7. Al-Khayri, J.M.; Sahana, G.R.; Nagella, P.; Joseph, B.V.; Alessa, F.M.; Al-Mssallem, M.Q. Flavonoids as potential anti-inflammatory molecules: A review. *Molecules* **2022**, *27*, 2901. [CrossRef]
8. Rakha, A.; Umar, N.; Rabail, R.; Butt, M.S.; Kieliszek, M.; Hassoun, A.; Aadil, R.M. Anti-inflammatory and anti-allergic potential of dietary flavonoids: A review. *Biomed. Pharmacother.* **2022**, *156*, 113945. [CrossRef]
9. Wang, X.; Cao, Y.; Chen, S.; Lin, J.; Yang, X.; Huang, D. Structure–activity relationship (SAR) of flavones on their anti-inflammatory activity in murine macrophages in culture through the NF-κB pathway and c-Src kinase receptor. *J. Agric. Food Chem.* **2022**, *70*, 8788–8798. [CrossRef]
10. Shen, N.; Wang, T.; Gan, Q.; Liu, S.; Wang, L.; Jin, B. Plant flavonoids: Classification, distribution, biosynthesis, and antioxidant activity. *Food Chem.* **2022**, *383*, 132531. [CrossRef]
11. Liu, Z.Q. What about the progress in the synthesis of flavonoid from 2020? *Eur. J. Med. Chem.* **2022**, *243*, 114671. [CrossRef] [PubMed]
12. Baker, W. 204. The synthesis of 5: 6-dihydroxyflavone and the structure of primetin. *J. Am. Chem. Soc.* **1939**, 956–961. [CrossRef]
13. Fan, H.; Liu, S.; Shen, W.; Kang, A.; Tan, J.; Li, L.; Liu, X.; Xu, C.; Xu, X.; Lai, Y. Identification of the absorbed components and metabolites of Xiao-Ai-Jie-Du decoction and their distribution in rats using ultra high-performance liquid chromatography/quadrupole time-of-flight mass spectrometry. *J. Pharm. Biomed. Anal.* **2020**, *179*, 112984. [CrossRef]
14. Park, S.; Kim, M.; Lin, Y.; Hong, M.; Nam, G.; Mieczkowski, A.; Kardos, J.; Lee, Y.H.; Lim, M.H. Designing multi-target-directed flavonoids: A strategic approach to Alzheimer's disease. *Chem. Sci.* **2023**, *14*, 9293–9305. [CrossRef]
15. Lombardo, E.; Sabellico, C.; Hájek, J.; Staňková, V.; Filipský, T.; Balducci, V.; De Vito, P.; Leone, S.; Bavavea, E.I.; Silvestri, I.P. Protection of cells against oxidative stress by nanomolar levels of hydroxyflavones indicates a new type of intracellular antioxidant mechanism. *PLoS ONE* **2013**, *8*, e60796. [CrossRef]
16. Behl, T.; Rana, T.; Alotaibi, G.H.; Shamsuzzaman, M.; Naqvi, M.; Sehgal, A.; Singh, S.; Sharma, N.; Almoshari, Y.; Abdellatif, A.A. Polyphenols inhibiting MAPK signalling pathway mediated oxidative stress and inflammation in depression. *Biomed. Pharmacother.* **2022**, *146*, 112545. [CrossRef] [PubMed]
17. Wang, X.; Cao, Y.; Chen, S.; Yang, X.; Bian, J.; Huang, D. Antioxidant and anti-inflammatory effects of 6, 3′, 4′-and 7, 3′, 4′-Trihydroxyflavone on 2D and 3D RAW264. 7 Models. *Antioxidants* **2023**, *12*, 204. [CrossRef]
18. Li, G.; Ding, K.; Qiao, Y.; Zhang, L.; Zheng, L.; Pan, T.; Zhang, L. Flavonoids regulate inflammation and oxidative stress in cancer. *Molecules* **2020**, *25*, 5628. [CrossRef]
19. Shamsudin, N.F.; Ahmed, Q.U.; Mahmood, S.; Shah, S.A.A.; Sarian, M.N.; Khattak, M.M.A.K.; Khatib, A.; Sabere, A.S.M.; Yusoff, Y.M.; Latip, J. Flavonoids as antidiabetic and anti-inflammatory agents: A review on structural activity relationship-based studies and meta-analysis. *Int. J. Mol. Sci.* **2022**, *23*, 12605. [CrossRef]
20. Lundberg, J.O.; Weitzberg, E. Nitric oxide signaling in health and disease. *Cell* **2022**, *185*, 2853–2878. [CrossRef]
21. Smith, J.A.; Das, A.; Ray, S.K.; Banik, N.L. Role of pro-inflammatory cytokines released from microglia in neurodegenerative diseases. *Brain Res. Bull.* **2012**, *87*, 10–20. [CrossRef] [PubMed]
22. Zhang, H.; Dhalla, N.S. The role of pro-Inflammatory cytokines in the pathogenesis of cardiovascular disease. *Int. J. Mol. Sci.* **2024**, *25*, 1082. [CrossRef] [PubMed]
23. Seruga, B.; Zhang, H.; Bernstein, L.J.; Tannock, I.F. Cytokines and their relationship to the symptoms and outcome of cancer. *Nat. Rev. Cancer.* **2008**, *8*, 887–899. [CrossRef] [PubMed]
24. Ciesielska, A.; Matyjek, M.; Kwiatkowska, K. TLR4 and CD14 trafficking and its influence on LPS-induced pro-inflammatory signaling. *Cell. Mol. Life Sci.* **2021**, *78*, 1233–1261. [CrossRef]
25. Behl, T.; Upadhyay, T.; Singh, S.; Chigurupati, S.; Alsubayiel, A.M.; Mani, V.; Vargas-De-La-Cruz, C.; Uivarosan, D.; Bustea, C.; Sava, C. Polyphenols targeting MAPK mediated oxidative stress and inflammation in rheumatoid arthritis. *Molecules* **2021**, *26*, 6570. [CrossRef]

26. Paracha, R.Z.; Ahmad, J.; Ali, A.; Hussain, R.; Niazi, U.; Tareen, S.H.K.; Aslam, B. Formal modelling of toll like receptor 4 and JAK/STAT signalling pathways: Insight into the roles of SOCS-1, interferon-β and proinflammatory cytokines in sepsis. *PLoS ONE* **2014**, *9*, e108466. [CrossRef]
27. Zinatizadeh, M.R.; Schock, B.; Chalbatani, G.M.; Zarandi, P.K.; Jalali, S.A.; Miri, S.R. The Nuclear Factor Kappa B (NF-kB) signaling in cancer development and immune diseases. *Genes Dis.* **2021**, *8*, 287–297. [CrossRef]
28. Ardito, F.; Giuliani, M.; Perrone, D.; Troiano, G.; Lo Muzio, L. The crucial role of protein phosphorylation in cell signaling and its use as targeted therapy. *Int. J. Mol. Med.* **2017**, *40*, 271–280. [CrossRef] [PubMed]
29. Tirichen, I.I.; Yaigoub, I.I.; Xu, W.; Wu, C.; Li, R.; Li, Y. Mitochondrial reactive oxygen species and their contribution in chronic kidney disease progression through oxidative stress. *Front. Physiol.* **2021**, *12*, 627837. [CrossRef]
30. Araki, E.; Nishikawa, T. Oxidative stress: A cause and therapeutic target of diabetic complications. *J. Diabetes Investig.* **2010**, *1*, 90–96. [CrossRef]
31. Bou-Teen, D.; Kaludercic, N.; Weissman, D.; Turan, B.; Maack, C.; Di Lisa, F.; Ruiz-Meana, M. Mitochondrial ROS and mitochondria-targeted antioxidants in the aged heart. *Free Radic. Biol. Med.* **2021**, *167*, 109–124. [CrossRef]
32. Casey, A.M.; Murphy, M.P. Uncovering the source of mitochondrial superoxide in pro-inflammatory macrophages: Insights from immunometabolism. *Biochim. Biophys. Acta-Mol. Basis Dis* **2022**, *1868*, 166481. [CrossRef] [PubMed]
33. Zorov, D.B.; Juhaszova, M.; Sollott, S.J. Mitochondrial reactive oxygen species (ROS) and ROS-induced ROS release. *Physiol. Rev.* **2014**, *94*, 909–950. [CrossRef]
34. Son, Y.; Cheong, Y.K.; Kim, N.-H.; Chung, H.T.; Kang, D.G.; Pae, H.O. Mitogen-activated protein kinases and reactive oxygen species: How can ROS activate MAPK pathways? *J. Signal Transduct.* **2011**, *2011*, 792639. [CrossRef] [PubMed]
35. Liu, J.; Wang, F.; Luo, F. The role of JAK/STAT pathway in fibrotic diseases: Molecular and cellular mechanisms. *Biomolecules* **2023**, *13*, 119. [CrossRef] [PubMed]
36. Lingappan, K. NF-κB in oxidative stress. *Curr. Opin. Toxicol.* **2018**, *7*, 81–86. [CrossRef]
37. Matsuo, M.; Sasaki, N.; Saga, K.; Kaneko, T. Cytotoxicity of flavonoids toward cultured normal human cells. *Biol. Pharm. Bull.* **2005**, *28*, 253–259. [CrossRef]
38. Zhai, K.; Mazurakova, A.; Koklesova, L.; Kubatka, P.; Büsselberg, D. Flavonoids synergistically enhance the anti-glioblastoma effects of chemotherapeutic drugs. *Biomolecules* **2021**, *11*, 1841. [CrossRef]
39. Eren-Guzelgun, B.; Ince, E.; Gurer-Orhan, H. In vitro antioxidant/prooxidant effects of combined use of flavonoids. *Nat. Prod. Res.* **2018**, *32*, 1446–1450. [CrossRef]
40. To, K.K.; Cho, W. Flavonoids overcome drug resistance to cancer chemotherapy by epigenetically modulating multiple mechanisms. *Curr. Cancer Drug Targets* **2021**, *21*, 289–305. [CrossRef]
41. More, G.K.; Makola, R.T. In-vitro analysis of free radical scavenging activities and suppression of LPS-induced ROS production in macrophage cells by Solanum sisymbriifolium extracts. *Sci. Rep.* **2020**, *10*, 6493. [CrossRef] [PubMed]
42. Jiang, C.; Zhang, J.; Xie, H.; Guan, H.; Li, R.; Chen, C.; Dong, H.; Zhou, Y.; Zhang, W. Baicalein suppresses lipopolysaccharide-induced acute lung injury by regulating Drp1-dependent mitochondrial fission of macrophages. *Biomed. Pharmacother.* **2022**, *145*, 112408. [CrossRef] [PubMed]

Disclaimer/Publisher's Note: The statements, opinions and data contained in all publications are solely those of the individual author(s) and contributor(s) and not of MDPI and/or the editor(s). MDPI and/or the editor(s) disclaim responsibility for any injury to people or property resulting from any ideas, methods, instructions or products referred to in the content.

International Journal of *Molecular Sciences*

MDPI

Article

Polyphenols from *Maackia amurensis* Heartwood Protect Neuronal Cells from Oxidative Stress and Prevent Herpetic Infection

Darya V. Tarbeeva [1,*], Evgeny A. Pislyagin [1], Ekaterina S. Menchinskaya [1], Dmitrii V. Berdyshev [1], Natalya V. Krylova [2], Olga V. Iunikhina [2], Anatoliy I. Kalinovskiy [1], Mikhail Y. Shchelkanov [2], Natalia P. Mishchenko [1], Dmitry L. Aminin [1] and Sergey A. Fedoreyev [1]

[1] G.B. Elyakov Pacific Institute of Bioorganic Chemistry, Far-Eastern Branch of the Russian Academy of Science, 690022 Vladivostok, Russia; pislyagin_ea@piboc.dvo.ru (E.A.P.); menchinskaya_es@piboc.dvo.ru (E.S.M.); berdyshev@piboc.dvo.ru (D.V.B.); kaaniw@piboc.dvo.ru (A.I.K.); mischenkonp@mail.ru (N.P.M.); daminin@piboc.dvo.ru (D.L.A.); fedoreev-s@mail.ru (S.A.F.)

[2] G.P. Somov Institute of Epidemiology and Microbiology, Rospotrebnadzor, 690087 Vladivostok, Russia; krylovanatalya@gmail.com (N.V.K.); olga_iun@inbox.ru (O.V.I.); adorob@mail.ru (M.Y.S.)

* Correspondence: tarbeeva1988@mail.ru; Tel.: +7-9143446441

Abstract: Here, we continued the investigation of anti-HSV-1 activity and neuroprotective potential of 14 polyphenolic compounds isolated from *Maackia amurensis* heartwood. We determined the absolute configurations of asymmetric centers in scirpusin A (**13**) and maackiazin (**10**) as *7R,8R* and $1''S,2''S$, respectively. We showed that dimeric stilbens maackin (**9**) and scirpusin A (**13**) possessed the highest anti-HSV-1 activity among polyphenols **1–14**. We also studied the effect of polyphenols **9** and **13** on the early stages of HSV-1 infection. Direct interaction with the virus (virucidal activity) was the main mechanism of the antiviral activity of these compounds. The neuroprotective potential of polyphenolic compounds from *M. amurensis* was studied using models of 6-hydroxydopamine (6-OHDA)-and paraquat (PQ)-induced neurotoxicity. A dimeric stilbene scirpusin A (**13**) and a flavonoid liquiritigenin (**6**) were shown to be the most active compounds among the tested polyphenols. These compounds significantly increased the viability of 6-OHDA-and PQ-treated Neuro-2a cells, elevated mitochondrial membrane potential and reduced the intracellular ROS level. We also found that scirpusin A (**13**), liquiritigenin (**6**) and retusin (**3**) considerably increased the percentage of live Neuro-2a cells and decreased the number of early apoptotic cells. Scirpusin A (**13**) was the most promising compound possessing both anti-HSV-1 activity and neuroprotective potential.

Keywords: polyphenolic compounds; HSV-1; virucidal activity; neuroprotective activity; ROS

1. Introduction

The leguminous woody plant *Maackia amurensis* Rupr et Maxim is widespread in the far east of the Russian Federation, particularly in Primorsky Krai. *M. amurensis* heartwood is a rich source of isoflavones, pterocarpans, flavanones, isoflavans, and isoflavanones. The heartwood of this plant contains not only isoflavonoids widely distributed in the Leguminosae such as genistein, daidzein, formononetin, tectorigenin, but also monomeric and dimeric stilbens, as well as rare isoflavonostilbenes and stilbenolignans [1]. These polyphenols constitute the polyphenolic complex of *M. amurensis* heartwood. In contrast to the heartwood, the roots of this plant produce isoflavone glycosides and prenylated flavanones [2,3].

The polyphenolic complex from *M. amurensis* heartwood possessed significant hepatoprotective activity and was registered in the Russian Federation as a drug named Maksar® [1]. In addition to hepatoprotective properties, Maksar® also possessed antithrombogenic, antiplatelet, antitumor, and antioxidant properties [4–6].

The therapeutic efficacy of Maksar® was shown when administrated to patients with chronic hepatitis including viral hepatitis B and C [7]. The authors suggested that the antiviral effect of Maksar® was due to the activation of immune response. A recent report revealed the anti-SARS-CoV-2 activity of Maksar® [8]. This effect can be due to the ability of polyphenols from Maksar® maackolin, maackin, and scirpusin A to bind to the main Mpro protease of SARS-CoV-2 predicted using molecular docking approach [9,10].

In addition to anti-SARS-CoV-2 activity, we found that Maksar® inhibited the replication of herpes simplex virus type 1 (HSV-1) and enterovirus B (ECHO-1) [11]. Herpesviruses can penetrate into the central nervous system and play a role in the pathogenesis of neurodegenerative disorders [12,13]. HSV-1 was reported to cause an increase in the level of intracellular reactive oxygen species (ROS) and lipid peroxidation. The virus also reduced the content of glutathione, which plays a crucial role in the body's antioxidant defense [14,15]. Additionally, HSV-1 infection induced a remodeling of mitochondrial shape [16]. HSV-1 was shown to bind to mitochondrion-associated factors. These interactions altered mitochondrial dynamics in neurons, thereby facilitating viral replication and pathogenesis [17]. There is a clear connection between herpes infection and the subsequent development of neurodegenerative diseases, which is largely due to the virus-induced accumulation of reactive oxygen species in neurons [18]. Impaired mitochondrial function results in elevated ROS levels and subsequently leads to the development of neurodegenerative diseases [19–21]. The activation of inflammatory processes and host immune responses caused chronic damage resulting in alterations of neuronal function and viability such as oxidative stress, protein aggregation, deficient autophagic processes, and neuronal death [15]. Therefore, there is a need for biologically active compounds possessing both antiviral and neuroprotective properties for the treatment of herpetic infections and neurodegenerative disorders.

Maksar® was shown to possess antiviral and neuroprotective properties, but it was unclear which individual compounds were responsible for these activities [11,22]. Recently, we started the study of antiherpetic and neuroprotective properties of polyphenols from *M. amurensis* in order to identify the most active compounds [22]. Here, we continue our investigation of anti-HSV-1 activity and neuroprotective potential of isoflavonoids and stilbenes from *M. amurensis*.

2. Results

2.1. Polyphenolic Compounds from M. amurensis Heartwood

We isolated 14 previously known compounds from *M. amurensis* heartwood: isoflavonoids genistein (**2**), retusin (**3**), maackiain (**4**), tectorigenin (**5**), medicarpin (**7**), formononetin (**8**), (±)-3-hydroxyvestitone (**12**), flavonoid liquiritigenin (**6**), monomeric stilbenes resveratrol (**1**), piceatannol (**11**), dimeric stilbens maackin (**9**), scirpusin A (**13**), isoflavonostilbenes maackiazin (**10**), and maackiapicevestitol (**14**) (Figure 1). We used ^{1}H NMR and HPLC-MS techniques for structural identification of the isolated polyphenols.

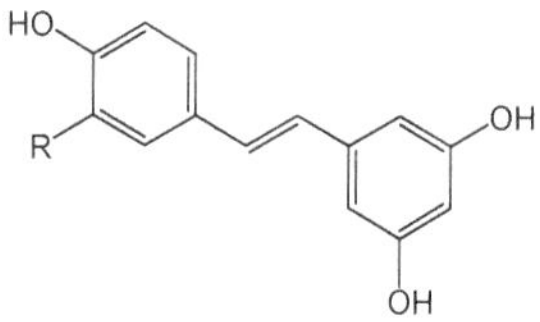

1 – R = H
11 – R = OH

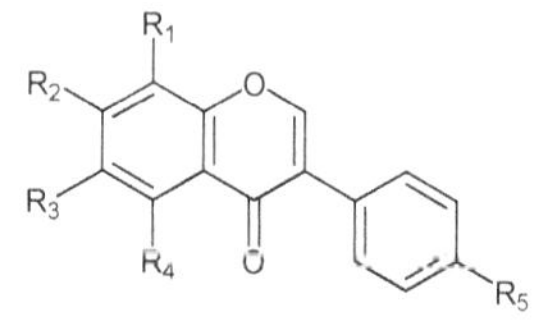

2 – $R_1 = R_3 = H$, $R_2 = R_4 = R_5 = OH$
3 – $R_3 = R_4 = H$, $R_1 = R_2 = OH$, $R_5 = OCH_3$
5 – $R_1 = H$, $R_2 = R_4 = R_5 = OH$, $R_3 = OCH_3$
8 – $R_1 = R_3 = R_4 = H$, $R_2 = OH$, $R_5 = OCH_3$

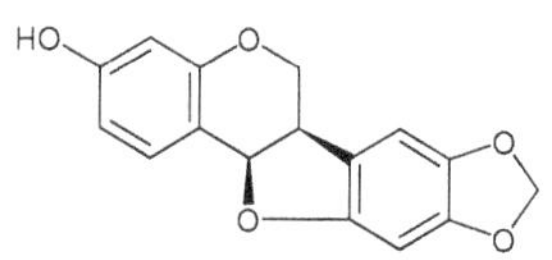

4

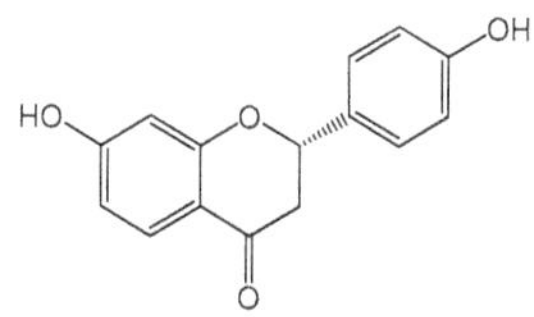

6

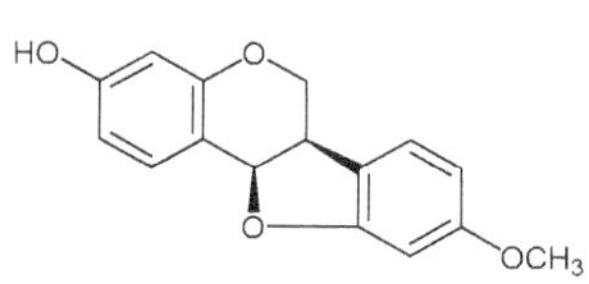

7

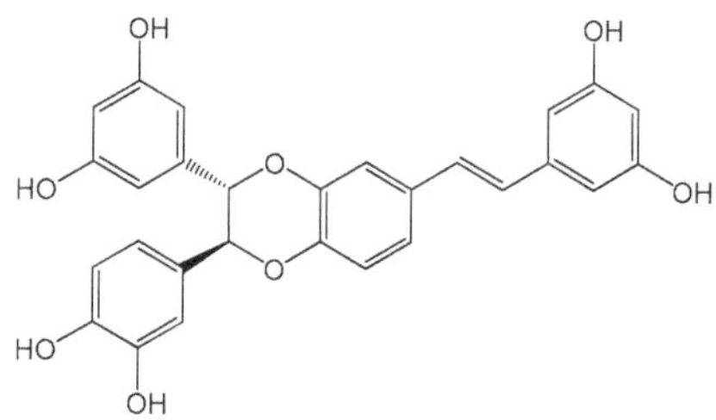

9

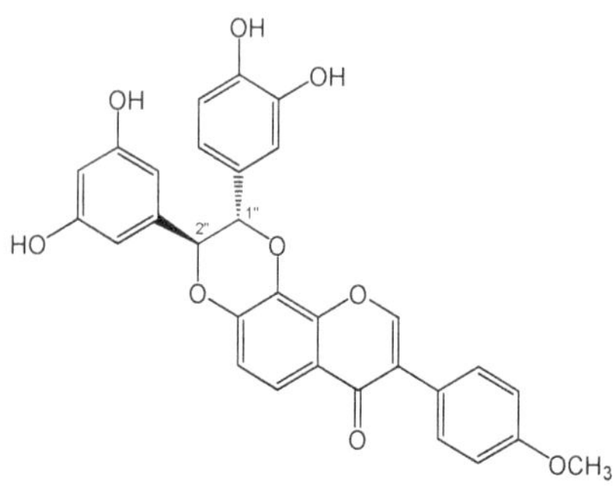

10

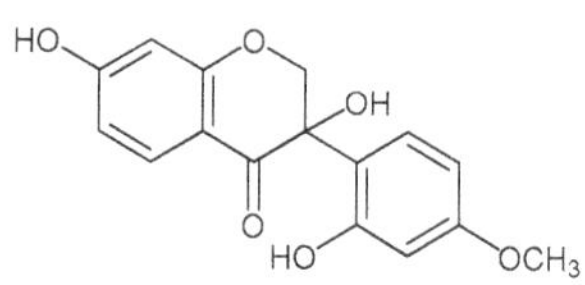

12

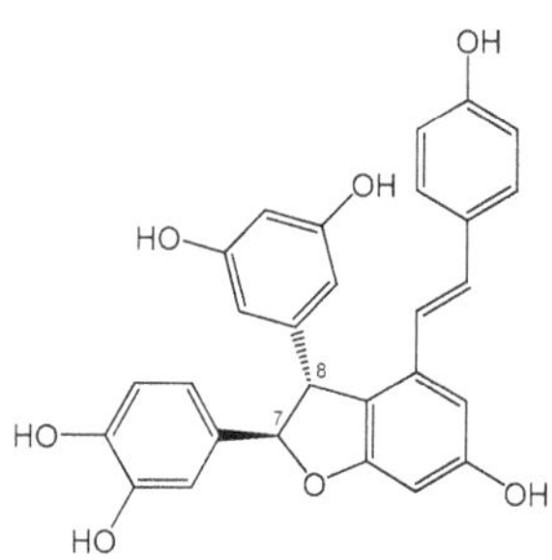

13

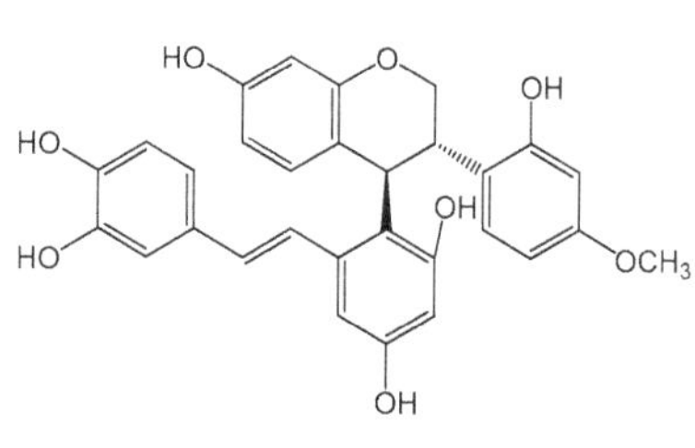

14

Figure 1. Polyphenolic compounds from *M. amurensis* heartwood.

2.2. Determination of Absolute Configurations for Maackiazin (10) and Scirpusin A (13)

We also determined the absolute configurations for scirpusin A (**13**) and maackiazin (**10**).

First of all, the vicinal spin–spin coupling constants J (H1″,H2″) were calculated using the B3LYP/6-311G(d)_PCM//B3LYP/6-311G(d)_PCM method for different conformations of 1″*R*,2″*S*-maackiazin (**10**) and 1″*S*,2″*S*-maackiazin (**10**). These significant conformations were preliminary selected during a conformational analysis performed at B3LYP/6-311G(d)_PCM level of theory (Supplementary data, File S1, Quantum modeling, Figure S1). The calculated mean values of J_{calc} (H1″,H2″) were 3.8 Hz and 7.8 Hz for 1″*R*,2″*S* and 1″*S*,2″*S*-maackiazin (**10**), respectively. The experimental value of J (H1″,H2″) was 7.9 Hz, which correlated well with the calculated value for 1″*S*,2″*S*-maackiazin (**10**) (Supplementary data, Table S1, Figure S1). In order to choose between 1″*S*,2″*S*-maackiazin (**10**) and 1″*R*,2″*R*-maackiazin (**10**) options, we compared the calculated ECD spectra for these stereoisomers with the experimental one (Figure 2).

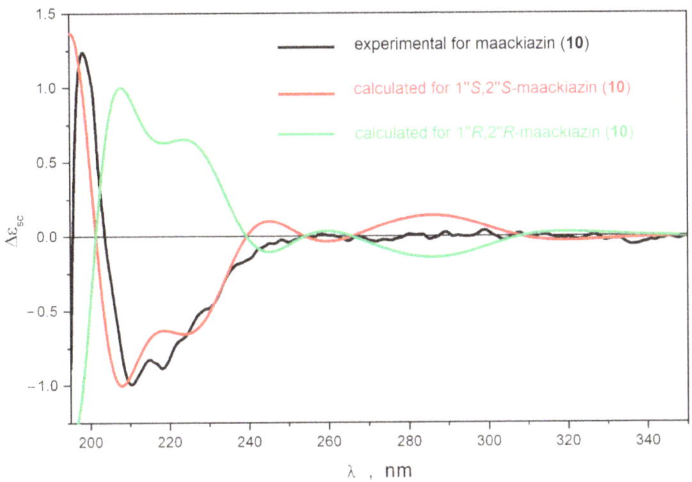

Figure 2. The comparison of the experimental and calculated ECD spectra for maackiazin (**10**).

The theoretical ECD spectra calculated for different conformations of 1″*S*,2″*S*-maackiazin (**10**) contained many bands at $\lambda \geq 240$ nm. We found that at $\lambda \geq 220$ nm the shapes of ECD spectra were strong functions of several large-amplitude motion (LAM) processes—internal rotations of anisol and polyphenol fragments around C-C bonds and the inversion of the 1,4-dioxane cycle. Thus, the internal rotation of the anisole substituent around C(1′)-C(3) bond resulted in the pronounced changes of the ECD spectrum at $\lambda \geq 220$ nm—many bands changed their intensity and even their sign. However, in the region $195 \leq \lambda \leq 220$ nm the positions and signs of the intensive positive and intensive negative bands did not change (negative band in the region $200 \leq \lambda \leq 240$ nm was a doublet). The inversion of 1,4-dioxane cycle led to the conformation, where both polyphenol substituents were in axial orientation relative to the molecular plane: $\theta 1(H1'',H2'') \equiv \angle H1''\text{-}C1''\text{-}C2''\text{-}H2'' \approx 81°$; $\theta 2 \equiv \angle C7\text{-}C1''\text{-}C2''\text{-}C8 \approx -162°$. For this conformation, the intensity of the high-frequency part of the negative band's doublet in the region $195 \leq \lambda \leq 220$ nm increased approximately twice, whereas the long-wave part disappeared. In the $\lambda \geq 250$ nm region, the broad positive band changed its sign. This conformation was minor, and its total portion was approximately 1%. The rotation of polyphenol substituents resulted in variations of intensities of bands in the $\lambda \leq 240$ nm region without changing their signs (Supplementary data, Figure S3). Thus, the shape of the ECD spectrum of 1″*S*,2″*S*-maackiazin (**10**) was stable during these LAM processes.

The data presented above confirmed that there were two characteristic bands in the ECD spectra of 1″S,2″S-maackiazin (**10**)—a positive band at $\lambda \approx 200$ nm and the high-frequency part of a negative band doublet in the region $195 \leq \lambda \leq 220$ nm. No other bands were characteristic due to the effective averaging of their intensities over LAM processes, which gave approximately zero. The comparison of the theoretical statistically averaged spectra, calculated for 1″S,2″S and 1″R,2″R stereoisomers of maackiazin (**10**) with the experimental ECD spectrum of **10**, thus, must be restricted to the analysis of their shapes at $\lambda \leq 250$. A good correspondence between calculated and experimental spectra was observed only for 1″S,2″S stereoisomer of **10** (Figure 2).

The absolute configuration of scirpusin A (**13**) was determined using the same approach. To distinguish between 7R,8R- and 7S,8R-scirpusin A (**13**) the J (H7,H8) vicinal constants were calculated and compared to the experimental value.

The performed conformational analysis of scirpusin A (**13**) showed that the potential energy profile for inversion of the dihydrofuran cycle in **13** depended on the configurations of its asymmetric centers. Thus, 7S,8R-scirpusin A (**13**) had only one stable conformation of the dihydrofuran cycle, whereas 7R,8R-scirpusin A (**13**)—two conformations. This difference in the topography of the potential energy surfaces led to differences in the predicted values of the vicinal spin–spin coupling constant. The value of J (H7,H8) was 9.2–9.4 Hz for 7S,8R-scirpusin A (**13**) and 6.3 Hz for 7R,8R-scirpusin A (**13**). The experimental value of 5.3 Hz correlated well enough with the last calculated value (Supplementary data, Table S2, Figure S2). Thus, the relative configuration of **13** was determined as 7R,8R or 7S,8S. Then, we calculated ECD spectra for these stereoisomers. We found that scirpusin A (**13**) was structurally flexible. Thus, the rotation of the p-vinylphenol substituent around C8′-C9′ was dynamically coupled to the inversion of the dihydrofuran cycle. Both these processes led to variations in the shape of the ECD spectrum, but the sign of the Cotton effect in the characteristic region at $\lambda \leq 240$ nm did not change. The position of the band at $\lambda \approx 200$ nm in experimental UV spectrum was chosen to be the reference for the determination of the UV shift $\Delta\lambda = +17$ nm (Supplementary data, Figure S4).

The calculated ECD spectra for 7R,8R-scirpusin A (**13**) and 7S,8S-scirpusin A (**13**) compared to the experimental spectrum for **13** are presented in Figure 3. A good correspondence between the calculated ECD spectrum for 7R,8R-scirpusin A (**13**) and the experimental spectrum confirmed the 7R,8R configuration of **13**.

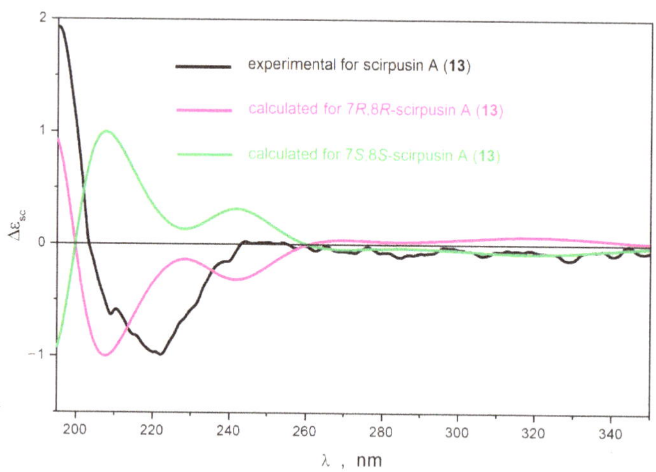

Figure 3. The comparison of experimental and calculated ECD spectra for scirpusin A (**13**).

The positive Cotton effect at 331 nm in the CD spectrum of liquiritigenin (**6**) confirmed its 2S configuration [23].

2.3. In Vitro Model of 6-OHDA Induced Neurotoxicity (Cell Viability Test)

Polyphenolic compounds **1–13** did not possess cytotoxic activity against Neuro-2a cells at a concentration up to 100 μM.

Neuro-2a cells were incubated with the studied polyphenolic compounds for 1 h. Then, 100 μM of 6-hydroxydopamine 6-OHDA were added. We determined the percentage of living Neuro-2a cells after treatment with 6-OHDA (Figure 4a–d). Liquiritigenin (**6**) and scirpusin A (**13**) at a concentration of 10 μM, as well as Maksar® (10 μg/mL), significantly increased the viability of 6-OHDA-treated Neuro-2a cells by 12%, 9%, and 11%, respectively (Figure 4b,d), whereas isoflavonoids **4**, **5**, **7**, **8**, maackiazin (**10**) and piceatannol (**11**) showed a weak effect on cell viability. Compounds **8** and **12** increased the viability of Neuro-2a cells at a concentration of 1 μM. Maackin (**9**) did not show significant activity in this test (Figure 4c), which was in accordance with the data obtained using a paraquat (PQ)-induced neurotoxicity model [22]. Resveratrol also did not cause any increase in the viability of 6-OHDA-treated Neuro-2a cells.

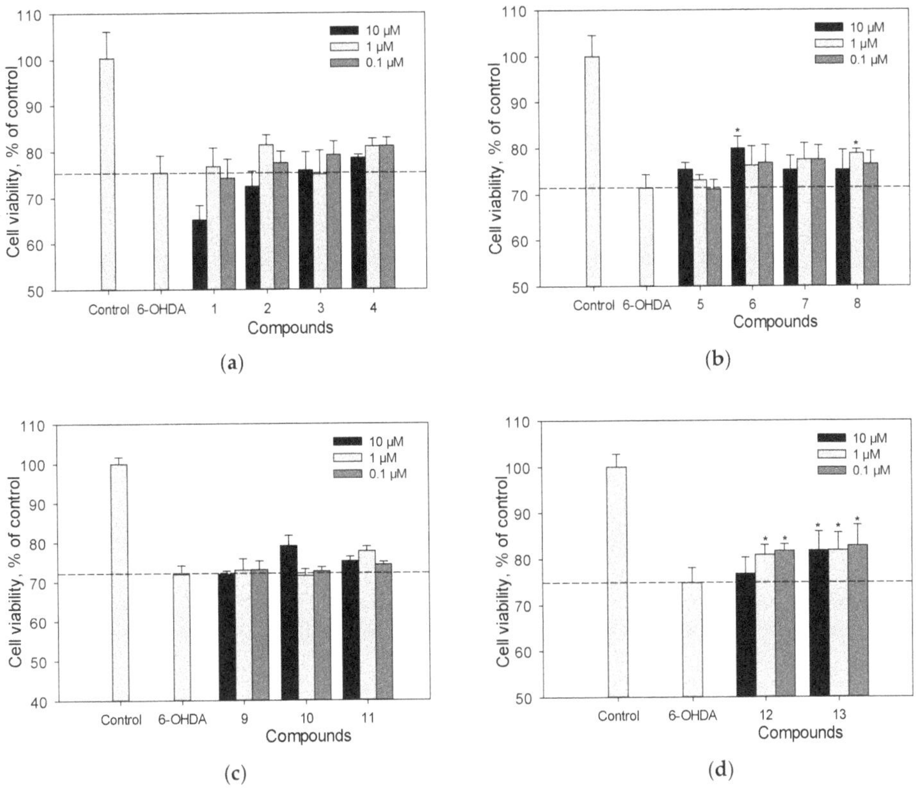

Figure 4. The effect of polyphenolic compounds isolated from *M. amurensis* heartwood on the viability of Neuro-2a cells treated with 6-OHDA (100 μM) (**a–d**). The viability of Neuro-2a cells treated with various compounds and 6-OHDA was measured by the MTT assay. Polyphenolic compounds isolated from *M. amurensis* heartwood were used at concentrations 10, 1, 0.1 μM. The data are presented as means ± SEM of three independent replicates. (*) indicate $p < 0.05$ versus 6-OHDA-treated cells.

2.4. Reactive Oxygen Species (ROS) Analysis in 6-OHDA-Treated Cells

We studied the effect of polyphenolic compounds set on ROS levels in 6-OHDA-treated Neuro-2a cells (Figure 5a–d). Scirpusin A (**13**), maackiasin (**10**), and liquiritigenin (**6**) possessed high activity in this test and at a concentration of 1 μM/mL they reduced ROS levels in 6-OHDA-treated Neuro-2a cells by 43%, 20%, and 15%, respectively (Figure 5a–d). Liquiritigenin (**6**) showed a dose-dependent effect (Figure 5b). Resveratrol (**1**), though it did not increase cell viability, at a concentration of 1 µM/mL it decreased ROS levels in 6-OHDA-treated cells by approximately 15%.

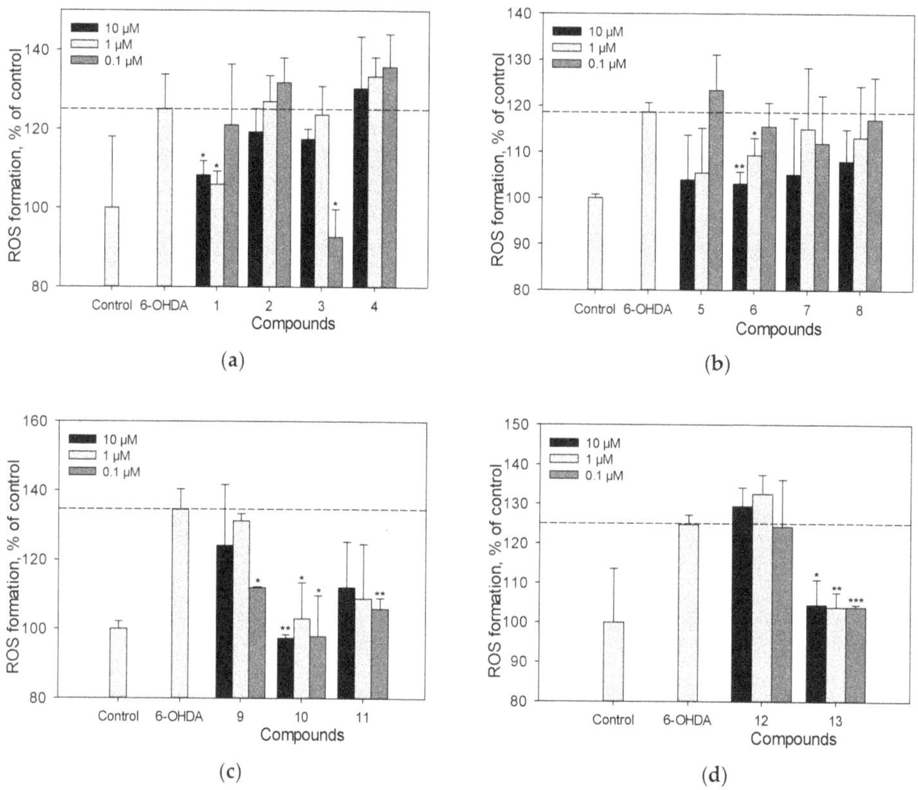

Figure 5. The effect of polyphenolic compounds isolated from *M. amurensis* heartwood on ROS levels in Neuro-2a cells treated with 6-OHDA (120 µM) (**a**–**d**). Polyphenolic compounds isolated from *M. amurensis* heartwood were used at concentrations 10, 1, 0.1 µM. The data are presented as means ± SEM of three independent replicates. (*), (**), and (***) indicate, respectively, $p < 0.05$, $p < 0.005$, and $p < 0.001$ versus 6-OHDA-treated cells.

2.5. Mitochondrial Membrane Potential (MMP) Detection

We determined the effect of polyphenolic compounds from *M. amurensis* on 6-OHDA-induced mitochondrial dysfunction in Neuro-2a cells (Figure 6a,b). The tetramethylrhodamine methyl (TMRM) fluorescence decreased by 14% after the treatment of Neuro-2a cells with 6-OHDA, which indicated that 6-OHDA caused depolarization of the mitochondrial membrane. We observed a significant effect of increasing mitochondrial membrane potential by 8–12% when Neuro-2a cells were incubated with resveratrol (**1**), piceatannol (**11**), liquiritigenin (**6**), medicarpin (**7**), and maackin (**9**) at a concentration of 10 µM (Figure 6a,b). Retusin (**3**) at a concentration of 10 µM showed weak activity in this test.

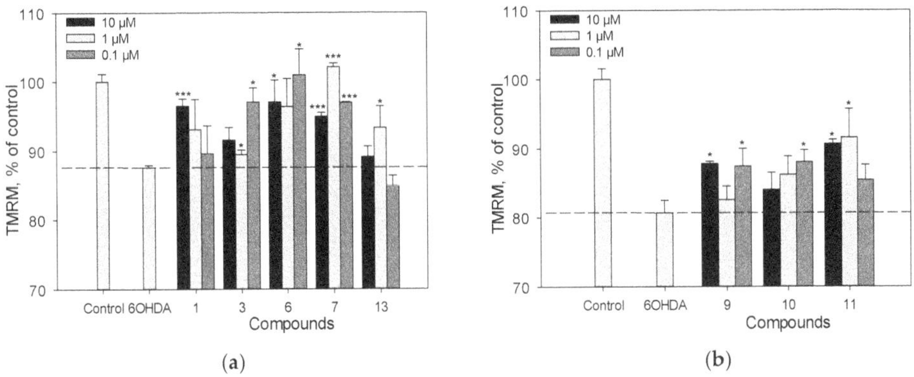

Figure 6. The effect of polyphenolic compounds isolated from *M. amurensis* heartwood on mitochondrial membrane potential in Neuro-2a cells treated with 6-OHDA (120 μM) (**a**,**b**). Polyphenolic compounds isolated from *M. amurensis* heartwood were used at concentrations 10, 1, 0.1 μM. The data are presented as means ± SEM of three independent replicates. (*) and (***) indicate, respectively, $p < 0.05$ and $p < 0.001$ versus 6-OHDA-treated cells.

2.6. Cell Viability and Reactive Oxygen Species (ROS) Analysis in PQ-Treated Cells

We previously studied the effect of stilbens **9**, **10**, **11**, and **13** on cell viability and ROS level in PQ-treated Neuro-2a cells. Here, we evaluated the neuroprotective potential against the toxic effect of PQ of stilbene **1**, flavonoid **6** and isoflavonoids **2–5**, **7**, **12** (Figure 7a–d). Resveratrol (**1**) and liquiritigenin (**6**) at a concentration of 10 μM demonstrated considerable activity and increased the viability of PQ-treated cells by 17% and 15%, respectively (Figure 7a–c). Liquiritigenin (**6**) also effectively reduced the ROS level in PQ-treated Neuro-2a cells (Figure 7d). Isoflavonoids genistein (**2**), medicarpin (**7**), and maackiain (**4**) at this concentration showed a much less significant effect and increased the cell viability by 5%, 7% and 2%, respectively (Figure 7a–c).

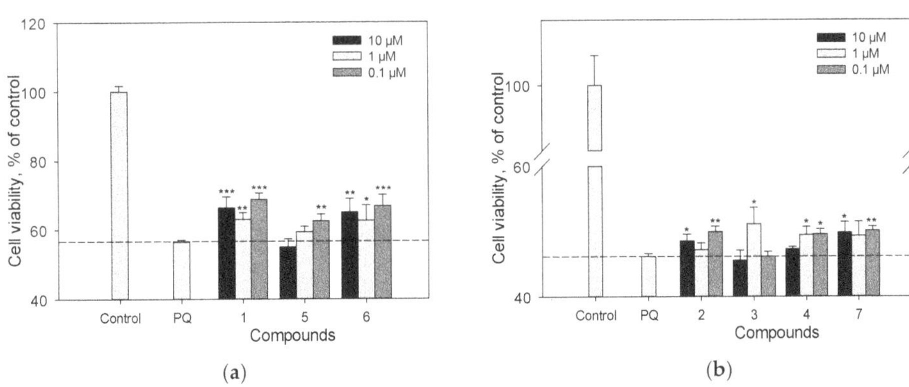

Figure 7. *Cont.*

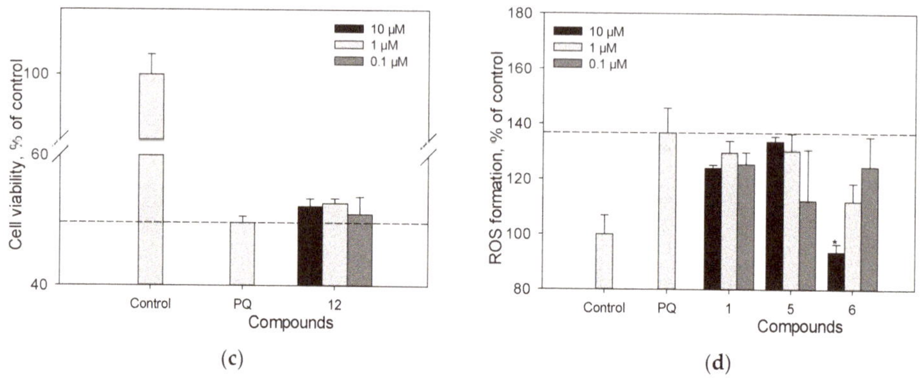

Figure 7. The effect of polyphenolic compounds isolated from *M. amurensis* heartwood on cell viability (**a–c**) and ROS levels (**d**) in Neuro-2a cells treated with PQ (1 mM). Polyphenolic compounds isolated from *M. amurensis* heartwood were used at concentrations 10, 1, 0.1 µM. The data are presented as means ± SEM of three independent replicates. (*), (**), and (***) indicate, respectively, $p < 0.05$, $p < 0.005$, and $p < 0.001$ versus PQ-treated cells.

2.7. Apoptosis

We investigated the effect of polyphenolic compounds from *M. amurensis* on apoptosis profile in 6-OHDA-treated Neuro-2a cells by flow cytometry using fluorescent annexin V conjugate as a fluorescent dye [24]. We observed significant changes in apoptosis profile in Neuro-2a cells after 6-OHDA-treatment. The percentages of early apoptotic and late apoptotic cells increased after 6-OHDA-treatment from 4.59 to 42.62% and from 4.3 to 12.15%, respectively (Table 1). Polyphenols **3**, **6** and **13** diminished the neurotoxic effect of 6-OHDA-treatment. These compounds considerably increased the percentage of live Neuro-2a cells and decreased the number of early apoptotic cells (Table 1, Figure 8).

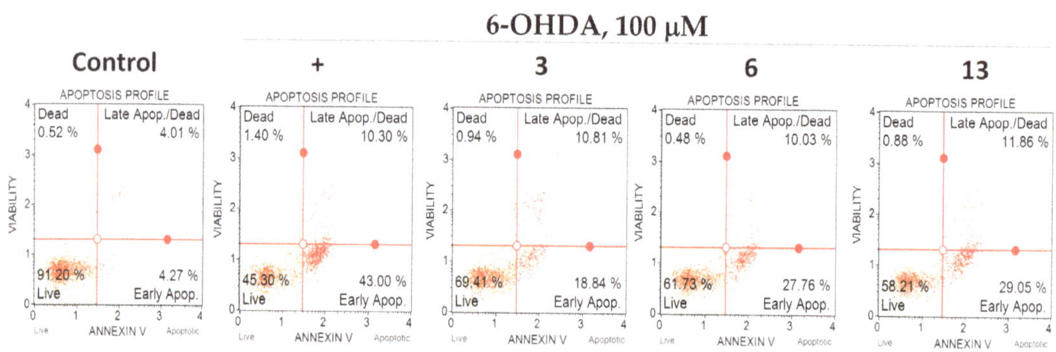

Figure 8. The effect of polyphenolic compounds isolated from *M. amurensis* heartwood on apoptosis profile in 6-OHDA-treated Neuro-2a cells.

Table 1. The effect of polyphenolic compounds from *M. amurensis* heartwood on apoptosis profile in 6-OHDA-treated Neuro-2a cells.

	Live	Early apop.	Late apop./Dead	Dead
Control	90.53 ± 0.68	4.59 ± 0.32	4.3 ± 0.29	0.63 ± 0.05
6-OHDA	44.14 ± 1.34	42.62 ± 2.3	12.15 ± 1.34	1.11 ± 0.37
3	67.89 ± 1.52 *	19.63 ± 0.79 *	11.31 ± 0.49	1.18 ± 0.24
6	61.83 ± 0.09 *	27.67 ± 1.09 *	10.89 ± 0.86	0.61 ± 0.13
13	56.90 ± 1.32 *	30.18 ± 1.13 *	12.2 ± 0.34	0.73 ± 0.15

* $p \leq 0.05$ compared to 6-OHDA-treated Neuro-2a cells.

2.8. Virucidal Activity of Polyphenolic Compounds from M. amurensis (CPE Inhibition Assay)

We determined the cytotoxicity of polyphenolic compounds **1–14** against Vero cells using the MTT assay. The 50% cytotoxic concentrations (CC_{50}) of the studied compounds were in the range of 120–300 μg/mL (Table 2).

Table 2. Virucidal activity of polyphenolic compounds from *M. amurensis* heartwood (CPE assay) against HSV-1. Values are given as the means ± SD of three or more independent experiments. Acyclovir® was used as the positive control.

Compounds	CC_{50}		IC_{50}		SI
	μg/mL	μM	μg/mL	μM	
1	123.0 ± 14.8	539.5 ± 64.9	9.6 ± 1.4	42.1 ± 6.1	12.8 ± 1.9
2	133.8 ± 20.1	495.6 ± 74.4	16.7 ± 2.3	61.9 ± 8.5	8.0 ± 1.0
4	152.1 ± 22.8	535.6 ± 80.3	28.3 ± 3.7	99.6 ± 13.0	5.4 ± 0.7
7	155.0 ± 23.2	574.1 ± 85.9	74.5 ± 11.1	275.9 ± 41.1	2.1 ± 0.3
9	269.6 ± 35.0	554.7 ± 72.0	2.4 ± 0.3	4.9 ± 0.6	112.3 ± 13.5
10	156.4 ± 18.8	297.3 ± 35.7	6.6 ± 0.9	12.5 ± 1.7	23.7 ± 3.1
11	250.2 ± 30.0	1025.4 ± 123.0	6.7 ± 0.9	27.4 ± 3.7	37.3 ± 5.2
13	199.4 ± 23.9	424.3 ± 50.1	2.6 ± 0.3	5.5 ± 0.6	76.7 ± 9.9
14	162.4 ± 17.9	316.0 ± 34.8	8.6 ± 1.2	16.7 ± 2.3	18.9 ± 2.8
Acyclovir	>1000	NA *	NA *	NA *	NA *

* NA—not active.

We previously reported that stilbenes **9–11**, **13**, and **14** possessed anti-HSV-1 activity when added to Vero cells simultaneously with the virus [22]. This method of exposure of the virus and cells to polyphenols was used to study the ability of the compounds to prevent the penetration of virus to cells. Here, we investigated the ability of polyphenols **1–14** to directly interact with the viral particles (HSV-1 was pretreated with the studied compounds) [25–27].

The antiviral activity of polyphenols from *M. amurensis* against HSV-1 was assessed using the cytopathic effect inhibition (CPE) assay. To study the inhibitory effect of polyphenolic compounds directly on the HSV-1 virus (direct virucidal activity), the monolayer of Vero cells grown on 96-well plates (1×10^4 cells/well) was infected with the virus (100 TCD_{50}/mL), pretreated with different concentrations of polyphenolic compounds. The antiviral activity of compounds **1–14** was studied at concentrations below CC_{50} values (Table 2, Figure S5). Maackin (**9**) and scirpusin A (**13**) inhibited viral replication more effectively than the other studied polyphenols. (IC_{50} values were 2.4 and 2.6 μg/mL, and SI values were 112.3 and 76.3, respectively). Polyphenolic compounds **1**, **7**, **10**, **11**, **14** also showed significant antiviral activity with IC_{50} values ranging from 6.7 to 9.6 μg/mL. Genistein (**2**) and maackiain (**4**) showed moderate activity with SI values of 8.0 and 5.4 and IC_{50} values of 16.7 and 28.3 μg/mL, respectively (Table 2). Polyphenols **3**, **5**, **6**, **8** and **12** did not show virucidal activity against HSV-1.

2.9. Virucidal Activity of Polyphenolic Compounds from *M. amurensis* (RT-PCR Assay)

The anti-HSV-1 activity of polyphenolic compounds from *M. amurensis* was also studied using the real-time PCR (RT-PCR) technique (Table 3, Figure S6). Vero cells grown on 96-well plates were infected with the virus (100 $TCID_{50}$/mL), pretreated with polyphenolic compounds at concentrations 0.1, 1 and 10 µg/mL. After 72 h of cell incubation, viral DNA was extracted from the supernatants, and the relative level of HSV-1 DNA was determined using RT-PCR. The effect of polyphenols on the relative level of viral DNA was assessed using the $2^{-\Delta Ct}$ method and reported as fold reduction compared to virus control.

Table 3. Virucidal activity of polyphenolic compounds from *M. amurensis* heartwood (RT-PCR assay) against HSV-1.

Compounds	Concentration µg/mL	µM	Ct	−ΔCt	$2^{-\Delta Ct}$	$\log_{10}$
	10		24.6 ± 2.9 *	−6.5 ± 0.8	0.0110 ± 0.0014	−1.9 ± 0.2 *
1	1		20.4 ± 2.8	−2.3 ± 0.2	0.203 ± 0.024	−0.7 ± 0.09
	0.1		18.1 ± 2.2	0	1.0	0
	10		22.4 ± 2.4 *	−4.3 ± 0.5	0.0508 ± 0.0061	−1.3 ± 0.2 *
2	1		18.4 ± 2.4	−0.30 ± 0.03	0.812 ± 0.110	−0.09 ± 0.01
	0.1		18.1 ± 2.2	0	1.0	0
	10		20.7 ± 2.3	−2.6 ± 0.3	0.165 ± 0.021	−0.8 ± 0.1
4	1		19.0 ± 2.1	−0.9 ± 0.1	0.536 ± 0.064	−0.30 ± 0.04
	0.1		18.1 ± 2.2	0	1.0	0
	10		19.8 ± 2.4	−1.75 ± 0.20	0.297 ± 0.04	−0.50 ± 0.06
7	1		18.3 ± 2.2	−0.20 ± 0.02	0.870 ± 0.11	−0.060 ± 0.008
	0.1		18.1 ± 2.2	0	1.0	0
	10		33.3 ± 4.3 *	−15.2 ± 1.8	0.0000266 ± 0.000003	−4.6 ± 0.6 *
9	1		25.9 ± 2.8 *	−7.8 ± 0.9	0.00449 ± 0.0006	−2.3 ± 0.3 *
	0.1		18.6 ± 2.4	−0.50 ± 0.05	0.707 ± 0.092	−0.15 ± 0.02
	10		28.0 ± 3.9 *	−9.9 ± 1.1	0.00105 ± 0.0001	−2.9 ± 0.4 *
10	1		20.2 ± 2.6	−2.1 ± 0.3	0.233 ± 0.03	−0.6 ± 0.08
	0.1		18.4 ± 2.4	−0.300 ± 0.03	0.812 ± 0.11	−0.09 ± 0.01
	10		30.2 ± 3.9 *	−12.1 ± 1.4	0.000228 ± 0.0003	−3.6 ± 0.5 *
11	1		24.5 ± 3.4 *	−6.4 ± 0.8	0.0118 ± 0.0015	−1.9 ± 0.2 *
	0.1		18.4 ± 2.4	−0.30 ± 0.03	0.812 ± 0.11	−0.09 ± 0.01
	10		31.2 ± 3.7 *	−13.1 ± 1.6	0.000114 ± 0.00001	−3.9 ± 0.5 *
13	1		25.8 ± 3.3 *	−7.7 ± 0.8	0.00481 ± 0.0006	−2.3 ± 0.3 *
	0.1		18.6 ± 2.4	−0.50 ± 0.05	0.707 ± 0.092	−0.15 ± 0.02
	10		27.1 ± 3.8 *	−9.0 ± 1.2	0.00195 ± 0.00020	−2.7 ± 0.3 *
14	1		19.2 ± 2.5	−1.1 ± 0.1	0.466 ± 0.056	−0.33 ± 0.04
	0.1		18.4 ± 2.4	−0.30 ± 0.03	0.812 ± 0.110	−0.09 ± 0.01
Virus control (DMSO)			18.1 ± 2.2	0	1.0	0
Cell control (DMSO)			≥37.0			

* $p \le 0.05$ compared to virus control.

We found that several polyphenolic compounds from *M. amurensis* at a concentration of 10 µg/mL significantly inhibited the replication of HSV-1 (Table 3). Maackin (**9**) and scirpusin A (**13**) caused a maximum reduction in viral DNA (4.6 log10 and 3.9 log10, respectively; $p < 0.05$) at this concentration. Compounds **10**, **11** and **14** also significantly reduced the level of viral DNA. Compounds **1**, **2**, **4**, and **7** possessed moderate anti-HSV-1 activity. Only compounds **9**, **11**, **13** inhibited the replication of HSV-1 at a concentration 1 µg/mL (Table 3). However, we did not observe any significant reduction in viral load when HSV-1 was pretreated with polyphenols at a concentration of 0.1 µg/mL. Polyphenols **3**, **5**, **6**, **8** and **12** did not demonstrate virucidal activity against HSV-1.

2.10. The Effect of Polyphenols on the Early Stages of HSV-1 Infection

The study of the virucidal activity of polyphenolic compounds from *M. amurensis* revealed that maackin (**9**) and scirpusin A (**13**) exhibited the highest inhibitory activity against the HSV-1 virus among the tested compounds (Tables 2 and 3). In order to study the mechanism of action of these compounds, two additional schemes of treatment of the virus and Vero cells with polyphenols were applied: the compounds were added before virus infection (pretreatment of cells) and after penetration of the virus into host cells (treatment of infected cells). The fourth method of exposure where the virus and polyphenols were added to cells simultaneously (simultaneous treatment) was previously described by us in [22]. The anti-HSV-1 activity of polyphenols was assessed using the cytopathic effect inhibition (CPE) assay. The obtained results are presented in Table 4.

Table 4. The effect of polyphenols **9** and **13** on the early stages of HSV-1 infection.

Compounds	CC_{50} (μg/mL)	Pretreatment of Cells			Simultaneous Treatment *			Treatment of Infected Cells		
		IC_{50} (μg/mL)	IC_{50} (μM/mL)	SI	IC_{50} (μg/mL)	IC_{50} (μM/mL)	SI	IC_{50} (μg/mL)	IC_{50} (μM/mL)	SI
9	269.6 ± 35.0	70.3 ± 9.1	144.6 ± 18.7	3.8 ± 0.5	17.6 ± 2.3	36.2 ± 4.7	15.3 ± 1.9	24.5 ± 3.2	50.4 ± 6.6	11.0 ± 1.4
13	199.4 ± 23.9	47.2 ± 6.1	100.4 ± 13.0	4.2 ± 0.6	13.8 ± 1.6	29.4 ± 3.4	14.4 ± 1.7	20.2 ± 2.8	43.0 ± 6.0	9.8 ± 1.2
Acyclovir	>1000		NA		2.1 ± 0.3	9.3 ± 1.3	430	0.4 ± 0.05	1.8 ± 0.2	>2500

* These data were published by us in [22].

The treatment of Vero cells with maackin (**9**) and scirpusin A (**13**) before infection (pretreatment of cells) had a weak effect on HSV-1 replication (IC_{50} 70.3 and 47.2 μg/mL, SI 3.8 and 4.2, respectively) (Table 4). A moderate inhibition of virus replication was observed after simultaneous treatment of cells with the virus and the tested polyphenols [22]. The application of polyphenols after virus adsorption and penetration to cells (treatment of infected cells) also had a moderate effect on HSV-1 replication (Table 4).

Thus, we showed that the tested polyphenols, especially maackin (**9**) and scirpusin A (**13**), possessed significant anti-HSV-1 activity mainly due to their direct virucidal effect. These compounds also inhibited virus–cell interactions when added simultaneously with the initiation of viral infection or even after virus adsorption and penetration to cells.

3. Discussion

Here, we continued our investigation of anti-HSV-1 activity and neuroprotective potential of isoflavonoids and stilbenes present in Maksar®.

The neuroprotective potential of polyphenolic compounds from *M. amurensis* heartwood was studied using models of 6-OHDA-and PQ-induced neurotoxicity [28,29]. Being a structural analogue of dopamine, 6-OHDA is an endogenous neurotoxin. It selectively penetrates dopaminergic and noradrenergic neurons, accumulates in the cytosol, and changes into dihydrophenylacetic acid or oxidizes to form hydrogen peroxide and para-quinone, which leads to ROS formation and oxidative stress in neurons, followed by cell death [28]. A dimeric stilbene scirpusin A (**13**) and a flavonoid liquiritigenin (**6**) were shown to be the most active compounds among the tested polyphenols preventing ROS formation in neuronal cells. These compounds significantly increased the viability of 6-OHDA-treated Neuro-2a cells. This effect was due to the ability of these polyphenols to reduce the level of intracellular ROS (Figures 4 and 5). Liquiritigenin (**6**) also increased mitochondrial membrane potential (Figure 6). Although resveratrol (**1**) previously demonstrated a protective effect against several type of insults that have shown a correlation to Parkinson's disease (PD) pathogenesis, it did not increase the viability of 6-OHDA-treated Neuro-2a cells in our study [30]. However, resveratrol (**1**) caused a significant decrease in intracellular ROS level and an increase in mitochondrial membrane potential in 6-OHDA-treated Neuro-2a cells. Maackiazin (**10**) had weak effect on cell viability, but considerably reduced ROS level in 6-OHDA-treated Neuro-2a cells. Compound **3** considerably inhibited ROS formation only at a concentration of 0.1 μM, which can be due to the possible prooxidant effect of this compound. Notably, the mechanism of action of polyphenolic compounds is complex and multitarget. This may include possible interactions with enzymes or receptors in a cell, which result a "bell-shaped" or reverse dependence of the dose-response for some polyphenolic compounds.

In contrast to 6-OHDA, PQ is an exogenous neurotoxin. PQ effects the redox cycle and increases ROS production [28]. Here, we evaluated the neuroprotective potential against the toxic effect of PQ of stilbene resveratrol (**1**), flavonoid liquiritigenin (**6**) and isoflavonoids, genistein (**2**), retusin (**3**), maackiain (**4**), tectorigenin (**5**), medicarpin (**7**), and ($\pm$)-3-hydroxyvestitone (**12**) (Figure 7).

Resveratrol (**1**) and liquiritigenin (**6**) at a concentration of 10 μM demonstrated considerable activity and increased the viability of PQ-treated cells by 17% and 15%, respectively. These compounds had significant effect even at a concentration of 1 μM (Figure 7a–c). Liquiritigenin (**6**) also effectively reduced the ROS level in PQ-treated Neuro-2a cells (Figure 7d). Isoflavonoids genistein (**2**), medicarpin (**7**), and maackiain (**4**) at this concentration showed weak effect on cell viability (Figure 7a–c).

Here, we also investigated the effect of polyphenolic compounds from *M. amurensis* on apoptosis profile in 6-OHDA-treated Neuro-2a cells by flow cytometry. We observed significant changes in apoptosis profile in Neuro-2a cells after 6-OHDA-treatment. Polyphenols **3**, **6** and **13** diminished the neurotoxic effect of 6-OHDA. These compounds considerably increased the percentage of live Neuro-2a cells and decreased the number of early apoptotic cells (Table 1, Figure 8).

The CPE inhibition assay and RT-PCR showed that dimeric stilbens with *trans*-double bond maackin (**9**) and scirpusin A (**13**) possessed the highest anti-HSV-1 activity among polyphenols **1–14** (Tables 2 and 3). The study of the effect of polyphenols **9** and **13** on the early stages of HSV-1 infection revealed that the main mechanism of their antiviral activity was direct interaction with the virus (virucidal activity) (Table 4). This activity may be due to the ability of these polyphenols to interact with the surface glycoproteins of HSV-1 or lipid components of the viral envelope.

Another mechanism of the antiviral action of polyphenols with pronounced antioxidant properties is the reduction in HSV-1-induced ROS production in the cells [31]. In our study, stilbenes **1**, **9–11**, **13**, **14** demonstrated much more pronounced anti-HSV-1 activity compared to isoflavonoids (**2–5**, **7**, **8**, **12**) (Tables 2 and 3), which may be due to the fact that stilbenes from *M. amurensis* possessed higher antiradical and antioxidant activity than isoflavonoids [6,22]. Notably, promising anti-HSV-1 activity of maackin (**9**) and scirpusin A (**13**) correlated with high antiradical and antioxidant activity of these stilbenes.

4. Materials and Methods

4.1. Plant Material

We provided the information on how *M. amurensis* was collected in our previous report [22].

4.2. Extraction and Isolation

We extracted *M. amurensis* heartwood (500 g) twice with a mixture of $CHCl_3$–EtOH at a ratio of 3:1 (*v*/*v*) for 3 h (60 °C) as previously described [22]. The air-dried extract (15 g) was applied to a polyamide column (50–160 μm, Sigma-Aldrich, St. Louis, MI, USA) and eluted with hexane–$CHCl_3$ and $CHCl_3$–EtOH solution systems to obtain fractions 1–8 and 9–18, respectively. We gradually increased the portions of $CHCl_3$ and EtOH in the solution systems (hexane/$CHCl_3$, *v*/*v*: 1:0, 10:0, 8:1, 5:1, 2:1, 1:1, 1:2, $CHCl_3$; $CHCl_3$/EtOH, *v*/*v*: 1:0, 100:1, 50:1, 40:1, 30:1, 20:1, 10:1, 5:1, 2:1). The fractions containing polyphenolic compounds according to HPLC data were selected for further purification.

Fraction 9 (630 mg) was subsequently chromatographed on a silica gel column (40–63 μm, Sigma-Aldrich, St. Louis, MI, USA) twice to obtain pure compounds **3** (6.3 mg) and **4** (5.2 mg). Fraction 10 (895 mg) was also chromatographed on a silica gel column twice to obtain individual compounds **2** (25.8 mg), **5** (7.8 mg), **7** (8.3 mg), **8** (33.6 mg), and **12** (3.7 mg). Compounds **6** (5.8 mg) and **10** (7.3 mg) were isolated from fraction 11 (654 mg) by silica gel chromatography. Fraction 13 (721 mg) contained compounds **1** (10.1 mg), **9** (12.4 mg), **11** (17.2 mg), **13** (14.3 mg), and **14** (3.1 mg). These polyphenols were separated on a silica gel column and subsequently purified on a C-18 column (YMC gel ODS-A 75 μm, Kyoto, Japan).

4.3. General Experimental Procedures

The CD spectra were measured on a Chirascan-plus Quick Start CD Spectrometer (Applied Photophysics Limited, Leatherhead, UK) (acetonitrile, 20 °C). We recorded the ^{1}H NMR spectra in acetone on an NMR Bruker AVANCE III DRX-700 instrument (Bruker, Karlsruhe, Germany).

4.4. HPLC–PDA–MS

We analyzed the fractions and individual polyphenolic compounds using a HPLC–PDA–MS system consisting of a CBM-20A system controller (Shimadzu, Kyoto, Japan), two LC-20AD pumps (Shimadzu, Kyoto, Japan), a DGU-20A3 degasser (Shimadzu, Kyoto, Japan), SIL-20A autosampler (Shimadzu, Kyoto, Japan), an SPD-M20A UV–VIS photodiode array detector (Shimadzu, Kyoto, Japan), and a LCMS-2020 mass detector (Shimadzu, Kyoto, Japan). The compounds were analyzed on a Discovery XR-ODS column (15 × 2.1 mm i.d.; 3.0 μm particle size; Supelco Analytical, Bellefonte, PA, USA) at a flow rate of 0.3 mL/min. The column was thermostated at 40 °C. The mobile phase consisted of 1% aqueous acetic acid (A) and acetonitrile containing 1% of acetic acid (B). We programmed the following gradient steps: 25–35% B (0–6 min), 35–60% B (6–11 min), 60–90% B (11–14 min), 90–25% B (14–16 min), 25% B (16–20 min). The injection volume was 2 μL. The MS settings were as follows: electrospray ionization (ESI), negative and positive ion modes, 150–800 m/z scans, N_2 drying gas (10 L/min), nebulizer gas flow 1.5 L/min, interface voltage 3.5 kV, and detector voltage 1.2 kV. We acquired and processed the data using Shimadzu LCMS Solution software (v. 5.42, Kyoto, Japan).

4.5. Neuro-2a Cell Line and Culture Condition

We acquired the murine neuroblastoma cell line Neuro-2a from American Type Culture Collection (ATCC® CCL-131, Manassas, VA, USA). We cultured the cells in DMEM supplied with 10% fetal bovine serum (Biolot, St. Petersburg, Russia) and 1% penicillin/streptomycin (Biolot, St. Petersburg, Russia) as previously described [22].

We performed the MTT assay to evaluate the cytotoxic activity of polyphenolic compounds against Neuro-2a as previously described [22].

4.6. In Vitro Model of 6-OHDA and PQ-Induced Neurotoxicity (Cell Viability Test)

After 24 h of adhesion, Neuro-2a cells (5 × 10^4 cells/mL) were treated with polyphenolic compounds at concentrations of 0.1, 1 and 10 μM for 1 h. Then, 100 μM of 6-OHDA or 1 mM of PQ (Sigma-Aldrich, St. Louis, MO, USA) were added. Cells incubated without inductors or with inductors were used as positive and negative control, respectively. Cell viability was measured after 24 h using the MTT assay.

4.7. Reactive Oxygen Species (ROS) Analysis in 6-OHDA- and PQ-Treated Cells

After 24 h of adhesion, Neuro-2a cells (5 × 10^4 cells/mL) were incubated with polyphenolic compounds at concentrations of 0.1, 1, and 10 μM for 1 h. Then, 120 μM of 6-OHDA or 1 mM of PQ were added to each well, and cells were incubated for 1 h. To study the ROS formation, 20 μL of 2,7-dichlorodihydrofluorescein diacetate solution (H2DCF-DA, Sigma-Aldrich, St. Louis, MO, USA) were added to each well, so that the final concentration was 10 μM. Then, the microplate was incubated for 30 min at 37 °C.

4.8. Mitochondrial Membrane Potential (MMP) Detection

The cells were incubated for 1 h in a 96-well plate (1 × 10^4 cells/well) with polyphenolic compounds (0.1, 1 and 10 μM). Then, 6-OHDA (120 μM) was added, and cell suspension was incubated for 1 h. Cells incubated without 6-OHDA and compounds and with 6-OHDA alone were used as positive and negative control, respectively. The tetramethylrhodamine methyl (TMRM) (Sigma-Aldrich, St. Louis, MO, USA) solution (500 nM) was added to each well and cells were incubated for 30 min at 37 °C. The intensity of fluorescence was measured using a PHERAstar FSplate reader (BMG Labtech, Ortenberg, Germany) at λ_{ex} = 540 nm and λ_{em} = 590 nm. The data were processed by MARS Data Analysis v. 3.01R2 (BMG Labtech, Ortenberg, Germany).

4.9. Apoptosis

The cells were incubated for 1 h in a 6-well plate (5×10^4 cells/mL) with polyphenolic compounds (10 μM). Then, 6-OHDA (100 μM) was added, and cell suspension was incubated for 24 h. After incubation, the maintenance media was collected, and the cells were washed by cold PBS twice and incubated with trypsin-EDTA solution. The cell suspension was washed with cold PBS twice again, centrifugated and then used for apoptosis detection by Muse® Annexin V & Dead Cell Kit in accordance with the manufacturer's instructions (Luminex, Austin, TX, USA). The fluorescence was measured with Muse® Cell Analyzer (Luminex, Austin, TX, USA) and the data were processed by Muse 1.5 Analysis software (Luminex, Austin, TX, USA).

4.10. HSV-1 Virus and Vero Cell Culture

The HSV-1 strain L2 obtained from N.F. Gamaleya Federal Research Centre for Epidemiology and Microbiology (Moscow, Russia) was grown in Vero cells as previously described [22].

We prepared the stock solutions of the tested compounds (10 mg/mL) in DMSO (Sigma, St. Louis, MO, USA). For the MTT assay and further studies of anti-HSV-1 activity, we diluted the stock solutions with DMEM so that the final concentration of DMSO was 0.5%.

4.11. Cytotoxicity of the Tested Compounds against Vero Cells

We used MTT assay to determine the cytotoxic effect of the tested compounds against Vero cells. We performed the MTT assay as previously described [22].

4.12. Virucidal Activity of Polyphenolic Compounds from M. amurensis

We determined the direct inhibitory effect of polyphenolic compounds from *M. amurensis* heartwood on HSV-1 (virucidal effect). We mixed the virus at an infectious dose (100 $TCID_{50}$/mL) with polyphenolic compounds (from 0.1 to 100 μg/mL) at a ratio of 1:1 (*v*/*v*) and incubated for an hour at 37 °C. Then, we infected the monolayer of Vero cells grown on 96-well plates (1×10^4 cells/well) with HSV-1 pretreated with the studied compounds. After 1 h adsorption at 37 °C, the cells were washed with phosphate-buffered saline (PBS) and overlaid by the maintenance medium with 1% FBS. The plates were incubated for 72 h at 37 °C in a CO_2-incubator. The cytopathic effect (CPE) inhibition assay and real-time reverse transcription-PCR (RT-PCR) assay were used to evaluate virucidal effect of the tested compounds.

4.12.1. Cytopathic Effect (CPE) Inhibition Assay

We previously described this method in detail in [22]. We performed the MTT assay to evaluate the antiviral activity of polyphenolic compounds. The viral inhibition rate (IR, %) was calculated according to Equation (1) [32]:

$$\text{IR, \%} = \frac{A_{tv} - A_{cv}}{A_{cd} - A_{cv}} \times 100, \quad (1)$$

where A_{tv} is the absorbance of cells infected with HSV-1 pretreated with the studied compounds,

A_{cv} is the absorbance of the virus-infected cells, but not treated with the studied compounds,

A_{cd} is the absorbance of control cells (untreated and non-infected).

We calculated 50% inhibitory concentration (the concentration of the compound that reduced the virus-induced CPE by 50%, IC_{50}) using a regression analysis of the dose–response curve [33,34]. We calculated the selectivity index (SI) as the ratio of CC_{50} to IC_{50}. Experiments were repeated three times.

4.12.2. Real-Time Polymerase Chain Reaction (RT-PCR) Assay

We performed the RT-PCR assay as we previously described in detail in [22]. Briefly, HSV-1 DNA was extracted after incubation from cell supernatants by using the DNA-sorb-AM (K1-12-100-CE) nucleic acid extraction kit (AmpliSens®, Moscow, Russia) according to the manufacturer's instructions. We treated the supernatants with a lysis solution that contained guanidine chloride (chaotropic agent) in the presence of sorbent (silica particles). As the elution solution was added, the DNA was adsorbed on silica particles and separated from the sorbent particles by centrifugation.

We detected the DNA of HSV-1 in the obtained samples using the AmpliSens® HSV I, II-FRT-100F PCR kit (AmpliSens®, Moscow, Russia) on a real-time PCR instrument Rotor-Gene Q (Qiagen, Hilden, Nordrhein-Westfalen, Germany) according to the manufacturer's instructions. The PCR reaction conditions were the following: initial denaturation at 95 °C for 15 min; then 5 cycles at 95 °C for 5 s, at 60 °C for 20 s, at 72 °C for 15 s; then 40 cycles at 95 °C for 5 s, at 60 °C for 20 s, and at 72 °C for 15 s. A negative sample was used as the amplification control for each run. The threshold cycle number C_t, was measured as the PCR cycle, where the amount of the amplified target reached the threshold value; the C_t values ≥ 37 indicated the absence of the HSV-1 DNA in the samples. We assessed the anti-HSV-1 activity of the studied compounds by the reduction in the viral DNA levels using the $2^{-\Delta Ct}$ method [35] according to Equation (2):

$$\Delta C_t = C_{tcv} - C_{ttv} \tag{2}$$

where C_t is the threshold cycle number measured as the PCR cycle, where the amount of the amplified target reached the threshold value,

C_{ttv} is the average C_t value for the infected samples (the virus was pretreated with polyphenolic compounds),

C_{tcv} is the average C_t value for virus control.

4.13. *The Effect of Polyphenols on the Early Stages of HSV-1 Infection*

In order to more deeply study the anti-HSV-1 effect of polyphenolic compounds that exhibited high virucidal activity we applied several additional schemes.

4.13.1. Pretreatment of Cells with Polyphenolic Compounds

The monolayer of cells was pretreated with different concentrations of the studied compounds for 2 h at 37 °C. After washing, the cells were infected with 100 $TCID_{50}$/mL of the virus at 37 °C for 1 h. Then, unabsorbed virus was removed by washing with PBS (Sigma, Saint-Louis, MO, USA), and cells were incubated in the maintenance medium until CPE appeared.

4.13.2. Treatment of Virus-Infected Cells with Polyphenols

The monolayer of cells was infected with the virus (100 $TCID_{50}$/mL) at 37 °C for 1 h, and then the cells were washed with PBS (Sigma, Saint-Louis, MO, USA) and treated with different concentrations of the studied compounds and incubated until CPE appeared.

After the incubation, MTT assay was performed and viral inhibition rate (IR), 50% inhibitory concentration (IC_{50}) and selectivity index (SI) of these polyphenolic compounds were calculated.

4.14. *Statistical Analysis*

The statistical analysis was performed using Statistica 10.0 software (StatSoft, Inc., Tulsa, OK, USA). CC_{50} and IC_{50} values were calculated using linear regression analysis of the dose-response curve. Results are presented as the mean values and standard deviations of three or more independent experiments. Comparison of differences between the indicators of the control and experimental groups was carried out using the Wilcoxon test for related samples. The differences were considered statistically significant if the p-value did not exceed 0.05.

5. Conclusions

The polyphenolic compounds contribute to the antiviral and neuroprotective activity of Maksar®. Stilbens scirpucin A (**13**) and maackin (**9**) possessed the highest anti-HSV-1 activity among the tested polyphenols. Notably, these stilbenes were more active than resveratrol, which was previously shown to possess antiherpetic properties [36,37]. Liquiritigenin (**6**) and scirpucin A (**13**) were mainly responsible for the neuroprotective potential of Maksar®. Scirpusin A (**13**) was the most promising compound possessing both anti-HSV-1 activity and neuroprotective potential.

Thus, Maksar® can be a perspective drug for the treatment of herpetic infections and neuronal disorders.

Supplementary Materials: The following supporting information can be downloaded at: https://www.mdpi.com/article/10.3390/ijms25084142/s1.

Author Contributions: Conceptualization, D.V.T., S.A.F., E.A.P., D.L.A., N.V.K., M.Y.S. and N.P.M.; methodology, D.V.T., E.A.P., E.S.M., D.V.B., N.V.K., A.I.K. and O.V.I.; investigation, D.V.T., S.A.F., D.V.B., E.A.P., E.S.M., N.V.K. and O.V.I.; writing—original draft preparation, D.V.T.; writing—review and editing, D.V.T., S.A.F., E.A.P., E.S.M., N.V.K. and D.V.B.; data analysis and interpretation, D.V.T., S.A.F., E.A.P., E.S.M., D.V.B. and N.V.K.; supervision, S.A.F. All authors have read and agreed to the published version of the manuscript.

Funding: This research was funded by Russian Science Foundation, grant number 22-75-00084, https://rscf.ru/en/project/22-75-00084/ accessed on 1 April 2022.

Institutional Review Board Statement: Not applicable.

Informed Consent Statement: Not applicable.

Data Availability Statement: The data are contained within the article and Supplementary Materials (Reference [38] is cited in the Supplementary Materials).

Conflicts of Interest: The authors declare no conflicts of interest.

References

1. Fedoreev, S.A.; Kulesh, N.I.; Glebko, L.I.; Pokushalova, T.V.; Veselova, M.V.; Saratikov, A.S.; Vengerovskii, A.I.; Chuchalin, V.S. Maksar: A preparation based on Amur Maackia. *Pharm. Chem. J.* **2004**, *38*, 605–610. [CrossRef]
2. Kulesh, N.I.; Fedoreev, S.A.; Veselova, M.V.; Denisenko, V.A.; Grigorchuk, V.P. Isoflavone glycosides from root bark of *Maackia amurensis*. *Chem. Nat. Compd.* **2020**, *56*, 415–419. [CrossRef]
3. Veselova, M.V.; Fedoreyev, S.A.; Tarbeeva, D.V.; Kulesh, N.I.; Kalinovskiy, A.I.; Kuzmich, A.S.; Kim, N.Y.; Grigorchuk, V.P. Cytotoxic prenylated polyphenolic compounds from *Maackia amurensis* root bark. *Nat. Prod. Commun.* **2017**, *12*, 1037–1040. [CrossRef]
4. Plotnikova, A.M.; Shulgau, Z.T.; Plotnikova, T.M.; Aliev, O.I.; Kulesh, N.I.; Mischenko, N.P.; Fedoreyev, S.A. Antithrombogenic and Antiplatelet activities of extract from *Maackia amurensis* wood. *Bull. Exp. Biol. Med.* **2009**, *2*, 204–207. [CrossRef] [PubMed]
5. Fedoreev, S.A.; Kulesh, N.I.; Mishchenko, N.P.; Ermakova, S.P.; Zvjagintseva, T.N. Anticancer Drug. RU Patent 2414920 C1, 1 December 2009.
6. Utkina, N.K.; Kulesh, N.I. Antioxidant activity of polyphenols and polyphenol complex from the far-eastern tree *Maackia amurensis*. *Pharm. Chem. J.* **2012**, *46*, 488–491. [CrossRef]
7. Putilova, E.A.; Ivanis, V.A.; Skljar, L.F. Clinical and immunological effects Maksar in chronic viral hepatitis. *Fundam. Res.* **2011**, *9*, 484–487.
8. Krylova, N.V.; Iunikhina, O.V.; Fedoreev, S.A.; Pott, A.B.; Persiyanova, E.V.; Mishchenko, N.P.; Tarbeeva, D.V.; Shchelkanov, M.Y. Anti-SARS-CoV-2 Activity of a Polyphenolic Complex from *Maackia amurensis*. *Bull. Exp. Biol. Med.* **2023**, *176*, 187–190. [CrossRef]
9. Cetin, A. In silico studies on stilbenolignan analogues as SARS-CoV-2 Mpro inhibitors. *Chem. Phys. Lett.* **2021**, *771*, 138563. [CrossRef]

10. Elsbaey, M.; Ibrahim, M.A.A.; Bar, F.A.; Elgazar, A.A. Chemical constituents from coconut waste and their in silico evaluation as potential antiviral agents against SARS-CoV-2. *S. Afr. J. Bot.* **2021**, *141*, 278–289. [CrossRef]
11. Krylova, N.V.; Fedoreev, S.A.; Iunikhina, O.V.; Mishchenko, N.P.; Pott, A.B.; Tarbeeva, D.V.; Shchelkanov, M.Y. Antiviral Agent against Human Herpesvirus Type 1 and Enterovirus B. RU Patent 2798659 C1, 23 June 2023.
12. He, Q.; Liu, H.; Huang, C.; Wang, R.; Luo, M.; Lu, W. Herpes Simplex Virus 1-Induced Blood-Brain Barrier Damage Involves Apoptosis Associated with GM130-Mediated Golgi Stress. *Front. Mol. Neurosci.* **2020**, *13*, 2. [CrossRef]
13. Santana, S.; Sastre, I.; Recuero, M.; Bullido, M.J.; Aldudo, J. Oxidative stress enhance neurodegeneration markers induced by herpes simplex virus type 1 infection in human neuroblastoma cells. *PLoS ONE* **2013**, *8*, e75842. [CrossRef] [PubMed]
14. Kavouras, J.H.; Prandovszky, E.; Valyi-Nagy, K.; Kovacs, S.K.; Tiwari, V.; Kovacs, M.; Shukla, D.; Valyi-Nagy, T. Herpes simplex virus type 1 infection induces oxidative stress and the release of bioactive lipid peroxidation by-products in mouse P19 neural cell cultures. *J. Neurovirol.* **2007**, *13*, 416–425. [CrossRef] [PubMed]
15. De Chiara, G.; Marcocci, M.E.; Sgarbanti, R.; Civitelli, L.; Ripoli, C.; Piacentini, R.; Garaci, E.; Grassi, C.; Palamara, A.T. Infectious agents and neurodegeneration. *Mol. Neurobiol.* **2012**, *46*, 614–638. [CrossRef] [PubMed]
16. Berry, N.; Suspène, R.; Caval, V.; Khalfi, P.; Beauclair, G.; Rigaud, S.; Blanc, H.; Vignuzzi, M.; Wain-Hobson, S.; Vartanian, J.P. Herpes simplex virus type 1 infection disturbs the mitochondrial network, leading to type I interferon production through the RNA Polymerase III/RIG-I pathway. *mBio* **2021**, *12*, e0255721. [CrossRef] [PubMed]
17. Manivanh, R.; Mehrbach, J.; Charron, A.J.; Grassetti, A.; Cerón, S.; Taylor, S.A.; Cabrera, J.R.; Gerber, S.; Leib, D.A. Herpes simplex virus 1 ICP34.5 alters mitochondrial dynamics in neurons. *J. Virol.* **2020**, *94*, e01784-19. [CrossRef] [PubMed]
18. Limongi, D.; Baldelli, S. Redox imbalance and viral infections in neurodegenerative diseases. *Oxid. Med. Cell. Longev.* **2016**, *2016*, 6547248. [CrossRef]
19. Cymerys, J.; Chodkowski, M.; Slonska, A.; Krzyzowska, M.; Banbura, M.W. Disturbances of mitochondrial dynamics in cultured neurons infected with human herpesvirus type 1 and type 2. *J. Neurovirol.* **2019**, *25*, 765–782. [CrossRef] [PubMed]
20. Area-Gomez, E.; de Groof, A.; Bonilla, E.; Montesinos, J.; Tanji, K.; Boldogh, I.; Pon, L.; Schon, E.A. A key role for MAM in mediating mitochondrial dysfunction in Alzheimer disease. *Cell Death Dis.* **2018**, *9*, 335. [CrossRef] [PubMed]
21. Grunewald, A.; Kumar, K.R.; Sue, C.M. New insights into the complex role of mitochondria in Parkinson's disease. *Prog. Neurobiol.* **2019**, *177*, 73–93. [CrossRef]
22. Tarbeeva, D.V.; Berdyshev, D.V.; Pislyagin, E.A.; Menchinskaya, E.S.; Kim, N.Y.; Kalinovskiy, A.I.; Krylova, N.V.; Iunikhina, O.V.; Persiyanova, E.V.; Shchelkanov, M.Y.; et al. Neuroprotective and antiherpetic properties of polyphenolic compounds from *Maackia amurensis* heartwood. *Molecules* **2023**, *28*, 2593. [CrossRef]
23. Levai, A. Utilization of the chiroptical spectroscopies for the structure elucidation of flavonoids and related benzopyran derivatives. *Acta Chim. Slov.* **1998**, *45*, 267–284. [CrossRef]
24. Moreno-Celis, U.; López-Martínez, F.J.; Cervantes-Jiménez, R.; Ferríz-Martínez, R.A.; Blanco-Labra, A.; García-Gasca, T. Tepary Bean (*Phaseolus acutifolius*) Lectins Induce Apoptosis and Cell Arrest in G0/G1 by P53 (Ser46) Phosphorylation in Colon Cancer Cells. *Molecules* **2020**, *25*, 1021. [CrossRef] [PubMed]
25. Figueiredo, G.G.; Coronel, O.A.; Trabuco, A.C.; Bazán, D.E.; Russo, R.R.; Alvarenga, N.L.; Aquino, V.H. Steroidal saponins from the roots of *Solanum sisymbriifolium* Lam. (Solanaceae) have inhibitory activity against dengue virus and yellow fever virus. *Braz. J. Med. Biol. Res.* **2021**, *54*, e10240. [CrossRef]
26. Low, Z.X.; OuYong, B.M.; Hassandarvish, P.; Poh, C.L.; Ramanathan, B. Antiviral activity of silymarin and baicalein against dengue virus. *Sci. Rep.* **2021**, *11*, 21221. [CrossRef]
27. Fernandes, L.S.; da Silva, M.L.; Dias, R.S.; da S. Lucindo, M.S.; da Silva, Í.E.P.; Silva, C.C.; Teixeira, R.R.; de Paula, S.O. Evaluation of Antiviral Activity of Cyclic Ketones against Mayaro Virus. *Viruses* **2021**, *13*, 2123. [CrossRef]
28. Bove, J.; Prou, D.; Perier, C.; Przedborski, S. Toxin-induced models of Parkinson's disease. *NeuroRX* **2005**, *2*, 484–494. [CrossRef]
29. Bastias-Candia, S.; Zolezzi, J.M.; Inestrosa, N.C. Revisiting the paraquat-induced sporadic Parkinson's disease-like model. *Mol. Neurobiol.* **2019**, *56*, 1044–1055. [CrossRef]
30. Okawara, M.; Katsuki, H.; Kurimoto, E.; Shibata, H.; Kume, T.; Akaike, A. Resveratrol protects dopaminergic neurons in midbrain slice culture from multiple insults. *Biochem. Pharmacol.* **2007**, *73*, 550–560. [CrossRef] [PubMed]
31. Sebastiano, M.; Chastel, O.; de Thoisy, B.; Eens, M.; Costantini, D. Oxidative stress favours herpes virus infection in vertebrates: A meta-analysis. *Curr. Zool.* **2016**, *62*, 325–332. [CrossRef] [PubMed]
32. Ngan, L.T.M.; Jang, M.J.; Kwon, M.J.; Ahn, Y.J. Antiviral activity and possible mechanism of action of constituents identified in *Paeonia lactiflora* root toward human rhinoviruses. *PLoS ONE* **2015**, *10*, e0121629. [CrossRef]
33. Weislow, O.S.; Kiser, R.; Fine, D.L.; Bader, J.; Shoemaker, R.H.; Boyd, M.R. New soluble-formazan assay for HIV-1 cytopathic effects: Application to high-flux screening of synthetic and natural products for AIDS-antiviral activity. *J. Natl. Cancer Inst.* **1989**, *81*, 577–586. [CrossRef] [PubMed]
34. Shchelkanov, M.Y.; Sakhuriya, I.B.; Burunova, V.V.; Pavlova, T.V.; Kornilaeva, G.V.; Karamov, E.V.; Eremin, V.F. Dehydrogenase activity of infected cells and biological properties of various HIV-1 variants. *Biochemistry* **1999**, *64*, 513–519.
35. Bovard, D.; van der Toorn, M.; Schlage, W.K.; Constant, S.; Renggli, K.; Peitsch, M.C.; Hoeng, J. Iota-carrageenan extracted from red algae is a potent inhibitor of SARS-CoV-2 infection in reconstituted human airway epithelia. *Biochem. Biophys. Rep.* **2022**, *29*, 101187. [CrossRef] [PubMed]

36. Annunziata, G.; Maisto, M.; Schisano, C.; Ciampaglia, R.; Narciso, V.; Tenore, G.C.; Novellino, E. Resveratrol as a novel anti-herpes simplex virus nutraceutical agent: An overview. *Viruses* **2018**, *10*, 473. [CrossRef]
37. Mattio, L.M.; Catinella, G.; Pinto, A.; Dallavalle, S. Natural and nature-inspired stilbenoids as antiviral agents. *Eur. J. Med. Chem.* **2020**, *202*, 112541. [CrossRef]
38. Frisch, M.J.; Trucks, G.W.; Schlegel, H.B.; Scuseria, G.E.; Robb, M.A.; Cheeseman, J.R.; Scalmani, G.; Barone, V.; Petersson, G.A.; Nakatsuji, H.; et al. *Gaussian 16, Revision A.03*; Gaussian, Inc.: Wallingford, CT, USA, 2016.

Disclaimer/Publisher's Note: The statements, opinions and data contained in all publications are solely those of the individual author(s) and contributor(s) and not of MDPI and/or the editor(s). MDPI and/or the editor(s) disclaim responsibility for any injury to people or property resulting from any ideas, methods, instructions or products referred to in the content.

International Journal of
Molecular Sciences

MDPI

Article

Endoplasmic Reticulum Stress Induces Vasodilation in Liver Vessels That Is Not Mediated by Unfolded Protein Response

Sergejs Zavadskis [1], Anna Shiganyan [1], Andrea Müllebner [2], Johannes Oesterreicher [1], Wolfgang Holnthoner [1], Johanna Catharina Duvigneau [2] and Andrey V. Kozlov [1,*]

1 Ludwig Boltzmann Institute for Traumatology, The Research Center in Cooperation with AUVA, Austrian Cluster for Tissue Regeneration, Donaueschingenstraße 13, 1200 Vienna, Austria; sergejs.zavadskis@trauma.lbg.ac.at (S.Z.)

2 Department of Biological Sciences and Pathobiology, Institute of Medical Biochemistry, University of Veterinary Medicine, Veterinärplatz 1, 1210 Vienna, Austria

* Correspondence: andrey.kozlov@trauma.lbg.ac.at

Abstract: There is a growing body of evidence that ER stress and the unfolded protein response (UPR) play a key role in numerous diseases. Impaired liver perfusion and ER stress often accompany each other in liver diseases. However, the exact impact of ER stress and UPR on the hepatic perfusion is not fully understood. The aim of this study was to disclose the effect of ER stress and UPR on the size of liver vessels and on the levels of Ca^{2+} and nitric oxide (NO), critical regulators of vascular tonus. This study was carried out in precisely cut liver tissue slices. Confocal microscopy was used to create 3D images of vessels. NO levels were determined either using either laser scan microscopy (LSM) in cells or by NO-analyser in medium. Ca^{2+} levels were analysed by LSM. We show that tunicamycin, an inducer of ER stress, acts similarly with vasodilator acetylcholine. Both exert a similar effect on the NO and Ca^{2+} levels; both induce significant vasodilation. Notably, this vasodilative effect persisted despite individual inhibition of UPR pathways—ATF-6, PERK, and IRE1—despite confirming the activation of UPR. Experiments with HUVEC cells showed that elevated NO levels did not result from endothelial NO synthase (eNOS) activation. Our study suggests that tunicamycin-mediated ER stress induces liver vessel vasodilation in an NO-dependent manner, which is mediated by intracellular nitrodilator-activatable NO store (NANOS) in smooth muscle cells rather than by eNOS.

Keywords: endoplasmic reticulum stress; unfolded protein response; ex-vivo model; vascular tonus; Ca^{2+}; NO; nitrodilators

1. Introduction

The liver produces large amounts of secretory proteins in the endoplasmic reticulum (ER) of hepatocytes. Apart from the synthesis, folding, and secretion of proteins, the ER is an important site for the intercellular calcium homeostasis [1] and subsequent calcium dependent regulatory pathways. Conditions that disrupt ER homeostasis create a cellular state commonly referred to as 'ER stress'. There is a growing body of evidence that ER stress plays a key role in numerous physiological and pathological processes [2].

Induction of ER stress causes disturbances in protein synthesis, calcium homeostasis, and triggers the unfolded protein response (UPR), which is a cellular adaptive mechanism aiming at the recovery of ER function. The UPR has three branches, which are characterized by the elevation of stress proteins IRE1, ATF6, and PERK [3]. IRE1 catalyses the splicing of XBP1, but it also interacts with other branches. For instance, the inhibition of the PERK pathway can be involved in an activation of the IRE1 pathway and splicing of XBP [4,5], while ATF6 elevates splicing by increasing levels of XBP1 [6]. Recent advances in the understanding of the regulation of UPR signalling have shown that it interferes with numerous signalling pathways [7]. ER stress has been implicated in numerous different human diseases [2,8].

Particularly important is the role the ER stress plays in liver pathophysiology. The well-functioning ER is required in liver for synthesis of clotting factors, numerous other proteins for metabolism of drugs and toxins. Hepatocytes, the predominant cell type in the liver, are enriched in the ER and they are responsible for the majority of liver functions. Therefore, most liver disorders are accompanied by ER stress [9]. In our previous study, we observed that also haemorrhagic shock (HS) induces sustained ER stress in the rat liver [8]. The liver is one of the organs affected foremost and instantly by haemorrhagic shock, causing dramatic changes in blood flow [10]. Other pathologies that are associated with a compromised liver perfusion also induce ER stress. These pathologies include liver tumours, liver trauma, and liver cirrhosis [11]. The frequent appearance of perfusion disorders affecting the liver may be explained by the fact that it has a unique dual blood supply, which originates from the hepatic artery (25%) and the portal vein (75%). Although impaired liver perfusion and ER stress often accompany each other, the exact impact of ER stress and UPR on the hepatic perfusion is not fully understood [12].

It is known that ER stress is capable of influencing vascular homeostasis [13], which could play a pivotal role in influencing impaired organ perfusion upon shock. Several reports suggest that ER stress can be associated with hypertension and vascular remodelling [14]. It has been shown that the induction of ER stress in renal arteries reduces endothelium-derived relaxing factors and contracting factor (EDCF)-mediated contractions [15]. In the case of hypovolemia, blood flow is predominantly redirected to vital organs, such as liver, through a process known as 'centralized' circulation, which is believed to be achieved exclusively via peripheral vasoconstriction. The contribution of blood flow through vital organs is less well understood.

Taking into account the evidence described above, we assumed that induction of the ER stress could affect liver perfusion. It is very difficult to address this question in vivo, because of systemic influences and the inability to visualize biochemical changes in single vessels within entire organs like the liver. These aspects, however, can be explored in a model of precisely cut liver slices, which preserve both the liver's structure and cellular functions, but it is free of systemic influences.

The general aim of this study was to understand whether activation of ER stress in the liver could influence liver perfusion. Specifically, we examined how tunicamycin, an inducer of ER stress, and inhibitors of UPR branches impact the signalling molecules regulating vascular tonus (Ca^{2+} and nitric oxide) and the single vessel volumes in precisely cut rat liver slices.

2. Results

2.1. Tunicamycin Induces Protein Aggregation Characteristics for ER Stress

Firstly, we examined whether the liver tissue in our experimental model is capable of responding to the ER stress-inducing agent, tunicamycin. The liver tissue slices were stained with Thioflavin-T, a fluorescent probe, known for its selective binding to amyloid fibrils, which are indicative of protein aggregation [3], a typical sign for ER stress (Figure 1).

The data presented in Figure 1 clearly show that tunicamycin induces ER stress. In order to test whether the induction of ER stress activates UPR in our model, we determined an early marker of UPR, the XBP1 splicing. XBP1 is a segment of IRE1 branch of UPR, but its splicing can be modulated by other branches of UPR [16]. Due to this fact, we determined effect of different UPR branch inhibitors on the XBP1s.

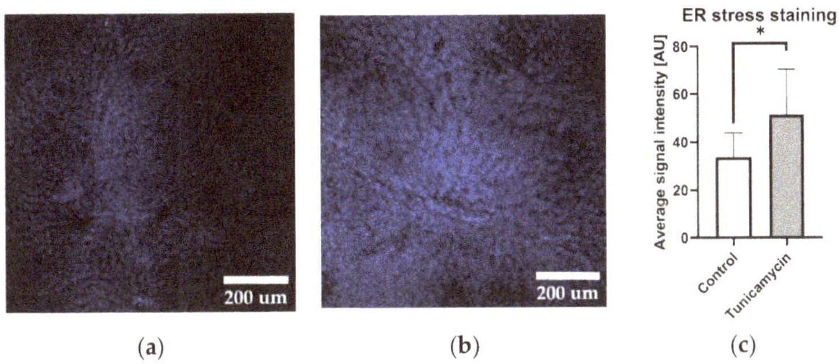

Figure 1. ER stress in vital liver tissue slices after 90 min visualized by staining with Thioflavin-T (10× magnification). (**a**) Typical fluorescence levels of a control tissue slice. (**b**) Representative fluorescence levels of liver tissue slices exposed to tunicamycin for 90 min. (**c**) Signal intensity quantification (whole image field) in control (treated with tunicamycin vehicle DMSO) and tunicamycin-treated samples. Unpaired *t*-test; data are presented as means ± SD; $n = 8$; * = $p < 0.05$.

2.2. Activation and Inhibition of UPR Pathways

To accomplish this, we employed novel inhibitors specific to distinct UPR pathways: MKC8866 targeting IRE1, GSK2606414 (GSK414) targeting PERK, and Exendin-4 (Exen4) targeting ATF-6. We observed that the increased XBP1 splicing induced by tunicamycin (Tun) was abolished by MKC8866 (an inhibitor of IRE1) and Exen4 (an inhibitor of ATF-6), while *G*SK414 augmented XBP1s, as depicted in Figure 2.

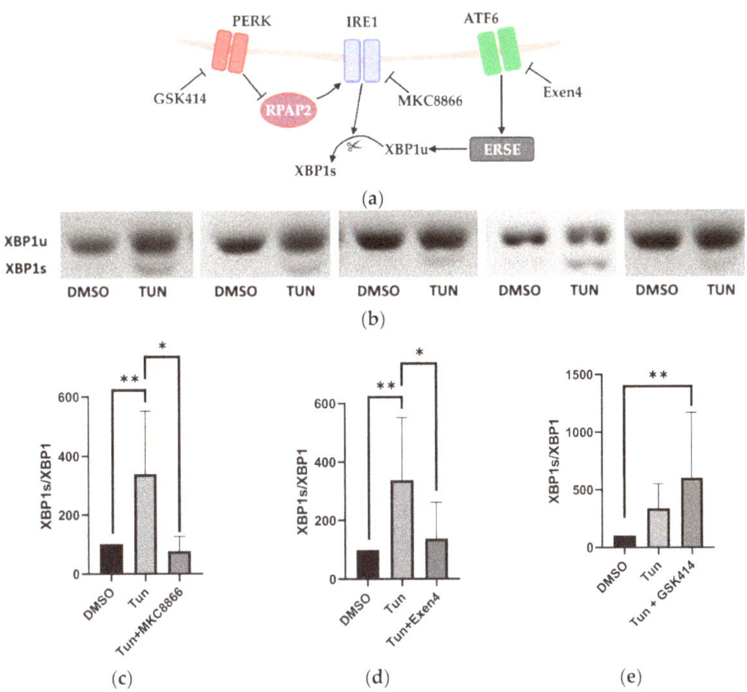

Figure 2. Determination of X-box binding protein 1 (XBP1) splicing as ratio between unspliced and spliced XBP1 mRNA in % of control in precision-cut tissue slices subjected to tunicamycin in combined with inhibitors targeting specific UPR pathways. (**a**) Scheme illustrating 3 UPR sensors with

respective inhibitors and interactions. (**b**) Representative images of XBP1 PCR products for slices treated with DMSO (vehicle of tunicamycin) and tunicamycin-treated slices. (**c**) Simultaneous administration of the IRE1 inhibitor MKC8866 with tunicamycin results in a substantial reduction in XBP1 indicated by a reduced ratio of spliced XBP1 (XBP1s) to the unspliced XBP1 (XBP1u) amount of PCR product. (**d**) Simultaneous application of the ATF6 inhibitor Exen4 with tunicamycin reduces XBP1 splicing. (**e**) Concurrent administration of the PERK inhibitor GSK414 with tunicamycin increases XBP1 splicing. The samples were taken 90 min after treatment. One-way ANOVA; data are presented as means ± SD excluding outliers DMSO n = 11; Tun n = 11; Tun + UPR inhibitor n = 4/8/5; outliers were excluded with 5% ROUT; * = $p < 0.05$; ** = $p < 0.005$.

Since ER stress disrupts the Ca^{2+} homeostasis, which is a key regulator of vascular tonus, we investigated intracellular levels of Ca^{2+} in response to tunicamycin and compared the effects with two well-known vasotropic agents, acetylcholine (Ach) and phenylephrine (Phn), a vasodilator and vasoconstrictor, respectively. To achieve this, we employed FLUO-4 AM calcium-sensitive staining.

2.3. Vascular Calcium Levels

As expected, the addition of the vasodilator Ach led to a decline in calcium levels. Likewise, the addition of the vasoconstrictor Phn resulted in an increase in calcium levels. The application of tunicamycin resulted in a reduction in calcium signal levels, similarly to that observed with acetylcholine, as illustrated in Figure 3.

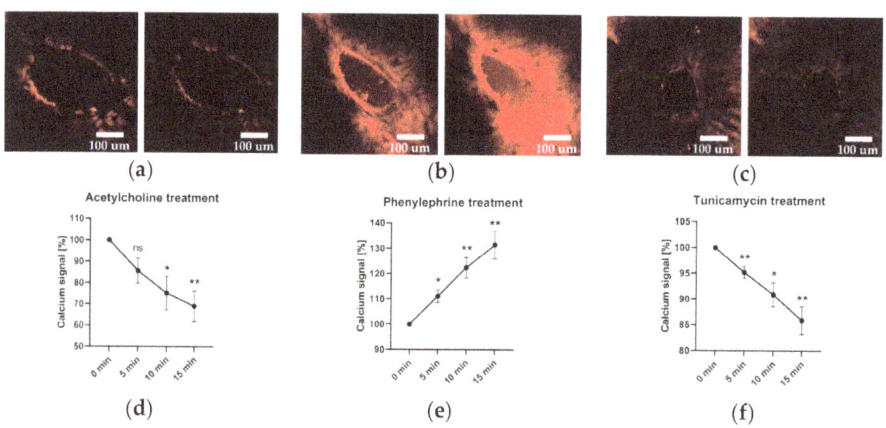

Figure 3. Effect of acetylcholine, phenylephrine, and tunicamycin on the intracellular calcium levels, visualized with FLUO-4 AM, f.c. 5 µM. The images are arranged in pairs, wherein the left image represents calcium levels immediately after treatment (0-min time point), while the corresponding right image represents the 15-min time point. The panels (**a**–**c**) are representative images showing the intracellular calcium levels upon treatment with acetylcholine, phenylephrine, and tunicamycin, respectively. Quantification of these changes are presented in panels (**d**–**f**) for acetylcholine, phenylephrine, and tunicamycin treatments, respectively. One-way ANOVA; data are presented as means ± SD; n = 4 biological replicates; n = 12 technical replicates; ns—not significant; * = $p < 0.05$; ** = $p < 0.005$.

Since the vasodilatory effects of acetylcholine are mediated by NO, we subsequently examined NO levels in liver slices treated with Ach, Phn, and tunicamycin. To accomplish this, we used DAF-FM, a nitric oxide-sensitive fluorescent dye. The images were taken 90 min after treatments.

2.4. Vascular Nitric Oxide Levels

The data presented in Figure 4 show that treatment with Ach and tunicamycin elevate the levels of NO in the vessel wall, while Phn does not change NO levels. These observations strongly suggest that tunicamycin modulates the intracellular Ca^{2+} via NO in a similar way as acetylcholine in our model. Therefore, we hypothesized that tunicamycin acts as a vasodilator. To validate this hypothesis, we assessed the effect of tunicamycin on changes of the vessel lumen (area).

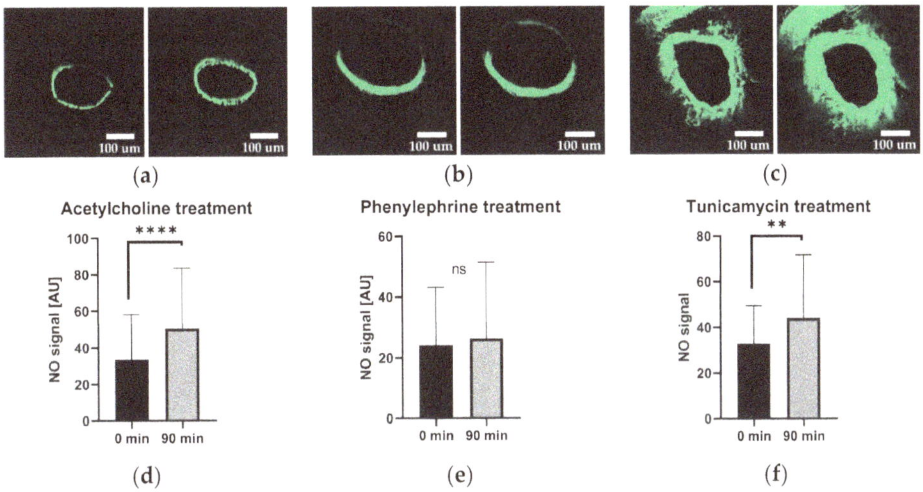

Figure 4. Effect of acetylcholine, phenylephrine, and tunicamycin on the intracellular nitric oxide levels, visualized with 10 μM DAF-FM. The representative images are arranged in pairs, wherein the left image depicts nitric oxide levels immediately after treatment (0-min time point), while the corresponding right image represents the 90-min time point, the scale bar represents 100 μm distance. (**a**) Representative images of NO levels following acetylcholine treatment. (**b**) Representative images of NO levels following phenylephrine treatment. (**c**) Representative images of NO levels following tunicamycin treatment. Quantification of signal changes are presented in panels (**d–f**) for acetylcholine, phenylephrine, and tunicamycin treatments, respectively. Statistical evaluation was carried out by Wilcoxon matched-pairs signed rank test (Shapiro–Wilk test not passed); data are presented as means ± SD; n = 5 biological replicates; n = 70 technical replicates; ns—not significant; ** = $p < 0.005$, **** = $p < 0.0001$.

2.5. Vessel Volume Changes in Response to Tunicamycin

Employing confocal microscopy, we captured 3D images of the vessels through 25 sections both prior to treatment, and at 45 min and 90 min after treatment, which rendered a three-dimensional image, as depicted in Figure 5a. The determination of changes in the vessel lumen was accomplished utilizing ImageJ. Only those vessel cuts which featured entire (or complete) vessel rings were taken into consideration; partial or damaged vessels were omitted from the analysis.

The data shown above suggest that tunicamycin has indeed a vasodilatory effect in liver vessels, which is mediated by an increase in intravascular NO generation. Increased NO generation upon tunicamycin treatment was additionally confirmed by elevated NO levels occurring in the incubation medium (Figure 6b). We next hypothesized that one specific branch of UPR triggers NO release, which mediates the vasodilatory effect of tunicamycin.

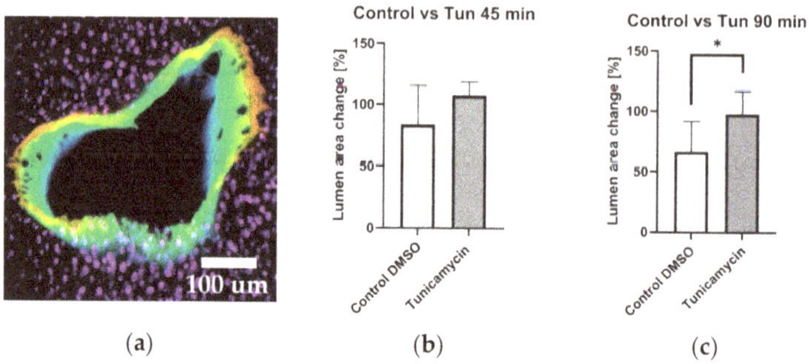

Figure 5. Estimation of the volumetric changes of vessels in response to tunicamycin. (**a**) Representative 3D image of an intact vessel with nuclear counterstain of surrounding hepatocytes. The liver slices were stained with DAF-FM and NucSpot650 Live for detection of NO and nuclei, respectively. (**b**) Effect of tunicamycin on blood vessel area after 45-min exposure to tunicamycin. (**c**) Effect of tunicamycin on blood vessel area after 90-min exposure to tunicamycin. Unpaired *t*-test; data are presented as means ± SD; $n = 6$; * = $p < 0.05$.

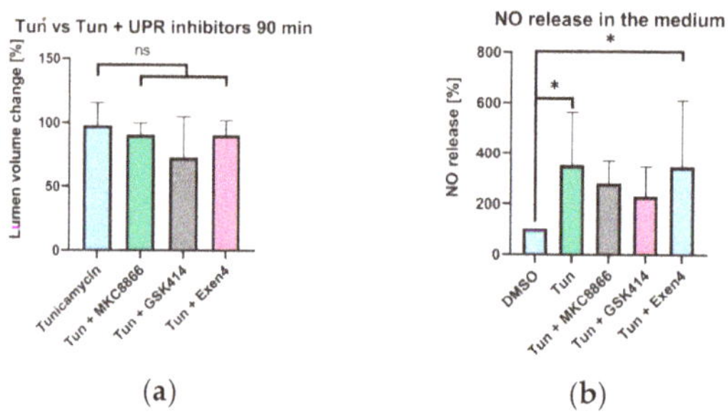

Figure 6. Effect of UPR inhibitors on the elevated lumen volume and NO generation induced by tunicamycin. (**a**,**b**) Effect of MKC8866, GSK414, and Exen4, the selective inhibitors of IRE1, PERK, and ATF6 pathways, respectively, on the vessel lumen volume (**a**) and on the levels of NO (**b**). $n = 6$. One-way ANOVA; data are presented as means ± SD; ns—not significant; * = $p < 0.05$.

2.6. Selective UPR Pathway Inhibition

To prove this hypothesis, we selectively inhibited various branches of the UPR and assessed their influence on the vasodilative effect of tunicamycin (Figure 6). We found that tunicamycin-induced vasodilation was not affected by selective UPR inhibitors, suggesting that it is not mediated by UPR (Figure 6a). Further, NO generation was also not modulated by UPR inhibitors (Figure 6b). Our data thus suggest it is the tunicamycin-induced ER stress that targets an NO-generation mechanism. In the vasculature several NO-generation mechanisms exist. The best known is the classical mechanism of Ach action, which involves NO generation by endothelial NOS (eNOS) of endothelial cells. In addition, several endogenous NO donors are located in smooth muscle cells, also capable of regulating vascular tonus.

Carefully analysing the images obtained in vessels treated with Ach and with tunicamycin, we observed that in the presence of tunicamycin, the NO-mediated fluorescence was present also in the deeper layers of the vessel walls, making the walls appear thicker upon treatment with tunicamycin compared to vessels treated with Ach (Figure A2). This suggests that NO may have been also generated in the smooth muscle layer of the vessel.

2.7. Experiments with Endothelial Cells

In order to clarify whether tunicamycin mediates vasodilation using the same mechanism as Ach, we performed experiments with cultured endothelial cells (HUVECs). Upon treating the cells with Ach or tunicamycin, we determined NO generation by fluorescent dye, DAF-FM. To dissect the contribution of eNOS we examined the effect of L-NAME an inhibitor of NOS. The abolished NO-related fluorescence of DAF-FM in the presence of L-NAME demonstrates that eNOS is indeed the source of NO release in HUVECs. Tunicamycin did not trigger NO release of the cells. In the contrary, we observed a strong decrease in the NO-levels in response to tunicamycin (Figure A1). This demonstrates that tunicamycin in contrast to Ach operates by a different NO-generation mechanism, possibly involving the part of the vessel located below the endothelium, such as the smooth muscle cells.

3. Discussion

In this study, we have shown that precisely cut liver slices (PCLSs) are a suitable model to study the functionality of hepatic vasculature and its response to the different treatments, such as the effects of vasodilators, vasoconstrictors, and activation of specific pathological processes such as induction of ER stress. The advantage of this model is that it allows the accessing of the local effects of treatments without systemic influences and enables to access specific morphological structures within the liver tissue.

We had shown that tunicamycin induces ER stress in the liver slice tissue using Thioflavin T (Figure 1a,b), which stains protein aggregates and amyloid fibrils. We did not observe any differences in the staining of liver tissue that would target specific structures; the staining was similar throughout the entire liver tissue, suggesting that tunicamycin induces ER stress in all types of liver cells. To determine whether or not ER stress was activating UPR in our model, we studied splicing of XBP1. XBP1 splicing is executed by the IRE1 branch, but can be additionally modulated by PERK and the ATF6 pathway. Our observation that inhibition of the PERK pathway resulted in an increased XBP1 splicing by trend (Figure 2) supports previous studies [5,16]. The PERK pathway typically activates in later stages, when the cell is not capable of recovering from ER stress. Once PERK-ATF4-CHOP is activated, it leads to apoptosis [17]. Further, addition of Exendin-4, predominantly inhibiting ATF6, has been shown to attenuate the XBP1 splicing [4], and we also found decreasing XBP1 splicing in our model. The data presented in Figure 2 suggest that tunicamycin is potentially activating all branches of UPR in our model.

Ca^{2+} and NO are two of the most important regulators of vascular tonus. In this study, we examined effect of Ach, a well-established vasodilator, Phn, a widely used vasoconstrictor, and tunicamycin in portal vessels of the liver. Ach and Phn were employed as references for changes in Ca^{2+} (Figure 3) and NO metabolism (Figure 4) upon induction of vasodilation and vasoconstriction, respectively. The results presented in the Figure 3 revealed that both Ach and tunicamycin caused a decrease in intracellular Ca^{2+} levels, while Phn caused an increase in Ca^{2+} levels (Figure 3). These data are in line with the known mechanisms of the action of Ach and Phn, which regulate vascular tonus by modulation of intracellular Ca^{2+} levels in smooth muscle cells. Phn directly interacts with smooth muscle cells, while Ach classically involves NO generated from eNOS of endothelial cells. NO in turn diffuses into smooth muscle cells and induces vasorelaxation via the NO/cGMP signalling cascade. The data shown in the Figure 3 prompted us to assume that tunicamycin-mediated induction of ER stress also modulates Ca^{2+} metabolism in a manner similar to Ach. Determination of NO levels in the vascular wall confirmed this assumption. We demonstrated that both Ach and tunicamycin increased intracellular NO levels, while Phn did not exhibit any detectable change in the NO levels (Figure 4). These observations further support our assumption that both Ach and tunicamycin operate via NO generation. An elevated NO generation was also confirmed by the increased concentrations of NO in the incubation medium upon treatment with tunicamycin (Figure 6b). The physiological effects of both Ach and Phn in our model are in line with observations using in-vivo findings [18,19]. The similarity in the actions of the ER stress inducer tunicamycin and Ach suggests that they both induce vasodilation mediated by NO and Ca^{2+}.

To quantify the change in vessel lumen and assess its response to tunicamycin, we employed laser scanning confocal microscopy to capture a comprehensive 3D image of the vessel and calculated the vessel area as described in the methods section. We observed that after 45 min of incubation with tunicamycin, the vessel area had a strong trend to increase (Figure 5a); this trend became significant after 90 min of incubation (Figure 5b). The findings illustrated in Figure 5 suggest that tunicamycin induces vasodilation in a NO-dependent manner. We assumed that this mechanism was activated by one of the branches of UPR, as previously suggested [20]. To test this assumption, we selectively inhibited various branches of the UPR and assessed their impact on the vasodilatory effect of tunicamycin and generation of NO. Given that the effects of tunicamycin were most pronounced at 90 min post-treatment, we evaluated effects of all inhibitors at this specific time point.

Surprisingly, none of the inhibitors exhibited a significant effect on the tunicamycin-mediated vasodilation (Figure 6a). Similarly, none of the inhibitors exerted a significant effect on the levels of NO released in the medium. Therefore, we could rule out UPR as mediator the vasotropic effects of tunicamycin. Instead, vasodilation may be elicited by the provoked ER stress itself. One possibility is that ER stress activates eNOS in endothelial cells, resulting in an increase in NO generation and subsequent vasodilation. This could be due to the release of Ca^{2+}. The activation of Ca^{2+}-dependent enzymes requires much lower levels of Ca^{2+} compared to the Ca^{2+} levels required for muscle contraction. That could be the reason why we did not see any increase in Ca^{2+} levels in response to Ach and tunicamycin.

To prove this assumption, we performed experiments with a HUVEC culture (Figure A1). The cells were treated with the same concentration of tunicamycin and incubated for 90 min. To assess the contribution of eNOS, we treated cells with the NOS inhibitor L-NAME. However, instead of the expected increase in the levels of NO, we observed a strong decrease in the NO levels, which was further completely abolished by L-NAME. We therefore can exclude involvement of eNOS generating NO. Nonetheless, the decrease in NO generation in response to tunicamycin is in line with previously reported data on the inhibition of NOS upon ER stress conditions recently reported [21].

Therefore, we questioned by which mechanism tunicamycin increased the NO generation in our portal liver vessels, which we determined using two independent methods, by NO specific fluorescent dye DAF-FM in the vessel walls and by chemiluminescence as a measure for NO products present in the incubation medium. Further, we show that this NO must have another source than the endothelial NOS. Recently, an alternative pathway of NO generation, which can induce vasodilation, was described by Liu et al. [22]. This pathway operates directly in smooth muscle cells. The authors describe a pool able to release vasodilatory NO under pathological conditions. Considering that induction of ER stress represents a stressful condition, it is well possible that tunicamycin activated this mechanism in our model as well. Indeed, when comparing the LSM images in the presence of tunicamycin and Ach, the vessel walls tended to be thicker upon treatment with tunicamycin compared to Ach (Figure A2). Therefore, our data suggest an involvement of smooth muscle cells in response to tunicamycin. We conclude that tunicamycin operates as vasodilator using NO. The mechanism by which NO is released is, however, different to Ach, as it does not involve eNOS. Instead, our findings point towards an involvement of the underlying smooth muscle cells, apparently capable of increasing NO generation in response to ER stress.

4. Materials and Methods

4.1. Animals and Respective Materials

The ex-vivo experiments were performed on Sprague-Dawley rats (390–500 g; 8–12 weeks, ♂; Janvier Labs, Le Genest-Saint-Isle, France) which were housed in pairs under controlled standard animal laboratory conditions at least for 7 days for adaptation purposes before experiments with free access to standard laboratory rodent diet and water. The rats with higher weight were used in priority. The animals were anesthetized by inhalation of 8% Sevoflurane mixture (Vapor 2000, Dräger, Austria) for at least 12 min until a complete loss of paw pinch reflex and immediately decapitated afterwards (DCAP-M, World Precision Instruments, Sarasota, FL, USA).

4.2. Human Umbilical Vein Endothelial Cell Isolation and Culture

Human umbilical vein endothelial cells (HUVECs) were isolated as described previously [23] under approval of the local ethics committee of the state of Upper Austria (ethics committee vote #200, 12/05/200). Cells were cultured in EGM-2 (Lonza, Walkersville, MD, USA) supplemented with 5% foetal calf serum (FCS) at 37 °C and 5% CO_2 and used in passage 6.

4.3. Reagents

The reagents were sourced from the following suppliers:

- DAF-FM Diacetate (Cat. Nr. D23842, Thermo Fischer, Waltham, MA, USA)
- FLUO-4 AM (Cat. Nr. F14201, Thermo Fischer, Waltham, MA, USA).
- Thioflavin-T (Cat. Nr. T3516-5g, f.c. 50 μM, Thermo Fischer, Waltham, MA, USA).
- Acetylcholine (Cat. Nr. AA6625-10MG, f.c 300 μM, Merck/Sigma-Aldrich, Darmstadt, Germany).
- Phenylephrine (Cat. Nr. P1240000, f.c 300 μM, Merck/Sigma-Aldrich, Darmstadt, Germany).
- Tunicamycin (Cat. Nr. Cay11445-10, f.c. 200 μg/mL, Cayman Chemical, Ann Arbor, MA, USA).

4.4. Preparation Methods

Freshly prepared 6 mm 200 μm thick rat liver slices were produced as described in a previous publication by our group [24] and afterwards stained in DPBS according to the respective protocols supplied by the fluorescent probe manufacturers. Liver slices were incubated for 90 min using tunicamycin 200 μg/mL or DMSO 1% vol., and tunicamycinin combination with UPR inhibitors, 20 μM for MKC8866, 1 μM for GSK414, and 100 nM for Exendin-4.

4.5. Determination of XBP1 Splicing

Total RNA from liver slices shock frozen in liquid nitrogen was isolated using RNAspin Mini Kit (Cytiva, Marlborough, MA, USA) according to the manufacturer's instructions. The RNA was quantified spectrophotometrically at 280 nm and the purity was assessed by ratio of absorbance at 260/280 nm using a Tecan Spark plate reader (Tecan Group Ltd., Männedorf, Switzerland). cDNA was synthesized from 1 μg of total RNA using Superscript^{TM} II reverse transcriptase (200 U/reaction; Invitrogen; Carlsbad, CA, USA) and anchored oligo dT primers (3.5 μmol/L final concentration). XBP1 mRNA of unspliced (XBP1u) and spliced variant (XBP1s) were amplified by means of PCR. Each PCR reaction contained SYBR R green I$^{®}$ (0.5×, Sigma Aldrich, Vienna, Austria), iTaq^{TM} DNA polymerase^{TM} (25 U/L; Bio Rad, Hercules, CA, USA), oligonucleotide primers (250 nmol/L each, Invitrogen; Carlsbad, CA, USA), dNTP (200 μmol/L each), and $MgCl_2$ (3 mmol/L). Primer pairs for XBP1 were described elsewhere [25]. Randomly assigned no-reverse transcriptase controls corresponding to ~15% of all the samples investigated, a no-template control, and an internal standard, a pool generated of equal aliquots of cDNA of all samples investigated, were included in each measurement. Analyses were run in duplicate on a CFX96^{TM} real-time cycler (Bio-Rad, Hercules, CA, USA). The PCR products of XBP1u (196 bp) and XBP1s (170 bp) mRNA were separated on a 2% Agarose Gel, stained with peqGreen (PeqLAB, Erlangen, Germany) and visualized by UV transillumination at 300 nm. Densidometric analyses were performed using the public domain image analyses software Scion Image Version Alpha 4.0.2.3 (Scion Corp., Frederick, MD, USA). Results were used to calculate the ratio of XBP1s to XBP1u mRNA.

4.6. Determination of NO Levels

The NO levels in the medium were measured with Sievers nitric oxide NOA-280i analyser (Analytix, Boldon, UK). The major product of NO oxidation in the medium is nitrite (NO_2^-). A reducing agent (NaI) is used to convert and measure nitrite back to NO.

$$I^- + NO_2^- + 2H^+ \rightarrow NO + 0.5I_2 + H_2O$$

Afterwards, the NO was quantified based on the chemiluminescent reaction with ozone by the device.

4.7. Microscopic Imaging and Analysis

LSM imaging was performed with an inverted confocal microscope (LSM 510, Zeiss, Oberkochen, Germany) and 10× objective. The images were acquired in ZEN 2009 (version 6.0.303; Carl Zeiss, Oberkochen, Germany) program and processed with ImageJ (version 1.53t, U. S. National Institutes of Health, Bethesda, MD, USA) software.

DAF-FM was identified as a viable option for unambiguously labelling vasculature within living tissue, as reported in the literature [26]. Our study successfully discerned a specific liver vascular component, namely the portal triad, as illustrated in Figure 7.

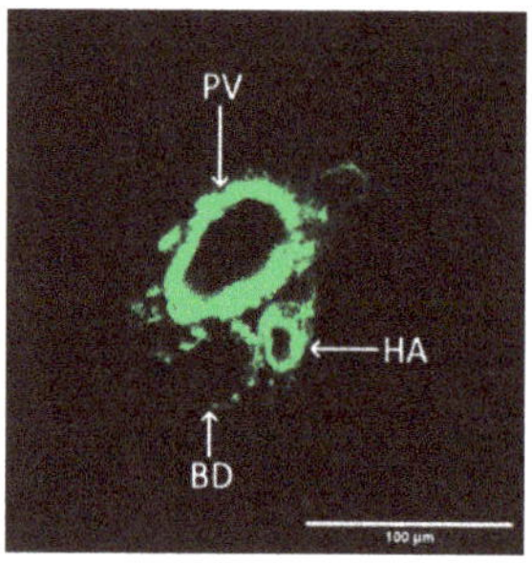

Figure 7. Representative image of liver vasculature depicting a portal triad, NO—sensitive staining. PV—portal vein, HA—hepatic artery, BD—bile duct.

The analysis of 3D images involved calculation of the vessel's volume using ImageJ software (version 1.53t) [27]. A specialized algorithm was developed, incorporating automated segmentation and computation of vessel lumen area for each image, with the objective of reducing bias originating from operator influence. In order to maintain alignment with our experimental schedule, we set a resolution of 512 × 512 pixels per scan, thereby preventing the image acquisition time from exceeding 10% of the intended experimental period.

The analysis of the vessel lumen area was performed with the following code:

```
run("Convert to Mask", "method = Huang background = Dark calculate black");
run("Despeckle", "stack");
close("ROI Manager");
run("Analyze Particles...", "size = 1500-Infinity summarize add slice");
roiManager("Select", 0);
run("Measure");
```

Afterwards, individual areas were multiplied with respective slice thickness for volume calculation.

It is crucial to highlight that only the undamaged sections of the vessels were considered for analysis, with those exhibiting outer damage resulting from cutting procedures deliberately excluded, as illustrated in Figure 8.

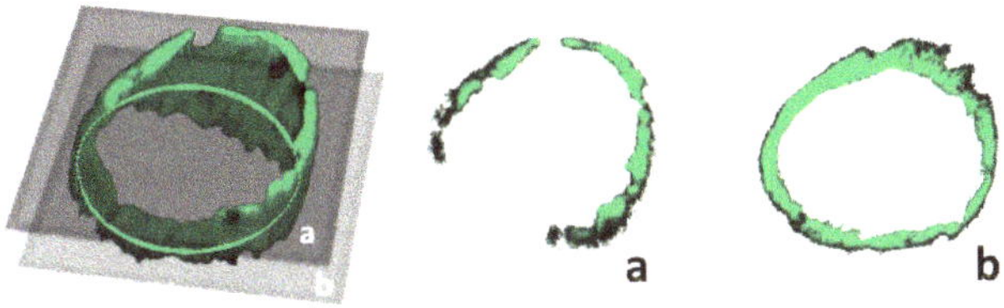

Figure 8. A complete 3D render of a portal vein. An example criterion for slice exclusion is images (**a**,**b**). (**b**) The top slice (1st out of 25) of the Z-stack render, excluded as lumen area calculation is not possible due to incomplete structure. (**a**) Representative image of an intact vessel suitable for lumen area calculation (15th out of 25).

5. Conclusions

In summary, the data obtained in this study suggest that the tunicamycin-mediated induction of ER stress results in a vasodilation of blood vessels in hepatic parenchyma. This vasodilation is NO dependent, but the source of NO is not eNOS. Our working hypothesis is that the mechanism underlying this effect is mediated by the nitrodilator-activatable intracellular NO store (NANOS), which was recently described [22]. The exact composition of this NO store is not yet clarified, but its components seem to be limited to a combination of nitrogen oxoanions such as nitrite (NO_2^-) and nitrate (NO_3^-) and S-nitrosothiols (SNOs) and that these substances are included in vesicle structures. Further studies are required to clarify mechanisms underlying the release of NO from NANOS. Although we do not know the exact mechanism of the NO release and the subsequent vasodilation, the data presented here can be important for understanding pathogenesis of diseases accompanied by circulatory failure, such as haemorrhagic shock.

Author Contributions: Conceptualization, A.V.K. and J.C.D.; methodology, A.V.K., J.C.D., A.M. and S.Z.; software, S.Z.; validation, S.Z., A.M., J.C.D. and A.V.K.; formal analysis, S.Z., A.V.K., A.M. and A.S.; investigation, S.Z., A.M., J.O., W.H. and A.S.; resources, A.V.K. and S.Z.; data curation, S.Z., A.M., A.V.K. and A.S.; writing—original draft preparation, S.Z.; writing—review and editing, A.V.K., A.M., J.C.D. and W.H.; visualization, S.Z., A.M. and A.V.K.; supervision, J.C.D. and A.V.K.; project administration, A.V.K.; funding acquisition, A.V.K. All authors have read and agreed to the published version of the manuscript.

Funding: This project is funded by the Austrian Research Promotion Agency (FFG). https://www.ffg.at/ (accessed on 1 October 2021). Project number #FO999887791.

Institutional Review Board Statement: Institutional veterinary review board approved this project as all the experiments were performed ex-vivo and did not require a special animal experiment permit.

Informed Consent Statement: Not applicable.

Data Availability Statement: The data presented in this study are available on request.

Acknowledgments: We would like to thank Karin Brenner, Astrid Hönigsberger, Gabrielle Leinfellner, and James Ferguson for assistance with handling the animals. We express our gratitude to Miriam Karas and Andreas Gasser for technical assistance. We thank Adelheid Weidinger for stimulating discussions.

Conflicts of Interest: The authors declare no conflicts of interest.

Appendix A

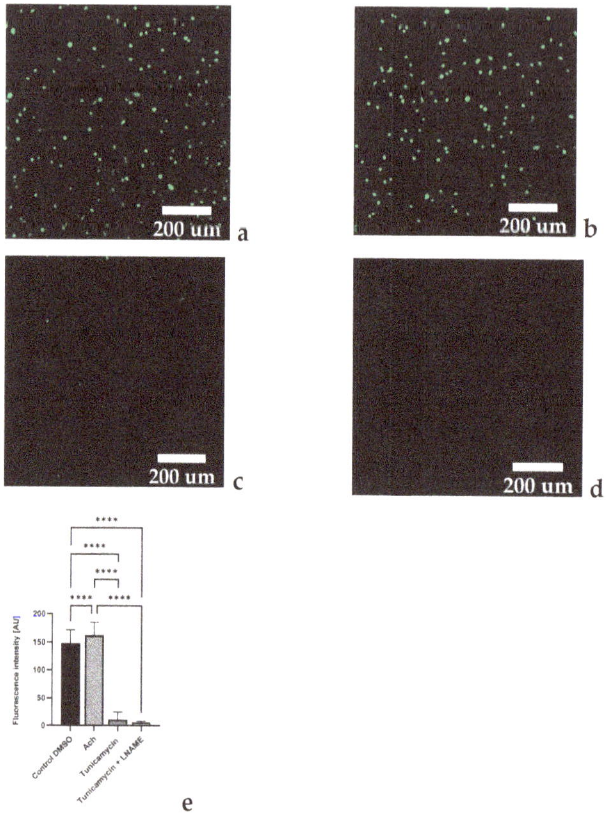

Figure A1. Representative images of fluorescent signal produced by DAF-FM NO-sensitive staining after 90 min treatment in endothelial cells, scale bar 200 μm. (**a**) Control group shows presence of NO signal. (**b**) Acetylcholine treated cell display an increase in NO signal. (**c**) Tunicamycin-treated group exhibits diminished NO signal. (**d**) Tunicamycin + L-NAME treated group exhibits almost complete elimination of NO signal. (**e**) Statistical evaluation of individual cell NO signal fluorescence values. One-way ANOVA; data are presented as means ± SD; n = 130; **** = $p < 0.0001$.

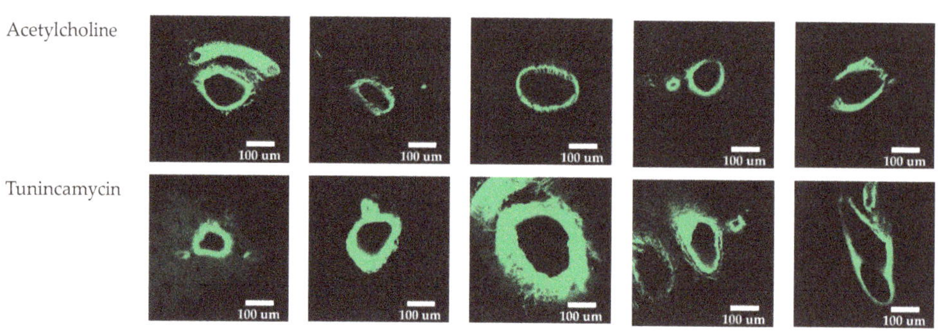

Figure A2. Images of vessels stained with NO-sensitive fluorescent dye DAF-FM and treated 90 min with either Ach or tunicamycin.

References

1. Braakman, I.; Bulleid, N.J. Protein folding and modification in the mammalian endoplasmic reticulum. *Annu. Rev. Biochem.* **2011**, *80*, 71–99. [CrossRef] [PubMed]
2. Almanza, A.; Carlesso, A.; Chintha, C.; Creedican, S.; Doultsinos, D.; Leuzzi, B.; Luís, A.; McCarthy, N.; Montibeller, L.; More, S.; et al. Endoplasmic reticulum stress signalling—From basic mechanisms to clinical applications. *FEBS J.* **2019**, *286*, 241–278. [CrossRef]
3. Sano, R.; Reed, J.C. ER stress-induced cell death mechanisms. *Biochim. Biophys. Acta* **2013**, *1833*, 3460–3470. [CrossRef] [PubMed]
4. Xue, C.; Lin, T.Y.; Chang, D.; Guo, Z. Thioflavin T as an amyloid dye: Fibril quantification, optimal concentration and effect on aggregation. *R. Soc. Open Sci.* **2017**, *4*, 160696. [CrossRef] [PubMed]
5. Lee, J.; Hong, S.-W.; Park, S.E.; Rhee, E.-J.; Park, C.-Y.; Oh, K.-W.; Park, S.-W.; Lee, W.-Y. Exendin-4 attenuates endoplasmic reticulum stress through a SIRT1-dependent mechanism. *Cell Stress Chaperones* **2014**, *19*, 649–656. [CrossRef] [PubMed]
6. Chang, T.-K.; Lawrence, D.A.; Lu, M.; Tan, J.; Harnoss, J.M.; Marsters, S.A.; Liu, P.; Sandoval, W.; Martin, S.E.; Ashkenazi, A. Coordination between Two Branches of the Unfolded Protein Response Determines Apoptotic Cell Fate. *Mol. Cell* **2018**, *71*, 629–636.e5. [CrossRef] [PubMed]
7. Hetz, C.; Zhang, K.; Kaufman, R.J. Mechanism, regulation and functions of the unfolded protein response. *Nat. Rev. Mol. Cell Biol.* **2020**, *21*, 421–438. [CrossRef] [PubMed]
8. Duvigneau, J.C.; Kozlov, A.V.; Zifko, C.; Postl, A.; Hartl, R.T.; Miller, I.; Gille, L.; Staniek, K.; Moldzio, R.; Gregor, W.; et al. Reperfusion does not induce oxidative stress but sustained endoplasmic reticulum stress in livers of rats subjected to traumatic-hemorrhagic shock. *Shock* **2010**, *33*, 289–298. [CrossRef] [PubMed]
9. Liu, X.; Green, R.M. Endoplasmic reticulum stress and liver diseases. *Liver Res.* **2019**, *3*, 55–64. [CrossRef]
10. Karmaniolou, I.I.; Theodoraki, K.A.; Orfanos, N.F.; Kostopanagiotou, G.G.; Smyrniotis, V.E.; Mylonas, A.I.; Arkadopoulos, N.F. Resuscitation after hemorrhagic shock: The effect on the liver—A review of experimental data. *J. Anesth.* **2013**, *27*, 447–460. [CrossRef]
11. Tian, J.-L.; Zhang, J.-S. Hepatic perfusion disorders: Etiopathogenesis and related diseases. *World J. Gastroenterol.* **2006**, *12*, 3265–3270. [CrossRef]
12. Shreya, S.; Grosset, C.F.; Jain, B.P. Unfolded Protein Response Signaling in Liver Disorders: A 2023 Updated Review. *Int. J. Mol. Sci.* **2023**, *24*, 14066. [CrossRef]
13. Kassan, M.; Galán, M.; Partyka, M.; Saifudeen, Z.; Henrion, D.; Trebak, M.; Matrougui, K. Endoplasmic reticulum stress is involved in cardiac damage and vascular endothelial dysfunction in hypertensive mice. *Arter. Thromb. Vasc. Biol.* **2012**, *32*, 1652–1661. [CrossRef]
14. Li, R.-L.; Zhuo, C.-L.; Yan, X.; Li, H.; Lin, L.; Li, L.-Y.; Jiang, Q.; Zhang, D.; Wang, X.-M.; Liu, L.-L.; et al. Irisin attenuates vascular remodeling in hypertensive mice induced by Ang II by suppressing Ca^{2+}-dependent endoplasmic reticulum stress in VSMCs. *Int. J. Biol. Sci.* **2024**, *20*, 680–700. [CrossRef]
15. Ando, M.; Matsumoto, T.; Taguchi, K.; Kobayashi, T. Decreased contraction induced by endothelium-derived contracting factor in prolonged treatment of rat renal artery with endoplasmic reticulum stress inducer. *Naunyn-Schmiedeberg's Arch. Pharmacol.* **2018**, *391*, 793–802. [CrossRef] [PubMed]
16. Tsuru, A.; Imai, Y.; Saito, M.; Kohno, K. Novel mechanism of enhancing IRE1α-XBP1 signalling via the PERK-ATF4 pathway. *Sci. Rep.* **2016**, *6*, 24217. [CrossRef]
17. Hu, H.; Tian, M.; Ding, C.; Yu, S. The C/EBP homologous protein (CHOP) transcription factor functions in endoplasmic reticulum stress-induced apoptosis and microbial infection. *Front. Immunol.* **2019**, *9*, 3083. [CrossRef] [PubMed]
18. Wilson, C.; Lee, M.D.; McCarron, J.G. Acetylcholine released by endothelial cells facilitates flow-mediated dilatation. *J. Physiol.* **2016**, *594*, 7267–7307. [CrossRef] [PubMed]
19. Lee, C.-H.; Poburko, D.; Sahota, P.; Sandhu, J.; Ruehlmann, D.O.; van Breemen, C. The mechanism of phenylephrine-mediated $[Ca^{2+}]$i oscillations underlying tonic contraction in the rabbit inferior vena cava. *J. Physiol.* **2001**, *534 Pt 3*, 641–650. [CrossRef]
20. Lenna, S.; Han, R.; Trojanowska, M. Endoplasmic reticulum stress and endothelial dysfunction. *IUBMB Life* **2014**, *66*, 530–537. [CrossRef]
21. Galán, M.; Kassan, M.; Kadowitz, P.J.; Trebak, M.; Belmadani, S.; Matrougui, K. Mechanism of endoplasmic reticulum stress-induced vascular endothelial dysfunction. *Biochim. Biophys. Acta (BBA)—Mol. Cell Res.* **2014**, *1843*, 1063–1075. [CrossRef] [PubMed]
22. Liu, T.; Schroeder, H.; Power, G.G.; Blood, A.B. A physiologically relevant role for NO stored in vascular smooth muscle cells: A novel theory of vascular NO signaling. *Redox Biol.* **2022**, *53*, 102327. [CrossRef] [PubMed]
23. Petzelbauer, P.; Bender, J.R.; Wilson, J.; Pober, J.S. Heterogeneity of dermal microvascular endothelial cell antigen expression and cytokine responsiveness in situ and in cell culture. *J. Immunol.* **1993**, *151*, 5062–5072. [CrossRef] [PubMed]
24. Weidinger, A.; Dungel, P.; Perlinger, M.; Singer, K.; Ghebes, C.; Duvigneau, J.C.; Müllebner, A.; Schäfer, U.; Redl, H.; Kozlov, A.V. Experimental data suggesting that inflammation mediated rat liver mitochondrial dysfunction results from secondary hypoxia rather than from direct effects of inflammatory mediators. *Front. Physiol.* **2013**, *4*, 138. [CrossRef] [PubMed]
25. Müllebner, A.; Herminghaus, A.; Miller, I.; Kames, M.; Luís, A.; Picker, O.; Bauer, I.; Kozlov, A.V.; Duvigneau, J.C. Tissue Damage, Not Infection, Triggers Hepatic Unfolded Protein Response in an Experimental Rat Peritonitis Model. *Front. Med.* **2022**, *9*, 785285. [CrossRef] [PubMed]

26. Paulo, L.L.; Cruz, J.C.; Zhuge, Z.; Carvalho-Galvão, A.; Brandão, M.C.; Diniz, T.F.; Haworth, S.M.; Athayde-Filho, P.F.; Lemos, V.S.; Lundberg, J.O.; et al. The novel organic mononitrate NDHP attenuates hypertension and endothelial dysfunction in hypertensive rats. *Redox Biol.* **2017**, *15*, 182–191. [CrossRef]
27. Schneider, C.A.; Rasband, W.S.; Eliceiri, K.W. NIH Image to ImageJ: 25 years of image analysis. *Nat. Methods* **2012**, *9*, 671–675. [CrossRef]

Disclaimer/Publisher's Note: The statements, opinions and data contained in all publications are solely those of the individual author(s) and contributor(s) and not of MDPI and/or the editor(s). MDPI and/or the editor(s) disclaim responsibility for any injury to people or property resulting from any ideas, methods, instructions or products referred to in the content.

International Journal of
Molecular Sciences

MDPI

Article

Radical-Generating Activity, Phagocytosis, and Mechanical Properties of Four Phenotypes of Human Macrophages

Shakir K. Suleimanov [1,2], Yuri M. Efremov [1], Timofey O. Klyucherev [1,2], Emin L. Salimov [3], Aligeydar A. Ragimov [3], Peter S. Timashev [1,4] and Irina I. Vlasova [1,*]

[1] Institute for Regenerative Medicine, I. M. Sechenov First Moscow State Medical University, 119991 Moscow, Russia; suleymanovef@gmail.com (S.K.S.); efremov_yu_m@staff.sechenov.ru (Y.M.E.); klyucherev_t_o@staff.sechenov.ru (T.O.K.); timashev_p_s@staff.sechenov.ru (P.S.T.)
[2] Laboratory of Clinical Smart Nanotechnologies, I. M. Sechenov First Moscow State Medical University, 119991 Moscow, Russia
[3] Laboratory Blood Transfusion Complex, I. M. Sechenov First Moscow State Medical University, 119991 Moscow, Russia; salimov_e_l@staff.sechenov.ru (E.L.S.); ragimov_a_a@staff.sechenov.ru (A.A.R.)
[4] World-Class Research Center "Digital Biodesign and Personalized Healthcare", I. M. Sechenov First Moscow State Medical University, 119991 Moscow, Russia
* Correspondence: iivlasova08@gmail.com or vlasova_i_i@staff.sechenov.ru

Abstract: Macrophages are the major players and orchestrators of inflammatory response. Expressed proteins and secreted cytokines have been well studied for two polar macrophage phenotypes—pro-inflammatory M1 and anti-inflammatory regenerative M2, but little is known about how the polarization modulates macrophage functions. In this study, we used biochemical and biophysical methods to compare the functional activity and mechanical properties of activated human macrophages differentiated from monocyte with GM-CSF (M0_GM) and M-CSF (M0_M) and polarized into M1 and M2 phenotypes, respectively. Unlike GM-CSF, which generates dormant cells with low activity, M-CSF confers functional activity on macrophages. M0_M and M2 macrophages had very similar functional characteristics—high reactive oxygen species (ROS) production level, and higher phagocytosis and survival compared to M1, while M1 macrophages showed the highest radical-generating activity but the lowest phagocytosis and survival among all phenotypes. All phenotypes decreased their height upon activation, but only M1 and M2 cells increased in stiffness, which can indicate a decrease in the migration ability of these cells and changes in their interactions with other cells. Our results demonstrated that while mechanical properties differ between M0 and polarized cells, all four phenotypes of monocyte-derived macrophages differ in their functional activities, namely in cytokine secretion, ROS production, and phagocytosis. Within the broad continuum of human macrophages obtained in experimental models and existing in vivo, there is a diversity of phenotypes with varying combinations of both markers and functional activities.

Keywords: myeloid-derived macrophages; reactive oxygen species; redox activity; phagocytosis; Young's modulus

1. Introduction

Macrophages are innate immune cells that play a crucial role in all phases of the immune response. They defend against pathogens in acute-phase response, participate in tissue repair, and orchestrate the entire process of regeneration.

Upon development of inflammation, monocytes migrate to an inflammatory site, where they undergo differentiation into macrophages as a result of exposure to tissue cytokines released by surrounding cells such as fibroblasts, endothelial cells, stromal cells, etc. [1–3]. Based on the signals received from the microenvironment, macrophages change the expression of genes, morphology and their specific functions in the processes called differentiation and polarization [4]. This process is mainly mediated by growth factors like granulocyte–macrophage colony-stimulating factor (GM-CSF) and macrophage colony-stimulating factor (M-CSF) [5]. Then, under the influence of a variety of biologically active compounds like cytokines and lipopolysaccharides (LPS), macrophages are polarized to various phenotypes. The traditional M1/M2 paradigm distinguishes two main macrophage subtypes depending on their phenotypic features, cytokine secretion, and transcriptional profiles: M1 (classically activated) and M2 (alternatively activated) macrophages [6–8]. LPS in combination with interferon-γ (IFN-γ) are the major inducers of macrophage polarization into M1 phenotype [3]. M1 macrophages participate in antigen presentation and produce reactive oxygen species (ROS), inflammatory factors (including interleukine-1β (IL-1β), IL-6, IL-12, and IL-23), and tumor necrosis factor TNF-α [9,10]. To differentiate macrophages to the precursor of M2 phenotype, the monocytes should be pre-incubated with M-CSF [11]. Then, polarization into the M2 state is accomplished by a number of inducers. Depending on the inducer, one of the four subtypes of M2 macrophages can be obtained. Incubation of macrophages with IL-4 results in the formation of M2a phenotype [12]. M2 macrophages demonstrate high phenotypic plasticity and anti-inflammatory and immunoregulatory activity; they contribute to the resolution of inflammation and regeneration due to the production of growth factors and anti-inflammatory cytokines IL-10, CCL18, and IL-1Ra. M2 macrophages express the mannose receptor CD206, known as a phagocytic receptor for pathogens, and CD163, a receptor for haptoglobin–hemoglobin complexes [13].

The polarization of M1 and M2 macrophages is a highly controlled process, and an imbalance of macrophage polarization is often associated with various diseases or inflammatory conditions, such as rheumatoid arthritis, atherosclerosis, cancer, and diabetes [2,12–14]. Balance between pro-inflammatory and anti-inflammatory macrophages is essential for the activated-to healing transition. Macrophage plasticity is considered a therapeutic target in different pathologies [15,16].

The radical-generating ability of macrophages is one of the most important activities of these cells both under normal physiological conditions and during inflammation. Under normal conditions, ROS play a crucial role in signal transduction, differentiation, gene expression, and apoptosis. NADPH oxidase (mainly NOX2) is the major source of ROS in macrophages [17]. A growing body of evidence points at mitochondria as essential sites of ROS formation, mainly due to electron leakage of the respiratory chain or to enzymes, such as monoamine oxidases [18]. Under physiological conditions, ROS production is important for the differentiation of human monocytes into M2 macrophages [19]. The levels of ROS and the production of extracellular H_2O_2 are lower in M2 macrophages compared to M1 cells due to a decreased expression of redox-active enzymes, most particularly NOX2 and NOX5 as well as higher levels of antioxidant enzymes [20].

In the beginning of inflammation, due to a high concentration gradient of GM-CSF and proinflammatory cytokines such as IFN-γ at the site of injury, a massive influx of monocytes from blood vessels gives rise to a population of M1 macrophages. M1 cells are activated at the site of inflammation and produce ROS and proinflammatory cytokines. The respiratory burst of activated macrophages is associated predominantly with NOX2, which produces a high level of the superoxide anion radical ($O_2^{\bullet-}$) [21]. Spontaneously or under the catalytic activity of superoxide dismutase (SOD), $O_2^{\bullet-}$ tunes into hydrogen peroxide. H_2O_2 reacts with iron in low-molecular complexes or in the active sites of enzymes with the production of ROS, thus initiating oxidative stress outside and inside the cells [22,23]. Upon development of an inflammatory response, the macrophage polarization is shifted into M2 phenotype. M2 macrophages exhibit multiple functions, including the secretion of anti-inflammatory cytokines, phagocytosis of apoptotic cells, and ROS production [3,9]. It is generally accepted that in conditions of acute inflammation, M1 macrophages have the highest radical-generating activity, creating a pro-oxidative environment and contributing to the killing and elimination of bacteria, whereas M2 macrophages remove damaged cells and matrix and are involved in tissue repair.

ROS production is closely related to another important function of macrophages—phagocytosis. After assembly on the membrane, NOX2 synthesizes the superoxide anion radicals inside the phagosome, which, along with proteolytic enzymes, ensures the destruction of a captured material [21]. Phagocytic activity is a characteristic of both M1 and M2 macrophages, which was shown in RAW 264.7 macrophages [24]. However, some studies demonstrated a decrease in the phagocytic activity of macrophages after polarization towards the M2 phenotype. IL-4-induced macrophage polarization towards an alternatively activated phenotype was demonstrated to cause a significant decrease in phagocytosis of N. meningitidis, but at the same time enhances the production of cathepsins and increases the acidity and proteolytic activity of phagosomes [25,26]. On the contrary, studies of human macrophages demonstrate that the level of E. coli phagocytosis in M2 macrophages is higher compared to M1 macrophages [27]. Phagocytic activity of macrophages may depend on the specific pathogen used in a study, which hinders making general conclusions on the results.

The redox activity of macrophages has been studied mainly using mouse macrophages, while human macrophages differ from mouse cells. For example, mouse M1 macrophages produce high amounts of NO, key storage (GSPT1), and transport (MRP1) proteins are engaged to protect professional killer cells from this cytotoxic agent [28]. On the contrary, despite the expression of iNOS, human macrophages do not produce a significant amount of nitric oxide, which indicates low activity of the enzyme [29–31]. At the same time, high activity of iNOS was reported under some pathological conditions [32]. In addition, arginase-1, the main marker of mouse M2 macrophages, is weakly expressed in human macrophages [7]. The clear dichotomy of polarization demonstrated for mouse macrophages is not as obvious in humans, and the separating line between M1-like and M2-like macrophages is rather represented by a continuum of phenotypes [9].

The study of the redox activity of human macrophages is important to understand the roles of M1 and M2 macrophages in the immune system. Using luminol-dependent chemiluminescence (CL), Lewis et al. showed similar radical-generating activity and found no differences in NOX2 expression in M1 and M2 human macrophage phenotypes either for THP-1-derived macrophages or monocyte-derived macrophages (MDM) [33]. The authors used high concentrations of phorbol 12,13-dibutyrate (10 μM) to activate macrophages. Nassif et al. demonstrated that the hypoglycemic agent metformin inhibits LPS-induced radical-generating M2 activity in human macrophages via the activation of adenosine monophosphate protein kinase [34].

Macrophage polarization and activation evokes changes in cytoskeleton structure and membrane composition, what results in changes of cell morphology and membrane stiffness. AFM is a powerful tool to study the cellular morphology and bio-mechanics of immune cells [35]. Activation of RAW264.7 macrophages by LPS (100 ng/mL) caused changes in cell morphology and membrane ultrastructure and increased the Young's modulus of cells, which was attributed to the redistribution of intracellular F-actin structures upon LPS treatment [36]. On the contrary, using colloidal force microscopy Leporatty et al. found that the stiffness of human macrophages decreased after cell stimulation by 10 μg/mL LPS [37]. The mechanical properties of macrophages are important in their interaction with the cell matrix and other cells involved in tissue regeneration and affect cell proliferation and migration [38].

In this study, we compare the functional activity and mechanical properties of four phenotypes of human macrophages with different phenotypic and secretory features. Monocytes were treated with GM-CSF (M0_GM) or M-CSF (M0_M) and then activated into M1 or M2 states, respectively. To simulate the conditions of the inflammatory site, we tested the functional activity and mechanics of macrophages in the presence of radical-generating inducers (low molecular phorbol-12-myristate-13-acetate (PMA, MW = 617 g/mol) and particulate opsonized zymosan (OZ)) and bacteria. Four phenotypes of macrophages differed in cytokine secretion, ROS generation, and ability to phagocytosis. For all macrophages studied, a similar decrease in viscoelastic parameter and height was observed upon activation, that evidenced cytoskeleton rearrangement, but only M1 and—to a lesser extent—M2 cells increased their rigidity. Changes in mechanical properties did not correlate with macrophage functional activity.

2. Results

2.1. Monocyte Differentiation and Macrophage Polarization

Macrophages were characterized by the expression of surface proteins, which serve as phenotype markers: CD86 indicated M1 phenotype, and CD206 indicated M2 phenotype. Macrophage functional response was determined by the secretion of cytokines—TNF-α for M1 and IL-10 for M2.

Treatment of monocytes with CSFs stimulates their differentiation into M0-macrophages. Incubation of macrophages with GM-CSF led to increased expression of CD86 and a shift of the anti-CD206-PE-Cy7 median fluorescence intensity (MFI) compared to cells treated with M-CSF. At the same time, low secretion of TNF-α and no secretion of IL-10 was observed for M0_GM cells, while cells incubated with M-CSF secreted a high amount of IL-10 (Figure 1). Further polarization of M0_GM macrophages into the M1 state (+LPS+IFN-γ) led to an increase in surface expression of CD86, an amplification of TNF-α production by about a hundredfold and secretion of a small amount of IL-10. IL-4-treated M2 macrophages were characterized by the highest expression of CD206 compared to other phenotypes and showed a slight increase in IL-10 production compared to their precursor M0_M macrophages (Figure 1). Another surface protein CD163 was not a good marker for M2 obtained after MDM treatment with IL-4, its expression was similar in four phenotypes of macrophages (Figure S1). CD163 was shown to be highly expressed on IL-10-treated MDM [13,39]. IL-1Ra can be considered as another marker of the M2a phenotype (Figure S1) [13]. Cellular morphology serves as an additional polarization marker. M2 macrophages predominantly exhibit an oblong shape, whereas M1 macrophages typically have a more rounded, spherical shape. (Figure S2) [20].

2.2. Effect of Growth Factors and Polarization Inducers on Macrophage Survival

In our experiments, the monocytes were seeded at the same number per well of the plate. However, by the sixth day of differentiation with CSF, a difference in cell numbers between the M0_GM and M0_M groups was already observed. The differences between cells increased after macrophage polarization into M1 and M2 phenotypes. To assess differences, the concentration of double-stranded DNA was measured using PicoGreen dye since the amount of DNA correlates with the number of cells. The relative difference between groups in each individual experiment, rather than the absolute values of DNA concentrations, was important. Data of four independent experiments are presented in Figure 2 as a heatmap reflecting levels of DNA (ng/mL) lower (blue shades) and higher (red shades) than expected values for each sample. The results showed that the group of macrophages incubated with M-CSF had a higher number of viable cells compared to macrophages incubated with GM-CSF. Moreover, the number of living cells decreased after M0_GM polarization to M1 state. The addition of IL-4 to M-CSF treated macrophages also resulted in a slight decrease in the number of viable cells compared to M0_M.

The results of DNA measurements were used to normalize the measured ROS parameters to the number of cells in a sample in order to compare the radical-generating activity of different phenotypes of macrophages.

2.3. ROS Generation by Activated Macrophages

Along with the secretion of cytokines, an important function of macrophages is ROS production. To simulate the conditions of the inflammatory site, macrophages were activated. We compared the radical-generating activity of different macrophage phenotypes after cell activation by substances with different mechanisms of action—opsonized zymosan and phorbol-12-myristate-13-acetate. Simultaneously with superoxide generation, activated macrophages release SOD3 from intracellular compartments, which catalyzes the dismutation of $O_2^{\bullet-}$ into hydrogen peroxide [40]. To measure H_2O_2, we used horseradish peroxidase (HRP) and two peroxidase substrates—luminol and 10-acetyl-3,7-dihydroxyphenoxazine (Amplex Red).

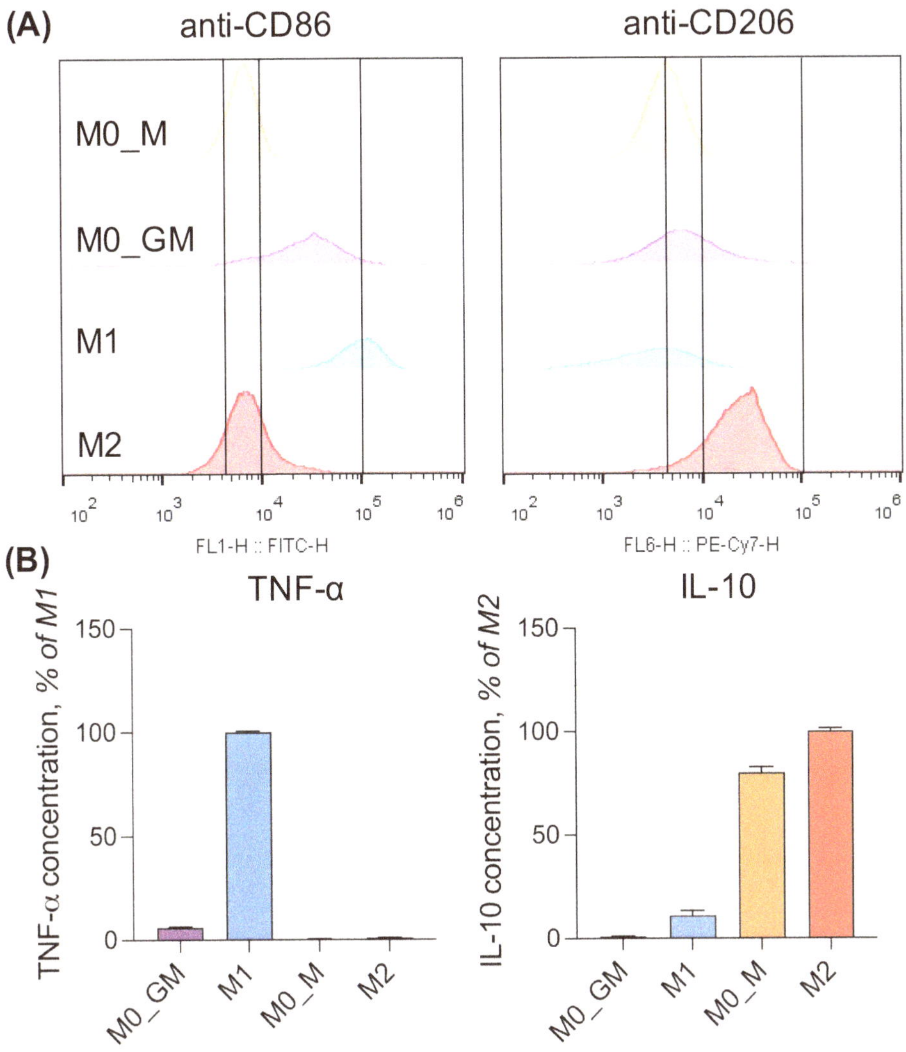

Figure 1. MDM polarization markers. (**A**) Flow cytometry histograms present fluorescence intensity of M0, M1, and M2 macrophages incubated with anti-CD86-FITC or anti-CD206-PE-Cy7 antibodies. Half of the cells cultured with GM-CSF (50 ng/mL) were treated with LPS (10 ng/mL) and IFN-γ (50 ng/mL); cells cultured with M-CSF (50 ng/mL) were treated with IL-4 (20 ng/mL). In untreated cells, the medium was changed (M0_GM and M0_M). In two days of incubation, macrophages were detached by accutase and stained with antibodies. (**B**) Amounts of cytokines TNF-α and IL-10 in conditioning media of macrophages were measured by ELISA.

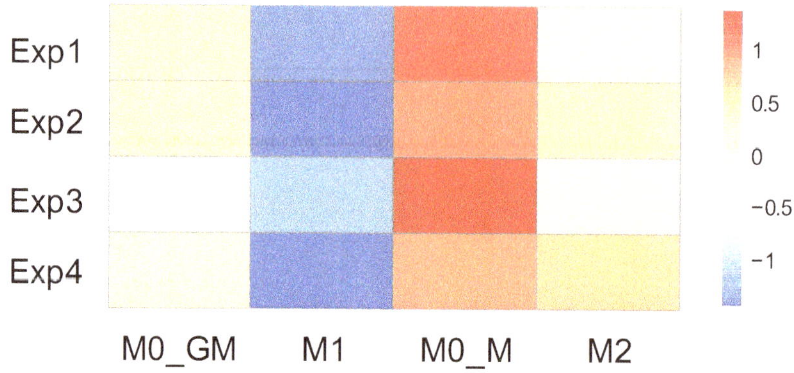

Figure 2. Effect of growth factors and polarizing inducers on the viability of human macrophages. Heatmap presents the amount of double-stranded DNA in a well after cell lysis. The results are presented as z-scores after z-standardization of each experiment.

2.3.1. Luminol-Dependent Chemiluminescence

Luminol is able to penetrate into cells. Therefore, using luminol, we measured extra- and intracellular ROS [41]. After activation with OZ, macrophages cultured in the presence of GM-CSF showed a very low level of luminol oxidation, while macrophages polarized to M1 state produced high levels of ROS: the luminol oxidation increased within 10–20 min, reached a maximum, and slowly decreased (Figure 3A). In the presence of PMA, M0_GM cells also demonstrated a slight increase in CL (Figure 3B). For both cellular activators, CL (area under CL curve) of M0_GM cells was approximately an order of magnitude lower than that for macrophages polarized in M1 phenotype (Figure 3C).

The addition of any of the activators to both types of macrophages differentiated in the presence of M-CSF caused CL comparable to the M1 level. Notably, M2 macrophages reached CL maximum significantly later than their precursor M0_M macrophages. Moreover, both M2 and M0_M macrophages activated with PMA reached CL intensity peak faster than those activated with OZ (Figures 3D,E and S3). Significant differences in the intensity of the luminescence peaks of M0_M and M2 macrophages for both activators were not found (Figure 3F).

Additionally, experiments were performed with macrophages detached by accutase; CL was measured in a cell suspension. The ratio of activities of different macrophage phenotypes in the suspensions was consistent with the results from experiments with adherent cells (Figure S4). We compared pro-inflammatory properties of detached M1 macrophages derived from MDM and THP-1 macrophages. Polarization of THP-1 macrophages into M1 state was evidenced by secretion of pro-inflammatory cytokines TNF-a and IL-1β, surface expression of CD86 and CD206 did not differ between groups (Figure S5). Both the ROS-generating activity and the cytokine secretion by M1 macrophages differentiated from leukemia cells were approximately 10 times lower than the activity of professional macrophages in M1 state (Figure S6). [33]. Taking into account the difference between the level of secreted cytokines [42,43], the direct comparison of MDM to THP-1 macrophages is not relevant.

2.3.2. Oxidation of Amplex Red by Horseradish Peroxidase in the Presence of Activated Macrophages

Contrary to luminol, Amplex Red cannot pass through the cell membrane. Using this reagent, we measured H_2O_2 outside the cells and in phagosomes. Activation of M0_GM by OZ resulted in a weak increase in Amplex Red oxidation and, accordingly, in an enhancement of fluorescence (Figure 4A). Notably, the response of M0_GM cells to the activator differed by an order of magnitude in different experiments. However, it was significantly weaker than the response of M1 cells (Figure 4C). Activation of M1 cells with OZ caused time-dependent amplification of fluorescence with a maximum at about 20 min. Similar to OZ, the fluorescence in the solution of PMA-activated M1 macrophages was significantly higher than that in the solution of precursor M0_GM (Figure 4B). For both activators, the wide scatter in the generated radical data of M0_GM group was observed (Figure 4C).

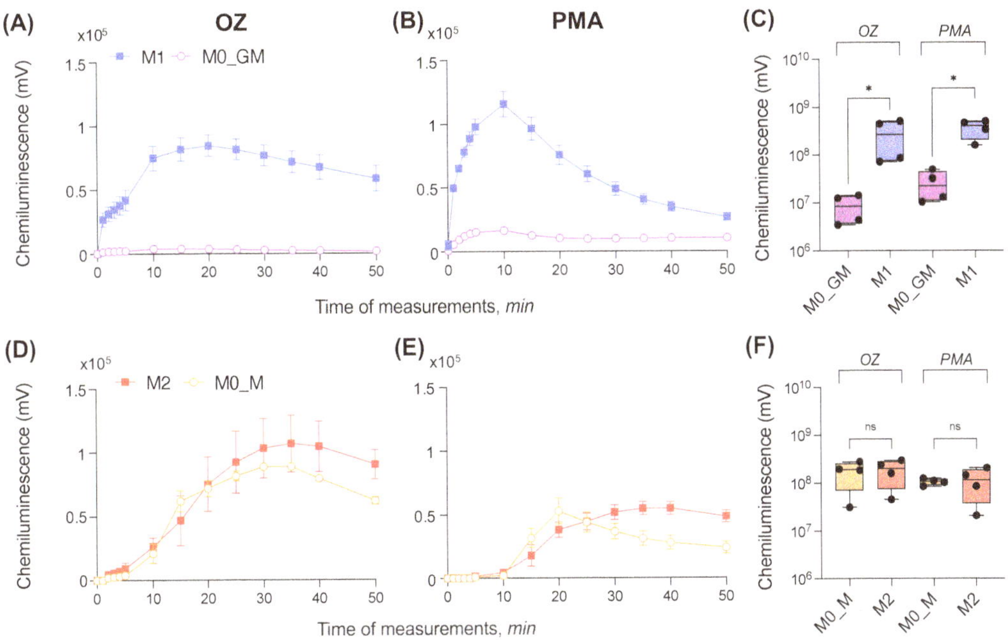

Figure 3. Radical-generating activity of human macrophages measured with luminol-dependent chemiluminescence. (**A–C**) Kinetics of CL in one independent experiment (**A**,**B**) and statistical comparison of ROS (**C**) for GM-CSF treated macrophages; (**D–F**) kinetics of CL in one independent experiment; (**D**,**E**) and statistical comparison of ROS level (**F**) for M-CSF-treated macrophages. Krebs–Ringer solution with $NaHCO_3$, $CaCl_2$, and 200 µM luminol was added to each well, and the plate was incubated at 37 °C for 5 min. Measurement of CL was started immediately after the addition of an activator: 0.3 mg/mL OZ (**A**,**D**) or 100 ng/mL PMA (**B**,**E**). Measurements were carried out at 37 °C until the CL maximum was reached for all samples. Each CL curve is the average of three curves obtained by measuring three technical replicates. Y-axis in logarithmic scale. $n = 4$, ^{ns} $p \geq 0.05$, * $p < 0.05$.

Both activators induced ROS generation by M0_M and M2 macrophages, increasing the fluorescence of resorufin in the medium (Figure 4D,E). Radical-generating activity of both types of cells treated with M-CSF was comparable to that of M1 phenotype, but the kinetics of accumulation of the fluorescent product was different—the maximum of fluorescence was reached no earlier than 30 min after the addition of any activator. Moreover, in contrast to CL measurements, the level of ROS production of M-CSF treated cells activated by PMA was statistically significantly lower if cells were polarized into the M2 phenotype (Figure 4F).

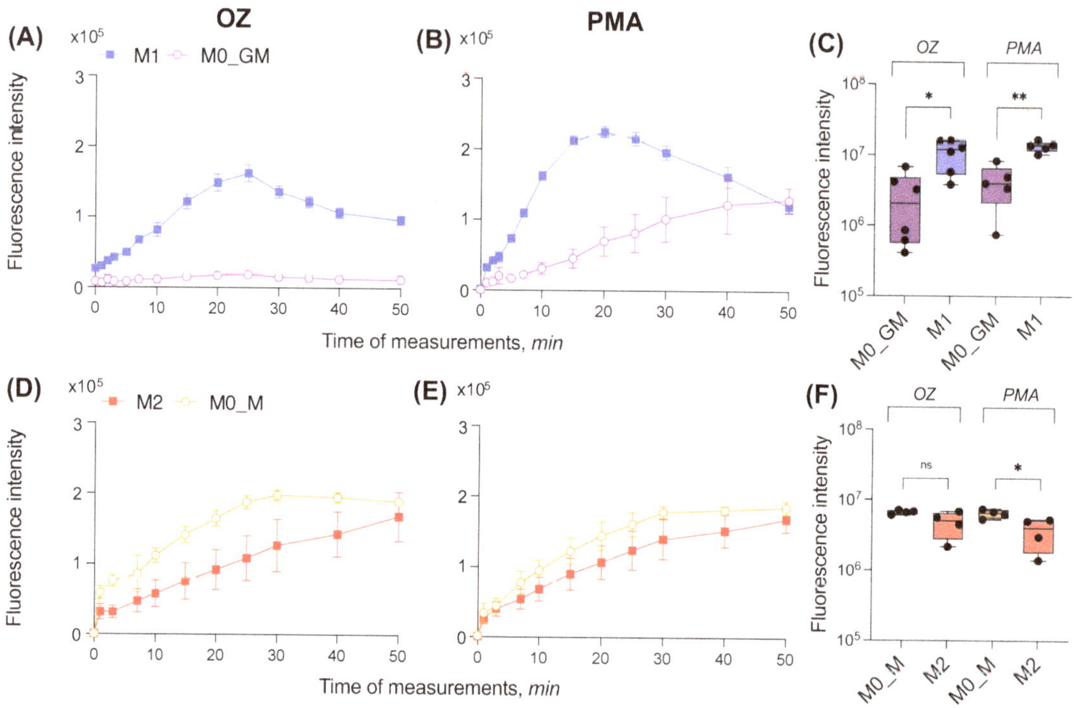

Figure 4. Radical-generated activity of human macrophages measured with Amplex red. (**A–C**) Kinetics of Amplex Red oxidation in one independent experiment (**A**,**B**) and statistical comparison of ROS generation (**C**) for GM-CSF treated macrophages; (**D–F**) Kinetics of Amplex Red oxidation in one independent experiment (**D**,**E**) and statistical comparison of ROS generation (**F**) for M-CSF treated macrophages. Krebs–Ringer solution with $NaHCO_3$, $CaCl_2$, and 10 μM of Amplex Red was added to each well, and the plate was incubated at 37 °C for 5 min. Measurement of fluorescence was started immediately after addition of an activator: 0.3 mg/mL OZ (**A**,**D**) or 100 ng/mL PMA (**B**,**E**). Measurements were carried out at 37 °C until the fluorescence maximum was reached for all samples. Each fluorescence curve is the average of three curves obtained by measuring three technical replicates. Y-axis in logarithmic scale. $n \geq 4$, ^{ns} $p \geq 0.05$, * $p < 0.05$, ** $p < 0.01$.

2.4. Comparison of ROS-Generating Ability of M1 and M2 Macrophages

ROS-generating activities (area under the curves) of M1 and M2 macrophage phenotypes are shown in Figure 5. Comparison of the level of ROS generation by macrophages activated by different inducers and measured by different methods revealed general trends: (1) M1 macrophages demonstrated the highest radical-generating activity; (2) a significant difference in the ROS-generated activity of macrophages M1 and M2 was observed upon activation by PMA, but not by the particulate agonist OZ (compare Figure 5A,C); (3) polarization of M0_GM macrophages into the M1 state amplified the ROS activity of cells, while polarization of M0_M macrophages into the M2 state does not change the level of total ROS production (CL) but slightly decreased the generation of H_2O_2 outside the cells (Amplex Red) (Figure 5B,D).

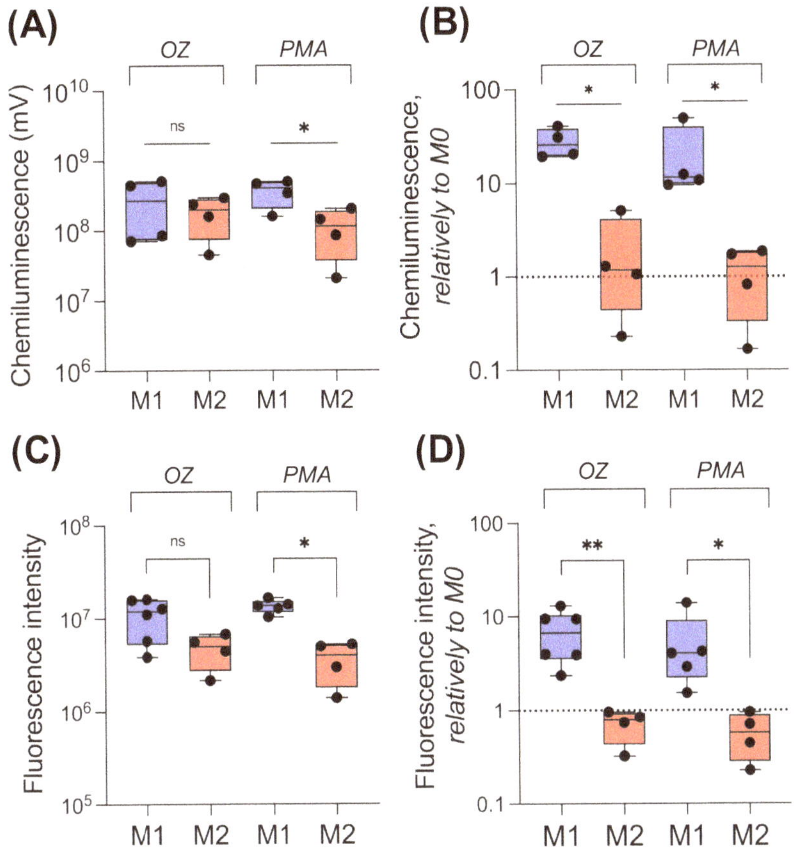

Figure 5. Comparison of radical-generating activity of human macrophages as assessed by CL (**A**,**B**) or Amplex Red oxidation (**C**,**D**). (**A**,**C**) represent the absolute values of the measured parameters; (**C**,**D**) represent the ratio of the measured parameter for M1 or M2 macrophages to the level of that parameter for the corresponding precursor (M0_GM or M0_M, respectively). $n \geq 4$, ^{ns} $p \geq 0.05$, * $p < 0.05$, ** $p < 0.01$.

2.5. Macrophage Phagocytosis

We employed flow cytometry and fluorescent microscopy to compare phagocytosis of inactivated FITC-labeled E. coli bioparticles by different phenotypes of human macrophages. Flow cytometry is a simple and precise method for quantifying cellular phagocytosis. First, we estimated the effect of different concentrations of bacteria on the phagocytic activity of macrophages. Figure 6A shows histograms of fluorescence distribution for different types of macrophages at various particle–cell ratios. The black line separates the phagocyting cells with high FITC MFI from non-phagocyting cells with low FITC MFI. To set the threshold separating non-phagocyting from phagocyting cells, we performed control experiments with macrophages, which were not incubated with BioParticles (Figure S7). Two parameters were measured for each histogram [27]: (1) the macrophage phagocytic affinity was calculated as the percentage of phagocytic cells; (2) the macrophage phagocytic capacity was estimated as the MFI of phagocytic cells. This parameter is indirectly proportional to the number of engulfed bacterial particles.

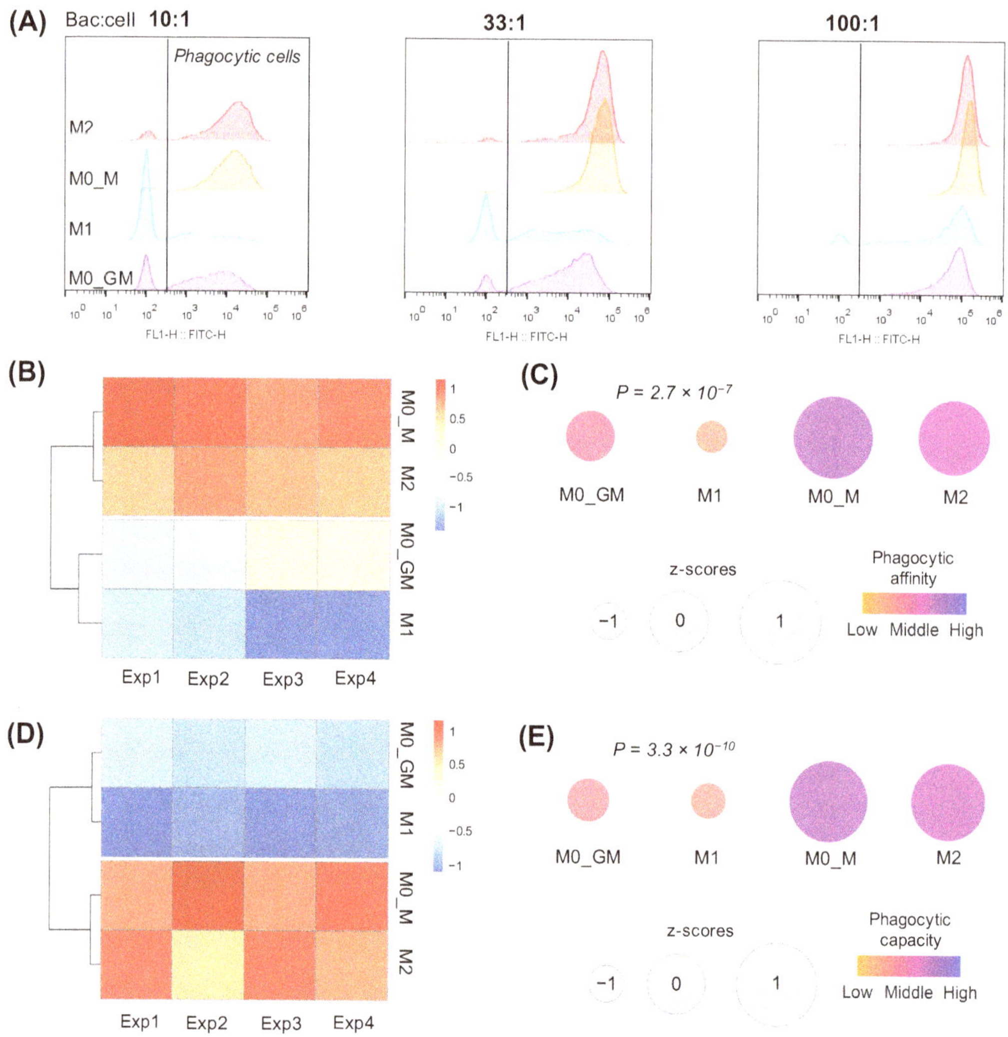

Figure 6. Phagocytosis activity of human macrophages assessed by flow cytometry. (**A**) Histograms of FITC fluorescence intensity distribution after incubation of different groups of macrophages with FITC-labeled E. coli bioparticles. The black vertical line separates the control group of macrophages not incubated with E. coli bioparticles (left) and the group of phagocytic cells (right). The ratio of bacteria to macrophage is indicated above the histograms. (**B**) Heatmap of the percentage of phagocyting macrophages in 4 biological experiments scaled to column z-score. (**C**) Average percentage of phagocyting cells in 4 groups of macrophages. Size and color represent z-score and phagocytic affinity, respectively. M1 versus M0_GM: $p = 0.00561$; M0_M versus M0_GM: $p < 0.001$; M2 versus M0_GM: $p < 0.001$; M0_M versus M1: $p < 0.001$; M2 versus M1: $p < 0.001$; M2 versus M0_M: $p = 0.303$, one-way ANOVA with post hoc Tukey test. (**D**) Heatmap of MFI, which reflects phagocytic capacity, in 4 groups of macrophages scaled to a column z-score. (**E**) Average median fluorescence intensity (MFI) in 4 groups of macrophages. Size and color represent z-score and phagocytic capacity, respectively. M1 versus M0_GM: $p = 0.0351$; M0_M versus M0_GM: $p < 0.001$; M2 versus M0_GM: $p < 0.001$; M0_M versus M1: $p < 0.001$; M2 versus M1: $p < 0.001$; M2 versus M0_M: $p = 0.4206$, one-way ANOVA with post hoc Tukey test).

Notably, the most significant differences in phagocytic affinity and capacity between groups were observed when bacterial particles were incubated with cells at a ratio of 10:1 (amount of bioparticle per cell) (Figure 6A). An increase in the ratio to 33:1 revealed the decrease in differences in phagocytic affinity, while variations in phagocytic capacity between groups of cells were more intense. When the ratio reached 100:1, only minor differences in MFI were detected between the groups. For further study, we chose the 10:1 ratio because it possessed the greatest differences between the measured parameters.

High variability of phagocytic affinity and capacity among macrophages from different donors did not allow us to perform statistical analysis between the experiments (Figure S8). At the same time, clear differences between groups of macrophages were observed within individual experiments. Thus, we performed a z-standardization of each experiment and used normalized values to analyze phagocytic activity. Figure 6 presents the statistical analysis of phagocytic affinity (B, C) and capacity (D, E) of different macrophage phenotypes. Comparison of both parameters revealed that (1) macrophages differentiated from M-CSF treated monocytes had significantly higher phagocytosis level as compared to GM-CSF treated cells; (2) M-CSF treated cells demonstrated a slight but statistically insignificant decrease in the percentage of phagocytic cells after polarization into M2 phenotype, while the phagocytic capacity remained almost the same for both groups of macrophages; (3) polarization of macrophages into M1 state substantially reduced the level of phagocytosis when compared to M0_GM phenotype. M1 macrophages have the lowest phagocytic activity among all phenotypes studied.

To estimate the impact of growth factors on phagocytic activity, we compared the phagocytosis of M1 macrophages generated after monocyte treatment with GM-CSF and M-CSF (M1 and M1_M, respectively). Analysis of both flow cytometry parameters revealed a significantly higher level of phagocytosis within the M1_M macrophage group compared to the M1_GM phenotype, thereby demonstrating that growth factors exert a significant influence on the functional activity of human macrophages (Figure S9A,B). In addition, we measured TNF-α and IL-6 secretion by M1_M and M1_GM macrophages to investigate the impact of growth factors on the secretory activity of macrophages. The level of both cytokines was significantly higher in the group of M1 macrophages differentiated with M-CSF (Figure S9C,D).

In addition to flow cytometry, fluorescent microscopy was employed to assess the phagocytic activity of macrophages at the 10:1 bioparticles-to-cell ratio. Images of fixed cells obtained with a fluorescent microscope were processed using image analysis software and the percentages of phagocytic cells for different macrophage phenotypes were compared (Figure 7A). The results were similar to the phagocytic affinity (percentage of phagocytic cells) obtained by flow cytometry (compare Figures 6B,C and 7B,C). Notably, a strong positive correlation was observed between the results of flow cytometry and fluorescence microscopy (Figure 7D).

To validate the results obtained with *E. coli* bioparticles, we have measured phagocytic capacity using FITC-conjugated OZ which is non-bacterial phagocytic agent . Fluorescent microscopy revealed that M1 macrophages had lower phagocytic capacity level of OZ particles, comparing to M2 macrophages (Figure S10).

2.6. Changes in Mechanical Properties of Macrophages after Activation

Upon activation, macrophages undergo changes in their morphology, which are attributed to alterations in the cytoskeleton and the molecular composition of the cell membrane. The latter changes are due to protein expression, activation of membrane proteins, as well as lipid and protein oxidation, among other factors. The morphological transformation of human macrophages during their activation by PMA is visually seen by both optical microscopy (Figure S11) and AFM (Figure 8).

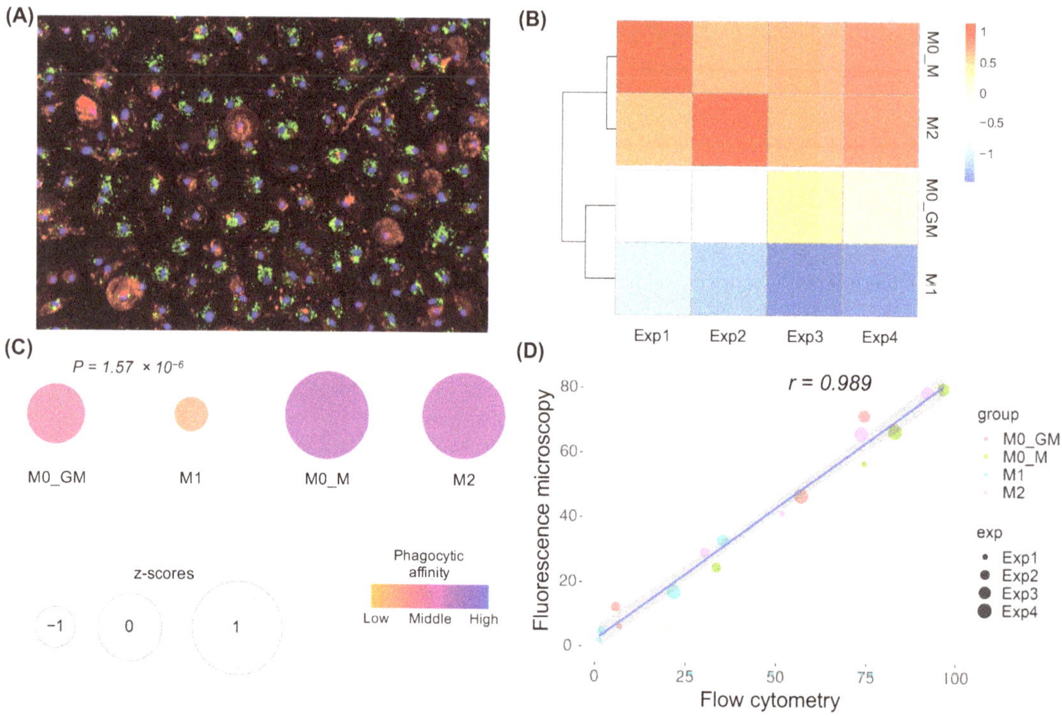

Figure 7. Analysis of phagocytosis activity of human macrophages assessed by fluorescence microscopy. (**A**) Fluorescence microscopy of fixed human macrophages after their incubation with *E. coli* bioparticles at a 10:1 ratio (20×). FITC green—*E. coli* bioparticles; red—phalloidin–rhodamine staining of the actin cytoskeleton; blue—DAPI, nuclear stain. (**B**) Heatmap of the percentage of phagocyting macrophages in 4 groups independent experiments scaled to column z-score. (**C**) Average percent of phagocyting cells in 4 groups of macrophages. Size and color represent z-score and phagocytic affinity, respectively. M1 versus M0_GM: $p = 0.0038$; M0_M versus M0_GM: $p = 0.0018$; M2 versus M0_GM: $p = 0.0027$; M0_M versus M1: $p < 0.001$; M2 versus M1: $p < 0.001$; M2 versus M0_M: $p = 0.9936$, one-way ANOVA with post hoc Tukey test. (**D**) Pearson correlation analysis of phagocytic affinity measured by flow cytometry and fluorescence microscopy. Label "group" is assigned to macrophage group, label "exp" is assigned to one of four biological replicates.

We compared the following parameters for the different macrophage phenotypes: cell height, stiffness (apparent Young's modulus), and the viscoelastic parameter power-law exponent (PLE) of the power-law rheology model, which indicates degree of closeness of the behavior of the sample to the liquid (PLE = 1) or solid (PLE = 0).

Firstly, we compared non-activated human macrophages. Various parameters of non-activated human macrophages exhibited variability depending on the cell donor, making it challenging to identify any differences between groups. Nevertheless, significant differences between macrophages were observed only for the cell height parameter. The cell height for both groups of macrophages incubated with M-CSF was significantly lower than that for GM-CSF macrophages, indicating a higher level of flattening (Figures 8 and S12).

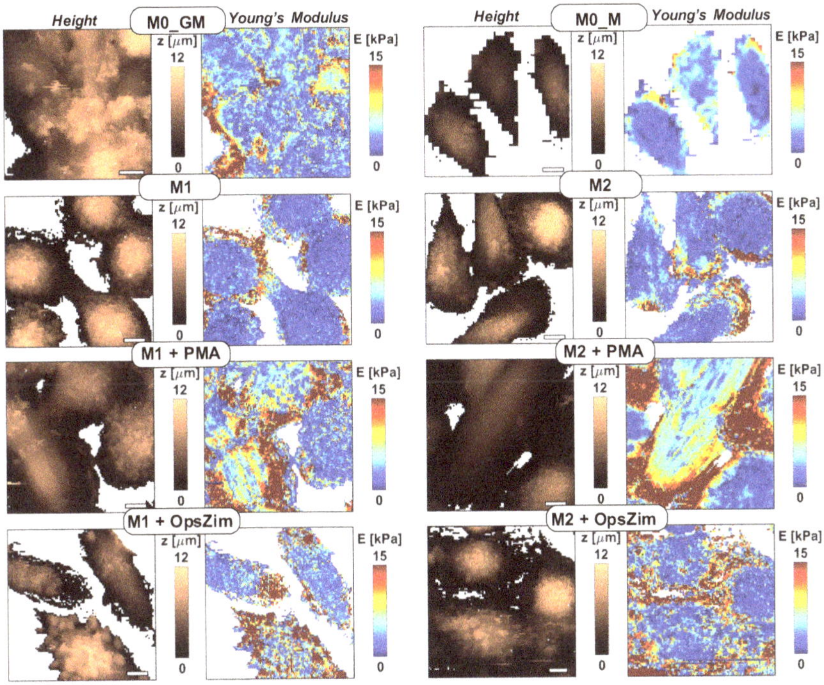

Figure 8. Representative AFM maps obtained on human macrophages untreated and treated with 100 nM PMA or 0.3 mg/mL OZ. Height and nanomechanical maps (distribution of Young's modulus) of the control and treated cells are presented, z (μm) and E (kPa). Individual particles of OZ on the surface of macrophages can be seen in topography images (bottom row, indicated with arrows). All images have the same color-coded scale for the height and modulus; the scale bar is 10 μm.

Activation of both M1 and M2 macrophages with PMA resulted in a significant increase in cell stiffness (Young's modulus). The greatest increase of the Young's modulus, by more than two times from (2.8 $\pm$ 1.09) to (6.92 $\pm$ 2.87) kPa, was observed for proinflammatory M1 macrophages. Significant changes in the Young's modulus were also registered in the group of anti-inflammatory M2 macrophages (Figure 9A). However, this parameter did not change significantly after the addition of PMA to precursor macrophages (M0_GM and M0_M). Addition of OZ also led to a significant increase in Young's modulus in both the M1 and M2 macrophage groups but was insignificant for their precursors (Figure 9B).

Activation of all four macrophage phenotypes with PMA resulted in a reduction of cell height, indicating their flattening (Figure 9). The highest level of flattening was observed for polarized macrophage phenotypes: 1.53- and 1.69-fold decrease of cell height for M1 and M2 macrophages, respectively (Figure 9C). This effect was least pronounced in the M0_GM macrophage group. On the other hand, activation of all macrophage groups with OZ had minimal impact on cell flattening except for a slight decrease observed in the M0_M group (Figure 9D). Particles of the activator can contribute to cell height. With AFM images, it was possible to observe individual particles of OZ on the surface of macrophages.

For all types of macrophages, a decrease in the PLE viscoelastic parameter characterizing the liquid–solid state was registered after activation by PMA (Figure 9E), which means transition to the solid state and an increase in pretension in the cytoskeleton, and usually occurs during cell spreading [44]. The fold of PLE decrease was the lowest for M0_GM cells. Activation of GM-CSF treated macrophages with OZ also led to a decrease in PLE, while M-CSF-treated macrophages demonstrated a slight increase (M0_M) or no change (M2) for this parameter (Figure 9F).

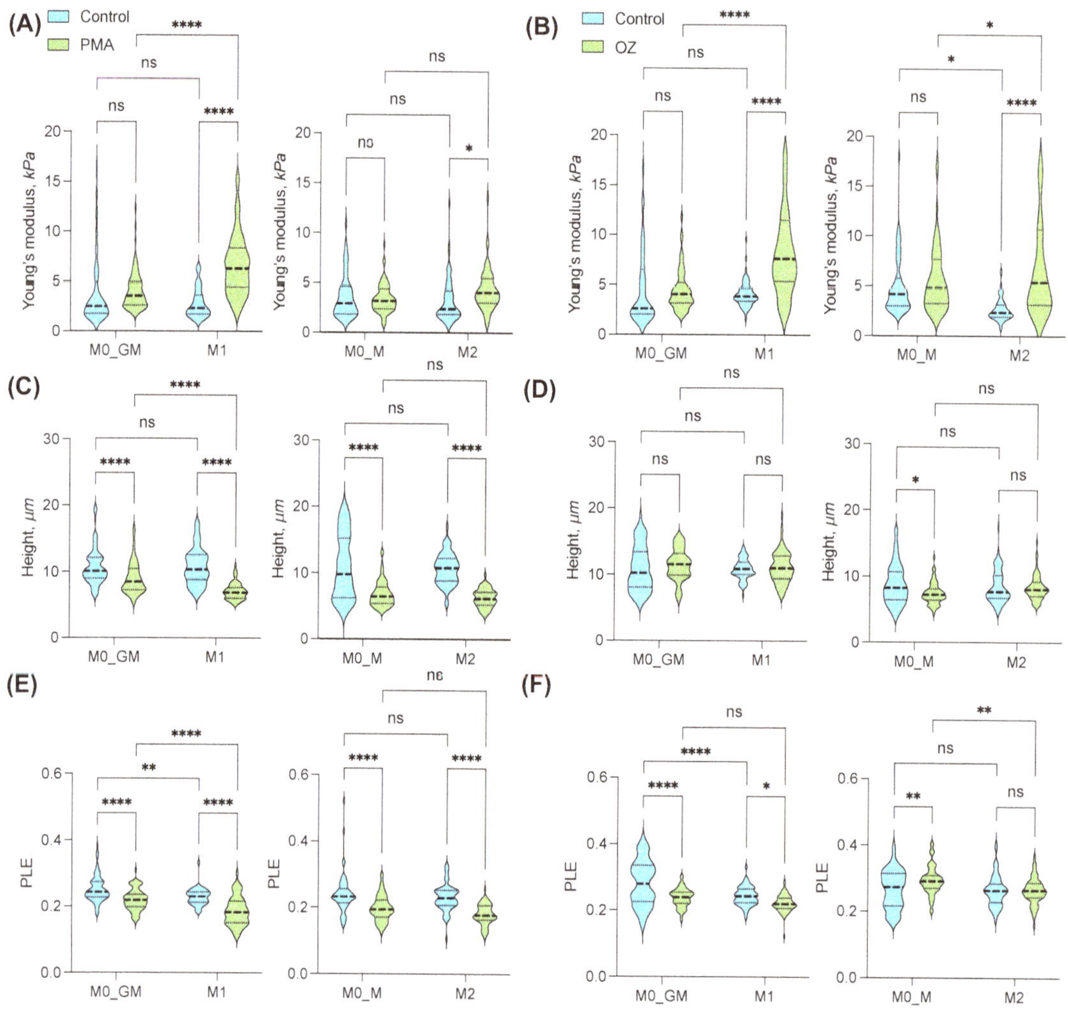

Figure 9. Young's modulus (**A**,**B**), height (**C**,**D**), and power law exponent (PLE) (**E**,**F**) of human macrophages activated with PMA (left) and OZ (right). Two-way analysis of variance with Tukey post hoc test was performed to compare differences between groups. The analysis includes the results of 2 independent experiments, in each of which the studied parameters were measured for a minimum of 20 individual cells. ^{ns} $p > 0.05$, * $p < 0.05$, ** $p < 0.01$, **** $p < 0.0001$.

3. Discussion

Macrophage activation is involved not only in immune response and tissue regeneration but also in the progression and outcome of many diseases, including autoimmune diseases, cancer, allergic disorders, and different types of infections [12,14]. Therefore, the study of specific macrophage activity is of especially importance [7]. Many studies have been devoted to comparing transcriptomes, proteomes, secreted cytokines, and expressed proteins of different types and phenotypes of macrophages [13,45–48]. Based on the results, conclusions were drawn about the difference in the functional activity of macrophages. However, macrophage functions are realized through a complex interplay between different proteins, mediators and membrane properties. ROS generation and phagocytosis are the major functions of phagocytes which require a further study.

As expected, functional activities of different types of macrophages differ. Among mouse macrophages, BMDMs were demonstrated to be significantly more phagocytic than other types of macrophages. Zajd et. al. have compared different functional and expression profiles of BMDM and peritoneal murine macrophages (pMAC). Despite some morphological similarities, BMDM have significantly higher phagocytic activity with both *E. coli* Bioparticles and Zymosan particles. ROS-generating activity measured by Amplex Red was almost the same for BMDM and pMAC [49]. A novel real-time imaging platform revealed the higher phagocytosis of BMDM towards different fluorescent pathogen particles as compared to mouse alveolar and peritoneal macrophages [50]. After treatment with fluorescent beads, BMDM displayed an increased rate of phagocytosis and phagosomal proteolysis but a similar oxidative burst compared to RAW 264.7 [51,52].

The majority of macrophage activity studies have been performed using mouse macrophages, while human and mouse macrophages differ significantly in size, immunometabolism, and differentiation stages [48,53,54]. Firstly, human macrophages are primarily derived from blood monocytes, whereas murine studies commonly use macrophages differentiated from bone marrow cells or tissue macrophages [55]. Secondly, markers used for characterizing murine macrophage polarization into M1 and M2 states, such as inducible NOS and arginase, do not serve as reliable markers for polarized human macrophages [7,30]. Lacey et al. (2012) compared the differences between the respective global gene expression profiles of MDM differentiated by GM-CSF or M-CSF with the differences between the respective profiles of murine bone marrow macrophages treated with the same CSF. They found that only 17% of genes regulated differently by these CSFs were common across the species [56].

As for human cells, two types of monocytes are used to obtain macrophages—blood monocytes or THP-1 cells, which is a human monocytic cell line derived from an acute monocytic leukemia patient. THP-1 cells are usually used as a control group for MDM; they are much easier and cheaper to culture, but their responses to activators differ substantially from the response of MDM [57]. Additionally, the markers and functional activity for both types of macrophages are highly dependent on a differentiation protocol [58,59]. While the radical-generating activity of THP-1 macrophages was demonstrated to be 20-fold lower than that of MDM [33], the phagocytic activity was comparable for two types of cells. The percentage of phagocytosis of *E. coli* BioParticles was the same for PMA-differentiated THP-1 cells and MDM [42], and the amount of particles consumed by a cell was significantly lower or comparable for two types of macrophages depending on the cell differentiation protocols and fluorescent particles used [42,43,59]. In our experiments, pro-inflammatory activity of THP-1-derived M1 macrophages was about 100 times lower than that of M1 MDM cells (Figure S6).

Upon activation, macrophages phagocytize foreign material and dead cells and produce reactive oxygen species that contribute, along with enzymes, to the digestion of trapped particles and cells. The primary source of ROS in activated macrophages is NADPH-oxidase which generates superoxide radicals. Concomitant release of SOD3 from intracellular compounds facilitates the dismutation of $O_2^{\bullet-}$ into hydrogen peroxide. H_2O_2 can pass through the cell membrane and diffuse over 100 μm in the extracellular space, affecting endocrine and paracrine redox-dependent signaling [40,60]. H_2O_2 turns via Fenton reaction into potent oxidizing agents, like hydroxy and hydroperoxyl radicals (OH• and OOH•), as well as into the corresponding lipid radicals. ROS generated by macrophages can damage not only foreign materials but also host tissues; therefore, one can expect that the ROS generating activity of anti-inflammatory regenerative M2 is significantly lower than the activity of M1 macrophages which produce proinflammatory environment during the acute phase of inflammation.

To compare the level of ROS generation by different macrophage phenotypes, we employed methods of chemiluminescence and fluorescent detection of $O_2^{\bullet-}/H_2O_2$. Advantages of these methods compared to fluorescent intracellular dyes measured by flow cytometry are as follows [20,61]: (1) cell detachment is not required, and (2) the kinetics of radical formation can be continuously monitored. We used substances with different mechanisms of macrophage activation. PMA is a potent activator of protein kinase C, and it is able to penetrate the cell membranes. The high-molecular-weight beta-glucan OZ activates cells through Toll-like type 2 and dectin-1 receptors [62]. Despite the different activation mechanisms, both inducers activated ROS production immediately after addition, and cell responses to both activators were very similar. Significant difference in ROS production and phagocytosis was observed between different macrophage phenotypes.

Exposure of monocytes to GM-CSF or M-CSF leads to different gene expression programs in cells [56], resulting in the appearance of intermediate states of M0 macrophages that are not M1 and M2 phenotypes but to varying degrees have markers and the functional activity of these cells (Figure 1). GM-CSF has low (if any) basal circulating levels that are elevated during immune/inflammatory reactions. GM-CSF is a driver of tissue inflammation, acting on different types of myeloid cells and promoting their differentiation/polarization, such as the maturation of the monocytes into macrophages [2,63]. Incubation of MDM with GM_CSF caused the expression of M1 marker CD86 on the cell surface, but did not activate any of the macrophage functions significantly (Figures 1, 3–7 and 10). Dormant M0_GM macrophages did not exist for a long time at the inflammatory site. GM-CSF appears almost simultaneously with proinflammatory cytokines and, possibly, LPS. Exposure of M0_GM cells to (+LPS,+IFNγ) initiates their polarization into M1 macrophages with the highest expression of the CD86 marker. The function of pro-inflammatory M1 macrophages is to create pro-oxidant environment, they can generate the high level of ROS immediately after activation by any inductor, secrete a high level of proinflammatory cytokine TNF-α but have the lowest ability to perform phagocytosis (Figure 10). With development of inflammation, short-term contact of M0_GM with apoptotic neutrophils can confer features of M2 cells on them [64].

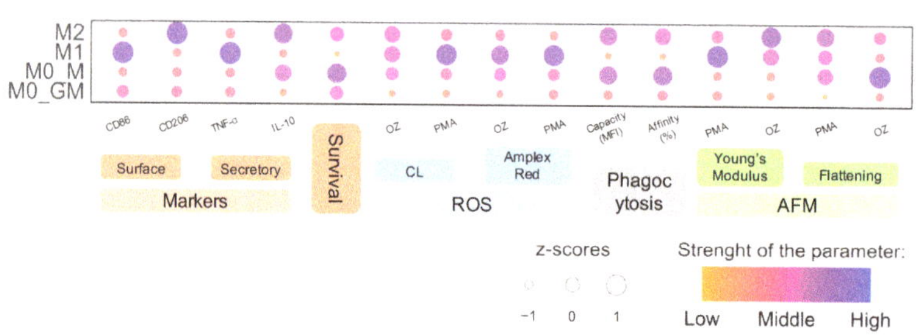

Figure 10. Comparative presentation of markers, functional activity, and mechanical properties of different phenotypes of MDM. Size and color represent z-score for each parameter measured.

Upon the development of inflammation, the concentration of GM-CSF decreases. M-CSF is constitutively produced by many types of cells including fibroblasts, endothelial cells, monocytes, macrophages, marrow stromal cells, etc., it can be detected in plasma at a concentration of ~10 ng/mL [5]. M-CSF controls the survival, proliferation, differentiation, and functions of monocytes and macrophages [2,65]. In our experiments, M-CSF not only differentiated monocytes but also polarized cells into a M2-like phenotype, which showed different types of M2 activity but did not express CD206 (Figures 1 and 3–7) [13]. These macrophages possessed the highest phagocytic ability and capacity among all phenotypes. M0_M macrophage treatment with cytokine IL-4 led to an anti-inflammatory macrophage phenotype M2. These cells secreted the highest level of IL-10 and produced radicals in an amount comparable to M1 and M0_M but somewhat more slowly (Figure S3), phagocytized bacteria comparably to M0_M, and expressed CD206 which is a marker of M2 phenotype (Figure 10). Lower ROS production by IL-4 generated M2 phenotype correlated with their highest T cell stimulatory capacity compared to other M2 phenotypes [66].

The effects of cytokines used to polarize macrophages on cell activity have been well documented [7,13,33,67]. M2 macrophages obtained after treatment of M0_M cells with

IL-4 or IL-13 differ not only in markers but also in radical-generating activity. Moreover, treatment of M1 with these cytokines alters cell activity to different degrees [68]. Following exposure to proinflammatory cytokines, M0_M macrophages can be polarized into the M1_M phenotype, which express the M1 markers but differs in the secretion of some cytokines [67]. M1_M had much higher phagocytic activity compared to M1 treated with GM-CSF (Figure S9A,B). The results demonstrate that macrophage treatment conditions should be analyzed before comparing the results of different studies.

As demonstrated earlier, activation of macrophages induces changes in their mechanical properties. After exposure to IFN-γ and LPS, an increase in relative elastic modulus was detected in RAW macrophages by optical magnetic twisting cytometry, [38]. An increase (of ~50%) in the Young's modulus of macrophages adhered to an ECM-coated substrate as compared to an uncoated glass substrate was demonstrated [69]. On the other hand, a decrease in the Young's modulus of human macrophages was observed upon stimulation with LPS [37]. The results of the latter study, however, can be affected by the high concentration of LPS (10 μg/mL), which induced cytotoxic effects, causing apoptosis and cell softening [70,71]. On the other hand, elasticity of macrophages determining by actin polymerization was demonstrated to affect innate macrophage functions including phagocytosis, secretion of cytokines, and transcriptomic profiles [38].

We analyzed the biomechanical properties of different macrophage phenotypes before and after activation and compared them to the functional activity of the cells. AFM provides a convenient method for the study of the biomechanical properties of the cell surface at the single nanoscale level [36,72]. For non-activated cells, M0_M and M2 macrophages display lower height (greater flattening) compared to M0_GM and M1, respectively. However, the substantial differences in the Young's modulus and PLE were not observed (Figure 9). When macrophages are activated, they stop migrating, start spreading out, and begin to perform one of their main functions—ROS generation. All macrophage phenotypes revealed a similar decrease in cell height after PMA activation, but there were no changes in cell height after the addition of OZ (Figure 9C,D). Most likely, the particulate agonist with a particle size of 1–3 μm [73], which is present in the medium at high concentrations, can bind to the cell membrane and distort the results of AFM measurements to some degree. The OZ particles attached to the cells can be seen in AFM topography images (Figure 8, bottom row).

Changes in the cytoskeleton and chemical changes in the cell membrane cause changes in the rigidity of macrophages, which are characterized by Young's modulus. In our experiments, the greatest increase in Young's modulus was observed for M1 cells. Notably, this phenotype had the highest radical-generating activity. A significant enhancement of stiffness was found in M2-cells. At the same time, Young's modulus did not change for their counterparts, M0_M and M0_GM, after PMA activation. A decrease in the viscoelastic parameter PLE, which characterizes the liquid–solid transition, was observed after PMA addition to all phenotypes of macrophages (Figure 9E,F). Decrease in PLE means that cell activation initiates pretension in the cytoskeleton, which usually results in macrophage spreading. PLE decrease was also observed for M0_GM and M1 macrophages after OZ addition, while it was insignificant for M0_M and M2.

4. Materials and Methods

4.1. Cell Culturing

Monocytes were isolated from peripheral blood mononuclear cells (PBMC). Blood was obtained from healthy volunteers. All donors have signed an informed consent form approved by the local ethics committee of Sechenov University (No. 07–17, 13 September 2017, Moscow, Russia).

Briefly, blood was mixed with PBS and layered onto Histopaque-1077 (Sigma Aldrich, Saint Louis, MO, USA) for density gradient centrifugation at 400 g for 40 min. After centrifugation, the PBMCs layer was collected and cells were washed twice with PBS and plated at a density of 7.5–8.5 $\times$ 10^5 PBMCs/cm^2. Cells were cultured in RPMI complete medium consisting of RPMI-1640 (Corning, Glendale, AZ, USA) supplemented with 10% autologous serum, penicillin (100 U/mL), and streptomycin (100 µg/mL) at 37 °C, 5% CO_2. After 2 h of incubation, non-adherent cells were washed with PBS, and fresh RPMI complete medium containing 50 ng/mL GM-CSF or 50 ng/mL M-CSF (SCI-store, Moscow, MOW, Russia) was added to adherent monocytes to differentiate them into M0 macrophages (M0_GM and M0_M, respectively). The medium was replaced with the fresh one on the 3rd day of culturing. Before choosing the parameters for macrophage treatment, we reviewed the literature and used the optimal ones to compare four different polar macrophage phenotypes. There is a consensus regarding the concentrations of CSFs and treatment times used for monocyte differentiation into macrophages, but cytokines and LPS used for MDM polarization differ 2–20 times (Table S1 in Supplementary Materials). We applied the middle concentration of cytokines IFN-γ and IL-4 but the lowest concentrations of LPS because LPS was demonstrated to cause dose-dependent apoptosis in macrophages differentiated from THP-1 cells [70]. On the sixth day of culturing, M0 macrophages differentiated with GM-CSF were polarized into M1 macrophages by treatment with 10 ng/mL LPS and IFN-γ 50 ng/mL to the medium, while M0 macrophages differentiated with M-CSF were polarized into M2 macrophages by treatment with IL-4 20 ng/mL [58]. The corresponding precursor M0_GM and M0_M macrophages were kept without any polarization inducers, and only the medium was replaced. After two days of incubation, four phenotypes of MDM were used for the experiments (Figure 11).

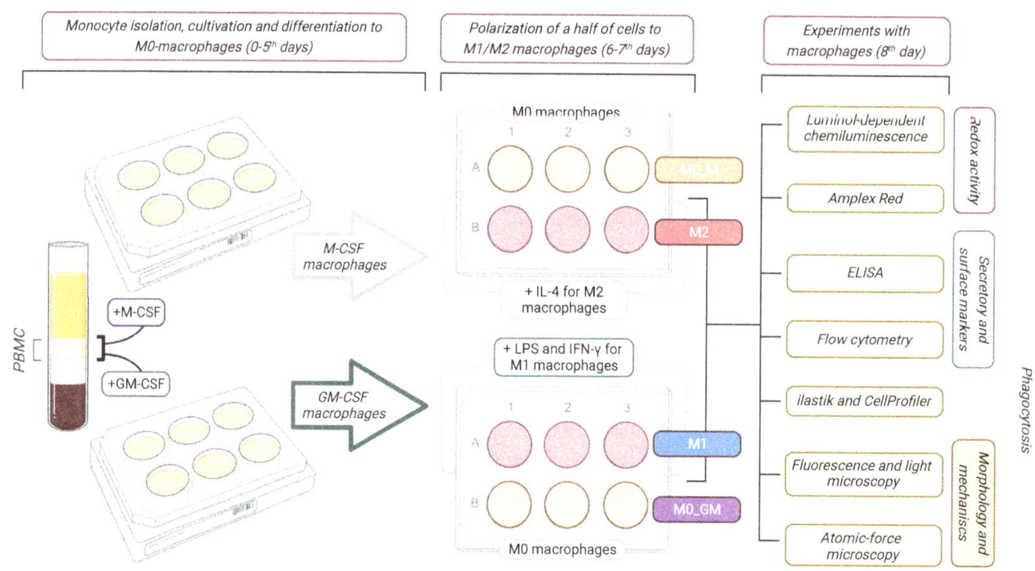

Figure 11. Schematic diagram of the experimental protocol.

4.2. Measurement of Surface Markers of Macrophage Polarization

On the 8th day of culturing, MDMs were washed three times with Versene (BioloT, Moscow, Russia) and detached with accutase solution (Sigma Aldrich, Saint Louis, MO, USA) for 5–10 min at 37 °C, 5% CO_2 to obtain single-cell suspensions and centrifuged at 300× g for 10 min. For better cell detachment, cell lifters (SPL Life sciences, Pocheon, Gyeonggi, Republic of Korea) were used. After centrifugation, cell pellets were washed twice with PBS to remove residual accutase, resuspended in flow buffer (PBS + 1% FBS), and aliquoted for subsequent staining with anti-CD206-PE-Cy7 and/or anti-CD86-FITC antibodies (Invitrogen, Waltham, MA, USA) in the dark at 4 °C. After 30 min of incubation, cells were washed in flow buffer and analyzed with Sony SH800 cell sorter (Sony

Biotechnology, San Jose, CA, USA). At least 10,000 events were recorded for each sample. Flow cytometry data visualization and analysis were performed with Sony Biotechnology software (version 1.8) and FlowJo (version 10).

4.3. Measurement of Secreted Markers of Macrophage Polarization

On the eighth day of culturing, supernatants were collected from MDM and centrifuged at 300× *g* for 10 min to exclude cell debris. ELISA kits for TNF-α and IL-6 determination ("Cytokine", St. Petersburg, SPB, Russia) were applied to evaluate the concentration of the M1 macrophage markers TNF-α and IL-6 according to the manufacturer's instructions. To measure the concentration of M2 macrophage marker IL-10, a human IL-10 ELISA kit (Sigma-Aldrich, USA) was used. Optical density was measured using a Multiskan™ FC Microplate Photometer (ThermoFisher, Waltham, MA, USA).

4.4. PicoGreen Assay

On the eighth day of culturing, the medium was removed and the attached cells were washed extensively with sterile PBS, the wells were filled with ddH_2O, and the plate was frozen at −80 °C until measurement. Before the PicoGreen assay, to break down cell membranes and release intracellular DNA, the plates were subjected to three freeze–thaw cycles. Subsequently, the solution from each well was transferred in triplicate to a fresh 96-well plate and the DNA concentration was measured using the Quant-iT PicoGreen dsDNA assay kit (Invitrogen, Waltham, MA, USA). Samples were incubated with PicoGreen for 15 min at room temperature and protected from light. The intensity of DNA fluorescence was measured using a spectrofluorometer Victor Nivo (Perkin Elmer, Shelton, CT, USA) with an excitation wavelength of 480 nm and an emission wavelength of 520 nm.

4.5. Measurements of ROS Production by Macrophages

The ability of macrophages to generate ROS was measured using luminol-enhanced CL and Amplex Red, which is converted in peroxidase reaction into fluorescent resorufin. Macrophages were cultured in 24-well plates so that at least three independent measurements were made for each group. The cells were washed twice with PBS to remove the culture medium and debris. After washing, 30 μg/mL horseradish peroxidase and 200 μM luminol were added to each well containing 500 μL of Krebs–Ringer solution with $NaHCO_3$ and $CaCl_2$ (pH 7.4) for CL measurement, and 60 μg/mL horseradish peroxidase and 10 μM Amplex Red (Sigma Aldrich, Saint Louis, MO, USA) were added for Amplex Red assay. To activate ROS production by macrophages, 300 μg/mL of OZ or 100 ng/mL of PMA were added. We selected concentrations of activators that are usually used to activate phagocytes [41,62,74]. To confirm the selected values, ROS-generating activity was measured at several concentration of activators (Figure S13). The intensity of emitted light, which is proportional to the amount of generated ROS, was recorded immediately after the addition of a macrophage activator by the use of a Victor Nivo multimode plate reader (PerkinElmer, Shelton, CT, USA). Gentle shaking was performed between measurement cycles. To compare the generation of ROS by cells over a certain period of time, the area under the CL or fluorescence curves was measured from the moment the activator was added to the selected time, which exceeded the light peak time. Additionally, we compared ROS generation by detached macrophages measuring CL in cell suspensions by the use of Lum-1200 (DISoft, Moscow, Russia). Macrophages were detached using accutase, resuspended in Krebs–Ringer solution, and transferred to a cuvette at a volume of 480 μL. After that, 400–500 μM luminol and 100–150 μg/mL HRP were added to each cuvette. OZ was added at 300 μg/mL concentration to activate ROS generation. All measurements were performed at 37 °C. For statistical analysis, the areas under the chemiluminescence or fluorescence curves were compared.

4.6. Phagocytosis Analysis

Macrophages were cultured in 35 mm dishes on microscopy slides for fluorescence microscopy or in 12-well culture plates for flow cytometry. Phagocytosis was studied using FITC-conjugated *E. coli* bioparticles (Molecular Probes, Eugene, OR, USA) on the eighth day of macrophage cultivation. Bioparticle suspension was thoroughly resuspended and mixed with complete RPMI medium and added to macrophages to achieve a required bacteria-to-cell ratio. Next, the bioparticles were incubated with cells for 1–2 h at 37 °C, 5% CO_2. After incubation, cells were washed three times with PBS to remove excess bacterial particles. Fluorescence Microscopy: Cells were incubated with ice-cold quenching solution to quench the fluorescence of external BioParticles. Fixation was carried out in 4% paraformaldehyde solution at 4 °C for 20 min. Next, the cells were permeabilized with 0.1% Triton X-100 solution for 5 min at room temperature. To stain the actin cytoskeleton of macrophages, phalloidin–rhodamine was used, and samples were incubated for 2 h in the dark, after which the cells were incubated for 20 min with DAPI to stain the nuclei. After each step, cells were washed three times with PBS. Cells were analyzed using an EVOS M5000 (Thermofisher Scientific, Waltham, MA, USA) fluorescence microscope. Two-class pixel classification with ilastik 1.4.0 was used to create probability maps and CellProfiler 4.2.5 was used to identify macrophages and measure phagocytic affinity. Flow cytometry: Cells were incubated with accutase for 5 min and detached using cell lifters. The obtained cell suspension was centrifuged at 300× *g* for 10 min. Cell pellets were filled with ice-cold quenching solution and incubated for 5 min. After centrifugation, cells were resuspended in fixation buffer (PBS with 0.5% FBS and 1% paraformaldehyde). After 20 min, cells were centrifuged, resuspended in flow buffer (PBS + 1% FBS), and analyzed on a Sony SH800 cell sorter (Sony Biotechnology, USA) using Sony Biotechnology software (version 1.8).

4.7. Atomic Force Microscopy

AFM measurements were conducted with a Bioscope Resolve AFM (Bruker, Santa Barbara, CA, USA) mounted on an Axio Observer inverted fluorescent microscope (Carl Zeiss, Oberkochen, Ostalbkreis, Germany). The microscope was equipped with a heated stage, and the sample temperature was kept constant at 37 °C. PeakForce QNM-Live Cell probes (PFQNM-LC-A-CAL, Bruker AFM Probes, Camarillo, CA, USA), short paddle-shaped cantilevers with a pre-calibrated spring constant (average value of 0.1 N/m) were used, the deflection sensitivity (nm/V) was calibrated from the thermal spectrum using the value of the spring constant [75]. The nanomechanical and topography maps were acquired in the fast force volume (FFV) mode with a map size from 20 × 20 to 80 × 80 µm and from 32 × 32 to 128 × 128 point measurements. The force curves (F-Z curves) had a vertical ramp distance of 3 µm, a vertical piezo speed of 183 µm/s, and a trigger force of 0.5–1 nN.

The numerical processing of the F-Z curves was conducted using Python 3.11.4 scripts (https://github.com/yu-efremov/ViscoIndent accessed on 1 January 2024) developed in the previous works [76,77] with the utilization of the Hertz's and Ting's models [78], for the elastic and viscoelastic processing, respectively. The Young's modulus (E) with the assumptions of the Hertz's theory, YM ("apparent" elastic modulus), was calculated from the approach part of the force curves:

$$F(\delta) = \frac{4\sqrt{R}}{3(1-\nu^2)} f_{BEC}(\delta) E \delta^{\frac{3}{2}} \tag{1}$$

where F is the force acting on the cantilever tip; δ is the indentation depth; ν is the Poisson's ratio of the sample (assumed to be time-independent and equal to 0.5); R is the radius of the indenter; $f_{BEC}(\delta)$ is the bottom-effect correction. The same curves were processed with the viscoelastic model:

$$F(t,\ \delta(t)) = \frac{4\sqrt{R}}{3(1-\nu^2)} \int_0^{t_1(t)} f_{BEC}(\delta) E(t-\xi) \frac{d\delta^{\frac{3}{2}}}{d\xi} d\xi \tag{2}$$

$$t_1(t) = \begin{cases} t_1(t) = t,\ 0 \le t \le t_m \\ \int_{t_1(t)}^{t} E(t-\xi)\frac{d\delta}{d\xi}d\xi = 0,\ t > t_m \end{cases} \tag{3}$$

where t is the time initiated at the contact; t_m is the duration of the approach phase; t_1 is the auxiliary function determined by Equation (3); ξ is the dummy time variable required for the integration; and $E(t)$ is the Young's relaxation modulus for the selected rheology model. Here we used the power law rheology (PLR) model (also known as a springpot in parallel with a dashpot) [79]:

$$E(t) = E_1 t^{-\alpha} + \eta\delta_D(t), \tag{4}$$

where E_1 is the relaxation modulus at t = 1 s (scale factor of the relaxation modulus); α is the power law exponent; η is the Newtonian viscous term (with Pa*s units); and $\delta_D(t)$ is the Dirac delta function. A larger α value means a larger amount of relaxation; materials exhibit a solid-like behavior at $\alpha = 0$ and a fluid-like behavior at $\alpha = 1$. The PLR model described by Equation (3) was successfully used for the description of cell mechanics in several previous studies [80].

The topography and local height were calculated from the force curves by the contact point position versus contact position over the substrate, the global tilt correction was performed if needed. We used the top 50% of each cell data set over a cell to define the central part, and the lower areas were discarded from the analysis, since the local properties there were highly affected by the high F-actin concentration at the periphery. From the cell datasets, the mean arithmetic values of YM and E_1, α, and η were used for further statistical comparison between the samples.

4.8. Statistical Analysis

Three technical replicates were measured for each biological sample. R programming language (R version 4.3.1) and GraphPad Prism 9.0.5 were used for statistical data analysis. For experiments on the evaluation of ROS generating capacity, the Mann–Whitney U test was used to compare groups. For phagocytosis experiments, deviations from the expected values were calculated as z-scores. Analysis of variance (ANOVA) with Tukey's post hoc test and Kruskal–Wallis one-way analysis of variance with Dunn's post hoc test was applied for multiple comparison. The Shapiro–Wilk test was performed to check if the sample fits the normal distribution. The correlation analysis was conducted using Pearson's correlation. p-values less than 0.05 were considered statistically significant.

5. Conclusions

Our study showed that three types of macrophage activity—specific cytokine secretion, ROS generation, and phagocytosis—were not identical for two of the four phenotypes studied. The high variability of the tested parameters evidences that the effects of CSFs and inducers on macrophage polarization differed for different donors. In spite of this variability, we demonstrated a general score of the activities of the macrophage phenotypes. The ability of macrophages to produce ROS can be ranked in the following order: M1 > M2 = M0_M >> M0-GM. The ability of different phenotypes for phagocytosis can be compared as follows: M2 $\geq$ M0_M > M0_GM > M1. Importantly, the functional activity of macrophage phenotypes is determined not only by polarization inducers but also by growth factors that are used for monocyte differentiation. Notably, M-CSF treated macrophages have significantly higher levels of phagocytosis compared to GM-CSF treated cells (Figures 6, 7 and S9) [27]. Other functions of macrophages, like antigen presentation, protein secretion, and extracellular trap formation, are expected to differ between macrophage phenotypes [81]. Mechanical properties of macrophages are responsible for cell migration and interaction with the extracellular matrix and other cells. Changes in M1 and M2 rigidity indicate another type of intracellular communication compared to their precursors M0_GM and M0_M. Considering the wide variety of different factors that can affect the polarization of macrophages after their differentiation by GM-CSF or M-CSF, such as cytokines, extracellular vesicles, lipid mediators, etc., one can assume a wide variety of different macrophage phenotypes, the functional activity of which is finely tuned to solve specific tasks of different phases of the inflammatory response.

Supplementary Materials: The following supporting information can be downloaded at: https://www.mdpi.com/article/10.3390/ijms25031860/s1.

Author Contributions: Methodology, data curation, investigation, and writing of original draft, S.K.S., Y.M.E., and T.O.K.; data curation, visualization, E.L.S.; resources, validation A.A.R.; project administration, funding acquisition, supervision, P.S.T.; conceptualization, methodology, validation, investigation, writing of original draft and editing, I.I.V. All authors have read and agreed to the published version of the manuscript.

Funding: This research was funded by grant from Russian Science Foundation (# 23-25-00497, rscf.ru/en/project/23-25-00497/).

Institutional Review Board Statement: The study was conducted in accordance with the Declaration of Helsinki and approved by the ethics committee of Sechenov University (No. 07–17, 13 September 2017, Moscow, Russia).

Informed Consent Statement: Informed consent was obtained from all subjects involved in the study.

Data Availability Statement: Data generated during the study and included in this article are available from the corresponding authors upon request.

Acknowledgments: We thank Maxim Balyasin (Peoples' Friendship University of Russia (RUDN University) for expert support in fluorescence microscopy image analysis. The study was carried out using the unique scientific facility Transgenebank.

Conflicts of Interest: The authors declare no conflicts of interest.

Abbreviations

AFM, atomic force microscopy; CL, chemiluminescence; HRP, horseradish peroxidase; MDM, monocyte-derived macrophages; M0_GM, GM-CSF-treated M0 macrophages; M0_M, M-CSF-treated M0 macrophages; M1 or M1_GM, GM-CSF-treated M1 macrophages; M1_M, M-CSF-treated M1 macrophages; M2, M-CSF-treated M2 macrophages; MFI, median fluorescence intensity; $O_2^{\bullet -}$, superoxide anion radical; OZ, opsonized zymosan; PBMC, peripheral blood mononuclear cells; PLE, power law exponent; PMA, phorbol-12-myristate-13-acetate; ROS, reactive oxygen species; SOD, superoxide dismutase.

References

1. Gordon, S.; Taylor, P.R. Monocyte and macrophage heterogeneity. *Nat. Rev. Immunol.* **2005**, *5*, 953–964. [CrossRef] [PubMed]
2. Hamilton, J.A. Colony-stimulating factors in inflammation and autoimmunity. *Nat. Rev. Immunol.* **2008**, *8*, 533–544. [CrossRef] [PubMed]
3. Italiani, P.; Boraschi, D. From Monocytes to M1/M2 Macrophages: Phenotypical vs. Functional Differentiation. *Front. Immunol.* **2014**, *5*, 514. [CrossRef] [PubMed]
4. Murray, P.J. Macrophage Polarization. *Annu. Rev. Physiol.* **2017**, *79*, 541–566. [CrossRef] [PubMed]
5. Ushach, I.; Zlotnik, A. Biological role of granulocyte macrophage colony-stimulating factor (GM-CSF) and macrophage colony-stimulating factor (M-CSF) on cells of the myeloid lineage. *J. Leukoc. Biol.* **2016**, *100*, 481–489. [CrossRef] [PubMed]
6. Locati, M.; Mantovani, A.; Sica, A. Macrophage activation and polarization as an adaptive component of innate immunity. *Adv. Immunol.* **2013**, *120*, 163–184. [CrossRef]
7. Murray, P.J.; Allen, J.E.; Biswas, S.K.; Fisher, E.A.; Gilroy, D.W.; Goerdt, S.; Gordon, S.; Hamilton, J.A.; Ivashkiv, L.B.; Lawrence, T.; et al. Macrophage activation and polarization: Nomenclature and experimental guidelines. *Immunity* **2014**, *41*, 14–20. [CrossRef]
8. Orekhov, A.N.; Orekhova, V.A.; Nikiforov, N.G.; Myasoedova, V.A.; Grechko, A.V.; Romanenko, E.B.; Dongwei, Z.; Chistiakov, D.A. Monocyte differentiation and macrophage polarization. *Vessel Plus* **2019**, *2019*, 10. [CrossRef]
9. Atri, C.; Guerfali, F.Z.; Laouini, D. Role of Human Macrophage Polarization in Inflammation during Infectious Diseases. *Int. J. Mol. Sci.* **2018**, *19*, 1801. [CrossRef]
10. Mills, C.D.; Ley, K. M1 and M2 macrophages: The chicken and the egg of immunity. *J. Innate Immun.* **2014**, *6*, 716–726. [CrossRef]
11. Hamidzadeh, K.; Belew, A.T.; El-Sayed, N.M.; Mosser, D.M. The transition of M-CSF-derived human macrophages to a growth-promoting phenotype. *Blood Adv.* **2020**, *4*, 5460–5472. [CrossRef] [PubMed]

12. Pérez, S.; Rius-Pérez, S. Macrophage Polarization and Reprogramming in Acute Inflammation: A Redox Perspective. *Antioxidants* **2022**, *11*, 1394. [CrossRef] [PubMed]
13. Wang, L.X.; Zhang, S.X.; Wu, H.J.; Rong, X.L.; Guo, J. M2b macrophage polarization and its roles in diseases. *J. Leukoc. Biol.* **2019**, *106*, 345–358. [CrossRef] [PubMed]
14. Mosser, D.M.; Edwards, J.P. Exploring the full spectrum of macrophage activation. *Nat. Rev. Immunol.* **2008**, *8*, 958–969. [CrossRef] [PubMed]
15. Mushenkova, N.V.; Nikiforov, N.G.; Shakhpazyan, N.K.; Orekhova, V.A.; Sadykhov, N.K.; Orekhov, A.N. Phenotype Diversity of Macrophages in Osteoarthritis: Implications for Development of Macrophage Modulating Therapies. *Int. J. Mol. Sci.* **2022**, *23*, 8381. [CrossRef] [PubMed]
16. Kadomoto, S.; Izumi, K.; Mizokami, A. Macrophage Polarity and Disease Control. *Int. J. Mol. Sci.* **2021**, *23*, 144. [CrossRef] [PubMed]
17. Herb, M.; Schramm, M. Functions of ROS in Macrophages and Antimicrobial Immunity. *Antioxidants* **2021**, *10*, 313. [CrossRef]
18. Canton, M.; Sánchez-Rodríguez, R.; Spera, I.; Venegas, F.C.; Favia, M.; Viola, A.; Castegna, A. Reactive Oxygen Species in Macrophages: Sources and Targets. *Front. Immunol.* **2021**, *12*, 734229. [CrossRef]
19. Zhang, Y.; Choksi, S.; Chen, K.; Pobezinskaya, Y.; Linnoila, I.; Liu, Z.G. ROS play a critical role in the differentiation of alternatively activated macrophages and the occurrence of tumor-associated macrophages. *Cell Res.* **2013**, *23*, 898–914. [CrossRef]
20. Griess, B.; Mir, S.; Datta, K.; Teoh-Fitzgerald, M. Scavenging reactive oxygen species selectively inhibits M2 macrophage polarization and their pro-tumorigenic function in part, via Stat3 suppression. *Free Radic. Biol. Med.* **2020**, *147*, 48–60. [CrossRef]
21. Nauseef, W.M. The phagocyte NOX2 NADPH oxidase in microbial killing and cell signaling. *Curr. Opin. Immunol.* **2019**, *60*, 130–140. [CrossRef] [PubMed]
22. Winterbourn, C.C. The biological chemistry of hydrogen peroxide. *Methods Enzymol.* **2013**, *528*, 3–25. [CrossRef] [PubMed]
23. Arnhold, J. Host-Derived Cytotoxic Agents in Chronic Inflammation and Disease Progression. *Int. J. Mol. Sci.* **2023**, *24*, 3016. [CrossRef] [PubMed]
24. Lam, R.S.; O'Brien-Simpson, N.M.; Holden, J.A.; Lenzo, J.C.; Fong, S.B.; Reynolds, E.C. Unprimed, M1 and M2 Macrophages Differentially Interact with Porphyromonas gingivalis. *PLoS ONE* **2016**, *11*, e0158629. [CrossRef]
25. Balce, D.R.; Li, B.; Allan, E.R.; Rybicka, J.M.; Krohn, R.M.; Yates, R.M. Alternative activation of macrophages by IL-4 enhances the proteolytic capacity of their phagosomes through synergistic mechanisms. *Blood* **2011**, *118*, 4199–4208. [CrossRef]
26. Varin, A.; Mukhopadhyay, S.; Herbein, G.; Gordon, S. Alternative activation of macrophages by IL-4 impairs phagocytosis of pathogens but potentiates microbial-induced signalling and cytokine secretion. *Blood* **2010**, *115*, 353–362. [CrossRef]
27. Schulz, D.; Severin, Y.; Zanotelli, V.R.T.; Bodenmiller, B. In-Depth Characterization of Monocyte-Derived Macrophages using a Mass Cytometry-Based Phagocytosis Assay. *Sci. Rep.* **2019**, *9*, 1925. [CrossRef] [PubMed]
28. Kovacevic, Z.; Sahni, S.; Lok, H.; Davies, M.J.; Wink, D.A.; Richardson, D.R. Regulation and control of nitric oxide (NO) in macrophages: Protecting the "professional killer cell" from its own cytotoxic arsenal via MRP1 and GSTP1. *Biochim. Biophys. Acta Gen. Subj.* **2017**, *1861*, 995–999. [CrossRef]
29. Schneemann, M.; Schoedon, G.; Hofer, S.; Blau, N.; Guerrero, L.; Schaffner, A. Nitric oxide synthase is not a constituent of the antimicrobial armature of human mononuclear phagocytes. *J. Infect. Dis.* **1993**, *167*, 1358–1363. [CrossRef]
30. Schneemann, M.; Schoeden, G. Macrophage biology and immunology: Man is not a mouse. *J. Leukoc. Biol.* **2007**, *81*, 579. [CrossRef]
31. Weinberg, J.B.; Misukonis, M.A.; Shami, P.J.; Mason, S.N.; Sauls, D.L.; Dittman, W.A.; Wood, E.R.; Smith, G.K.; McDonald, B.; Bachus, K.E. Human mononuclear phagocyte inducible nitric oxide synthase (iNOS): Analysis of iNOS mRNA, iNOS protein, biopterin, and nitric oxide production by blood monocytes and peritoneal macrophages. *Blood* **1995**, *86*, 1184–1195. [CrossRef] [PubMed]
32. Neves-Souza, P.C.; Azeredo, E.L.; Zagne, S.M.; Valls-de-Souza, R.; Reis, S.R.; Cerqueira, D.I.; Nogueira, R.M.; Kubelka, C.F. Inducible nitric oxide synthase (iNOS) expression in monocytes during acute Dengue Fever in patients and during in vitro infection. *BMC Infect. Dis.* **2005**, *5*, 1–12. [CrossRef] [PubMed]
33. Lewis, C.V.; Vinh, A.; Diep, H.; Samuel, C.S.; Drummond, G.R.; Kemp-Harper, B.K. Distinct Redox Signalling following Macrophage Activation Influences Profibrotic Activity. *J. Immunol. Res.* **2019**, *2019*, 1278301. [CrossRef]
34. Nassif, R.M.; Chalhoub, E.; Chedid, P.; Hurtado-Nedelec, M.; Raya, E.; Dang, P.M.C.; Jean-Claude, M.; El-Benna, J. Metformin Inhibits ROS Production by Human M2 Macrophages via the Activation of AMPK. *Biomedicines* **2022**, *10*, 319. [CrossRef]
35. Li, J.; Liu, Y.; Yuan, Y.; Huang, B. Applications of atomic force microscopy in immunology. *Front. Med.* **2021**, *15*, 43–52. [CrossRef]
36. Pi, J.; Li, T.; Liu, J.; Su, X.; Wang, R.; Yang, F.; Bai, H.; Jin, H.; Cai, J. Detection of lipopolysaccharide induced inflammatory responses in RAW264.7 macrophages using atomic force microscope. *Micron* **2014**, *65*, 1–9. [CrossRef] [PubMed]
37. Leporatti, S.; Gerth, A.; Köhler, G.; Kohlstrunk, B.; Hauschildt, S.; Donath, E. Elasticity and adhesion of resting and lipopolysaccharide-stimulated macrophages. *FEBS Lett.* **2006**, *580*, 450–454. [CrossRef]
38. Patel, N.R.; Bole, M.; Chen, C.; Hardin, C.C.; Kho, A.T.; Mih, J.; Deng, L.; Butler, J.; Tschumperlin, D.; Fredberg, J.J.; et al. Cell elasticity determines macrophage function. *PLoS ONE* **2012**, *7*, e41024. [CrossRef]
39. Thornton, S.; Tan, R.; Sproles, A.; Do, T.; Schick, J.; Grom, A.A.; DeLay, M.; Schulert, G.S. A Multiparameter Flow Cytometry Analysis Panel to Assess CD163 mRNA and Protein in Monocyte and Macrophage Populations in Hyperinflammatory Diseases. *J. Immunol.* **2019**, *202*, 1635–1643. [CrossRef]

40. Hu, L.; Zachariae, E.D.; Larsen, U.G.; Vilhardt, F.; Petersen, S.V. The dynamic uptake and release of SOD3 from intracellular stores in macrophages modulates the inflammatory response. *Redox. Biol.* **2019**, *26*, 101268. [CrossRef]
41. Jancinová, V.; Drábiková, K.; Nosál, R.; Racková, L.; Májeková, M.; Holománová, D. The combined luminol/isoluminol chemiluminescence method for differentiating between extracellular and intracellular oxidant production by neutrophils. *Redox. Rep.* **2006**, *11*, 110–116. [CrossRef]
42. Hoppenbrouwers, T.; Bastiaan Net, S.; Garssen, J.; Pellegrini, N.; Willemsen, L.E.; Wichers, H.J. Functional differences between primary monocyte-derived and THP-1 macrophages and their response to LCPUFAs. *PharmaNutrition* **2022**, *22*, 100322. [CrossRef]
43. Tedesco, S.; De Majo, F.; Kim, J.; Trenti, A.; Trevisi, L.; Fadini, G.P.; Bolego, C.; Zandstra, P.W.; Cignarella, A.; Vitiello, L. Convenience versus Biological Significance: Are PMA-Differentiated THP-1 Cells a Reliable Substitute for Blood-Derived Macrophages When Studying. *Front. Pharmacol.* **2018**, *9*, 71. [CrossRef]
44. Efremov, Y.M.; Velay-Lizancos, M.; Weaver, C.J.; Athamneh, A.I.; Zavattieri, P.D.; Suter, D.M.; Raman, A. Anisotropy vs isotropy in living cell indentation with AFM. *Sci. Rep.* **2019**, *9*, 5757. [CrossRef] [PubMed]
45. Spiller, K.L.; Wrona, E.A.; Romero-Torres, S.; Pallotta, I.; Graney, P.L.; Witherel, C.E.; Panicker, L.M.; Feldman, R.A.; Urbanska, A.M.; Santambrogio, L.; et al. Differential gene expression in human, murine, and cell line-derived macrophages upon polarization. *Exp. Cell Res.* **2016**, *347*, 1–13. [CrossRef] [PubMed]
46. Li, P.; Hao, Z.; Wu, J.; Ma, C.; Xu, Y.; Li, J.; Lan, R.; Zhu, B.; Ren, P.; Fan, D.; et al. Comparative Proteomic Analysis of Polarized Human THP-1 and Mouse RAW264.7 Macrophages. *Front. Immunol.* **2021**, *12*, 700009. [CrossRef] [PubMed]
47. Qie, J.; Liu, Y.; Wang, Y.; Zhang, F.; Qin, Z.; Tian, S.; Liu, M.; Li, K.; Shi, W.; Song, L.; et al. Integrated proteomic and transcriptomic landscape of macrophages in mouse tissues. *Nat. Commun.* **2022**, *13*, 7389. [CrossRef] [PubMed]
48. Han, J.; Gallerand, A.; Erlich, E.C.; Helmink, B.A.; Mair, I.; Li, X.; Eckhouse, S.R.; Dimou, F.M.; Shakhsheer, B.A.; Phelps, H.M.; et al. Human serous cavity macrophages and dendritic cells possess counterparts in the mouse with a distinct distribution between species. *Nat. Immunol.* **2024**, *25*, 155–165. [CrossRef] [PubMed]
49. Zajd, C.M.; Ziemba, A.M.; Miralles, G.M.; Nguyen, T.; Feustel, P.J.; Dunn, S.M.; Gilbert, R.J.; Lennartz, M.R. Bone Marrow-Derived and Elicited Peritoneal Macrophages Are Not Created Equal: The Questions Asked Dictate the Cell Type Used. *Front. Immunol.* **2020**, *11*, 269. [CrossRef] [PubMed]
50. Kapellos, T.S.; Taylor, L.; Lee, H.; Cowley, S.A.; James, W.S.; Iqbal, A.J.; Greaves, D.R. A novel real time imaging platform to quantify macrophage phagocytosis. *Biochem. Pharmacol.* **2016**, *116*, 107–119. [CrossRef] [PubMed]
51. Guo, M.; Härtlova, A.; Dill, B.D.; Prescott, A.R.; Gierliński, M.; Trost, M. High-resolution quantitative proteome analysis reveals substantial differences between phagosomes of RAW 264.7 and bone marrow derived macrophages. *Proteomics* **2015**, *15*, 3169–3174. [CrossRef]
52. Huang, N.N.; Becker, S.; Boularan, C.; Kamenyeva, O.; Vural, A.; Hwang, I.Y.; Shi, C.S.; Kehrl, J.H. Canonical and noncanonical g-protein signaling helps coordinate actin dynamics to promote macrophage phagocytosis of zymosan. *Mol. Cell. Biol.* **2014**, *34*, 4186–4199. [CrossRef]
53. Krombach, F.; Münzing, S.; Allmeling, A.M.; Gerlach, J.T.; Behr, J.; Dörger, M. Cell size of alveolar macrophages: An interspecies comparison. *Environ. Health Perspect.* **1997**, *105*, 1261–1263. [CrossRef]
54. Vijayan, V.; Pradhan, P.; Braud, L.; Fuchs, H.R.; Gueler, F.; Motterlini, R.; Foresti, R.; Immenschuh, S. Human and murine macrophages exhibit differential metabolic responses to lipopolysaccharide—A divergent role for glycolysis. *Redox. Biol.* **2019**, *22*, 101147. [CrossRef]
55. Sieweke, M.H.; Allen, J.E. Beyond stem cells: Self-renewal of differentiated macrophages. *Science* **2013**, *342*, 1242974. [CrossRef]
56. Lacey, D.C.; Achuthan, A.; Fleetwood, A.J.; Dinh, H.; Roiniotis, J.; Scholz, G.M.; Chang, M.W.; Beckman, S.K.; Cook, A.D.; Hamilton, J.A. Defining GM-CSF- and macrophage-CSF-dependent macrophage responses by in vitro models. *J. Immunol.* **2012**, *188*, 5752–5765. [CrossRef]
57. Shalhoub, V.; Faust, J.; Boyle, W.J.; Dunstan, C.R.; Kelley, M.; Kaufman, S.; Scully, S.; Van, G.; Lacey, D.L. Osteoprotegerin and osteoprotegerin ligand effects on osteoclast formation from human peripheral blood mononuclear cell precursors. *J. Cell. Biochem.* **1999**, *72*, 251–261. [CrossRef]
58. Unuvar Purcu, D.; Korkmaz, A.; Gunalp, S.; Helvaci, D.G.; Erdal, Y.; Dogan, Y.; Suner, A.; Wingender, G.; Sag, D. Effect of stimulation time on the expression of human macrophage polarization markers. *PLoS ONE* **2022**, *17*, e0265196. [CrossRef] [PubMed]
59. Daigneault, M.; Preston, J.A.; Marriott, H.M.; Whyte, M.K.; Dockrell, D.H. The identification of markers of macrophage differentiation in PMA-stimulated THP-1 cells and monocyte-derived macrophages. *PLoS ONE* **2010**, *5*, e8668. [CrossRef] [PubMed]
60. Ledo, A.; Fernandes, E.; Salvador, A.; Laranjinha, J.; Barbosa, R.M. In vivo hydrogen peroxide diffusivity in brain tissue supports volume signaling activity. *Redox. Biol.* **2022**, *50*, 102250. [CrossRef] [PubMed]
61. Müllebner, A.; Dorighello, G.G.; Kozlov, A.V.; Duvigneau, J.C. Interaction between Mitochondrial Reactive Oxygen Species, Heme Oxygenase, and Nitric Oxide Synthase Stimulates Phagocytosis in Macrophages. *Front. Med.* **2017**, *4*, 252. [CrossRef]
62. Gantner, B.N.; Simmons, R.M.; Canavera, S.J.; Akira, S.; Underhill, D.M. Collaborative induction of inflammatory responses by dectin-1 and Toll-like receptor 2. *J. Exp. Med.* **2003**, *197*, 1107–1117. [CrossRef] [PubMed]
63. Hamilton, J.A. GM-CSF in inflammation. *J. Exp. Med.* **2020**, *217*, e20190945. [CrossRef] [PubMed]

64. Chernykh, E.R.; Sakhno, L.V.; Shevela, E.Y.; Tikhonova, M.A.; Khonina, N.A.; Ostanin, A.A. Phenotypic and functional changes of GM-CSF differentiated human macrophages following exposure to apoptotic neutrophils. *Cell. Immunol.* **2018**, *331*, 93–99. [CrossRef] [PubMed]
65. Popova, A.; Kzhyshkowska, J.; Nurgazieva, D.; Goerdt, S.; Gratchev, A. Pro- and anti-inflammatory control of M-CSF-mediated macrophage differentiation. *Immunobiology* **2011**, *216*, 164–172. [CrossRef] [PubMed]
66. Kraaij, M.D.; Koekkoek, K.M.; van der Kooij, S.W.; Gelderman, K.A.; van Kooten, C. Subsets of human type 2 macrophages show differential capacity to produce reactive oxygen species. *Cell. Immunol.* **2013**, *284*, 1–8. [CrossRef] [PubMed]
67. Jaguin, M.; Houlbert, N.; Fardel, O.; Lecureur, V. Polarization profiles of human M-CSF-generated macrophages and comparison of M1-markers in classically activated macrophages from GM-CSF and M-CSF origin. *Cell. Immunol.* **2013**, *281*, 51–61. [CrossRef] [PubMed]
68. Scott, T.E.; Lewis, C.V.; Zhu, M.; Wang, C.; Samuel, C.S.; Drummond, G.R.; Kemp-Harper, B.K. IL-4 and IL-13 induce equivalent expression of traditional M2 markers and modulation of reactive oxygen species in human macrophages. *Sci. Rep.* **2023**, *13*, 19589. [CrossRef]
69. Souza, S.T.; Agra, L.C.; Santos, C.E.; Barreto, E.; Hickmann, J.M.; Fonseca, E.J. Macrophage adhesion on fibronectin evokes an increase in the elastic property of the cell membrane and cytoskeleton: An atomic force microscopy study. *Eur. Biophys. J.* **2014**, *43*, 573–579. [CrossRef]
70. Genin, M.; Clement, F.; Fattaccioli, A.; Raes, M.; Michiels, C. M1 and M2 macrophages derived from THP-1 cells differentially modulate the response of cancer cells to etoposide. *BMC Cancer* **2015**, *15*, 1–14. [CrossRef]
71. Murashko, A.V.; Frolova, A.A.; Akovantseva, A.A.; Kotova, S.L.; Timashev, P.S.; Efremov, Y.M. The cell softening as a universal indicator of cell damage during cytotoxic effects. *Biochim. Biophys. Acta Gen. Subj.* **2023**, *1867*, 130348. [CrossRef] [PubMed]
72. Zhou, G.; Zhang, B.; Tang, G.; Yu, X.F.; Galluzzi, M. Cells nanomechanics by atomic force microscopy: Focus on interactions at nanoscale. *Adv. Phys. X* **2021**, *6*, 1866668. [CrossRef]
73. Venkatachalam, G.; Arumugam, S.; Doble, M. Synthesis, Characterization, and Biological Activity of Aminated Zymosan. *ACS Omega* **2020**, *5*, 15973–15982. [CrossRef] [PubMed]
74. Vlasova, I.I.; Suleimanov, S.K.; Mikhalchik, E.V.; Urmantaeva, N.T.; Salimov, E.L.; Ragimov, A.A.; Khlebnikova, T.M.; Timashev, P.S. Redox-Activation of Neutrophils Induced by Pericardium Scaffolds. *Int. J. Mol. Sci.* **2022**, *23*, 15468. [CrossRef]
75. Schillers, H.; Rianna, C.; Schäpe, J.; Luque, T.; Doschke, H.; Wälte, M.; Uriarte, J.J.; Campillo, N.; Michanetzis, G.P.; Bobrowska, J. Standardized nanomechanical atomic force microscopy procedure (SNAP) for measuring soft and biological samples. *Scientific reports* **2017**, *1*, 5117. [CrossRef]
76. Efremov, Y.M.; Wang, W.H.; Hardy, S.D.; Geahlen, R.L.; Raman, A. Measuring nanoscale viscoelastic parameters of cells directly from AFM force-displacement curves. *Sci. Rep.* **2017**, *7*, 1541. [CrossRef]
77. Efremov, Y.M.; Shpichka, A.I.; Kotova, S.L.; Timashev, P.S. Viscoelastic mapping of cells based on fast force volume and PeakForce Tapping. *Soft Matter* **2019**, *15*, 5455–5463. [CrossRef] [PubMed]
78. Ting, T.C.T. The contact stresses between a rigid indenter and a viscoelastic half-space. *J. Appl. Mech.* **1966**, *4*, 845–854. [CrossRef]
79. Kollmannsberger, P.; Fabry, B. Active soft glassy rheology of adherent cells. *Soft Matter* **2009**, *9*, 1771–1774. [CrossRef]
80. Efremov, Y.M.; Okajima, T.; Raman, A. Measuring viscoelasticity of soft biological samples using atomic force microscopy. *Soft Matter* **2020**, *5*, 64–81. [CrossRef]
81. Smith, T.D.; Tse, M.J.; Read, E.L.; Liu, W.F. Regulation of macrophage polarization and plasticity by complex activation signals. *Integr. Biol.* **2016**, *16*, 946–955. [CrossRef] [PubMed]

Disclaimer/Publisher's Note: The statements, opinions and data contained in all publications are solely those of the individual author(s) and contributor(s) and not of MDPI and/or the editor(s). MDPI and/or the editor(s) disclaim responsibility for any injury to people or property resulting from any ideas, methods, instructions or products referred to in the content.

International Journal of
Molecular Sciences

MDPI

Article

Influence of Microbiota-Related Metabolites Associated with Inflammation and Sepsis on the Peroxidase Activity of Cyclooxygenase in Healthy Human Monocytes and Acute Monocytic Leukemia Cells

Natalia Beloborodova [1], Roman Fadeev [2] and Nadezhda Fedotcheva [1,2,*]

[1] Federal Research and Clinical Center of Intensive Care Medicine and Rehabilitology, 25-2 Petrovka St., 107031 Moscow, Russia; nvbeloborodova@yandex.ru
[2] Institute of Theoretical and Experimental Biophysics, Russian Academy of Sciences, 3, Institutskaya St., 142290 Pushchino, Russia; fadeevrs@gmail.com
* Correspondence: nfedotcheva@mail.ru

Abstract: The human microbiota produces metabolites that can enter the bloodstream and exert systemic effects on various functions in both healthy and pathological states. We have studied the participation of microbiota-related metabolites in bacterial infection by examining their influence on the activity of cyclooxygenase (COX) as a key enzyme of inflammation. The influence of aromatic microbial metabolites, derivatives of phenylalanine (phenylpropionic acid, PPA), tyrosine (4-hydroxyphenyllactic acid, HPLA), and tryptophan (indolacetic acids, IAA), the concentrations of which in the blood change notably during sepsis, was evaluated. Also, the effect of itaconic acid (ITA) was studied, which is formed in macrophages under the action of bacterial lipopolysaccharides (LPS) and appears in the blood in the early stages of infection. Metabiotic acetyl phosphate (AcP) as a strong acetylating agent was also tested. The activity of COX was measured via the TMPD oxidation colorimetric assay using the commercial pure enzyme, cultured healthy monocytes, and the human acute monocytic leukemia cell line THP-1. All metabolites in the concentration range of 100–500 μM lowered the activity of COX. The most pronounced inhibition was observed on the commercial pure enzyme, reaching up to 40% in the presence of AcP and 20–30% in the presence of the other metabolites. On cell lysates, the effect of metabolites was preserved, although it significantly decreased, probably due to their interaction with other targets subject to redox-dependent and acetylation processes. The possible contribution of the redox-dependent action of microbial metabolites was confirmed by assessing the activity of the enzyme in the presence of thiol reagents and in model conditions, when the COX-formed peroxy intermediate was replaced with *tert*-butyl hydroperoxide (TBH). The data show the involvement of the microbial metabolites in the regulation of COX activity, probably due to their influence on the peroxidase activity of the enzyme.

Keywords: cyclooxygenase; COX activity; monocytes; inflammation; microbial metabolites; 4-hydroxyphenyllactic acid; itaconic acid; tert-butyl hydroperoxide; thiol reagents; TMPD

1. Introduction

Currently, much attention is being given to the components of microbial cells, including lipopolysaccharides (LPS), that are detected by the innate host immune system throughout the body. LPS are a structural component of the cell wall of Gram-negative bacteria, but a number of other compounds, small molecules originating from microbes (SMOMs), are the products of the metabolism of living bacteria, and above all, the microbiota [1]. The human microbiota produces metabolites that may enter the bloodstream and exert systemic effects on various functions of the human organism in both healthy and pathological states. Microbial metabolites have been implicated in the regulation of the immune system, the central nervous system, metabolism, and epigenetic control [2–4]. They regulate the immune system via several ways, including receptor and metabolic signaling pathways [5]. It is believed that pathological conditions cause microbiota dysbiosis and disturb the production of microbial metabolites, which leads to the dysregulation of the immune system and metabolism [5,6]. Additionally, these relationships are bidirectional, since gut microbiota dysbiosis may provoke a high risk of the onset of chronic diseases.

Under conditions of dysbiosis, bacteria themselves may provide a proinflammatory effect through the influence of LPS, which are a part of the outer membrane of Gram-negative bacteria. In the case of an increased translocation of intestinal bacteria in the bloodstream, LPS trigger systemic inflammation and stimulate the synthesis of pro-inflammatory cytokines, namely interleukins IL-6, IL-1, and IL-27, as well as the tumor necrosis factor TNF-α [6]. In addition, LPS induce the biosynthesis of itaconic acid (ITA), which is also an important link in processes associated with macrophage-mediated inflammation [7]. It has recently been shown that itaconic acid lowers macrophage inflammation, diminishing the production of inflammatory agents, TNF-α, and the cyclooxygenase COX-2 [8], and ameliorates autoimmunity by regulating the imbalance of T cells via metabolic and epigenetic reprogramming [9].

Cyclooxygenase is one of the key participants in the inflammatory process. It enables both the inclusion of molecular oxygen into arachidonic acid to form the cyclic peroxy endoperoxide PGG2 and the subsequent reduction of PGG2 into the hydroxy endoperoxide prostaglandin H2 (PGH2), a precursor of all prostaglandins and thromboxanes. The formation of hydroxy endoperoxide occurs with the involvement of glutathione as a reducing cofactor. Prostaglandins play a key role in the generation of the acute inflammatory response; in addition, they promote chronic inflammation via participation in processes such as the amplification of cytokine signaling, the induction of the release of cytokines, and the infiltration of inflammatory cells at the inflamed site [10,11].

Some studies showed the involvement of microbial metabolites in the regulation of COX activity. Short-chain fatty acids were found to influence the immune response by stimulating the biosynthesis of prostaglandins in human monocytes, and their effect could increase in the presence of LPS. In addition, they inhibited the LPS-induced production of interleukin-10 (IL-10), TNF-α, and interferon-gamma (IFN-y), which indicates their ability to produce both pro- and anti-inflammatory effects [12]. It is assumed that the effect of short-chain fatty acids is related to their influence on the acetylation/deacetylation of transcription factors involved in the production of inflammatory cytokines [13,14]. It was found that microbial metabolites, in particular butyrate, propionate, D-lactate, and some phenolic acids, participate in the regulation of acetylation, inhibiting deacetylation reactions [15]. In addition, the intestinal phenolic acids identified in fecal water 3-phenylpropionic acid, 3-hydroxyphenylacetic acid, and 3-(4-hydroxyphenyl)-propionic acid decreased the COX-2 protein level in the human colon adenocarcinoma cell line HT-29 by up to 63% [16]. The participation of phenolic acids in the modulation of the immune response is also evidenced by our data showing their effect on neutrophils, the main producer of ROS in circulation. It was found that, among the phenolic acids tested, phenyllactate and 4-hydroxyphenyllactate most significantly decreased ROS production in PMA-activated rat neutrophils [17].

The basic conception of our investigations on this subject is that the metabolites of the microbiota of a healthy person have been implicated in the regulation of inflammation due to their ability to control the immune process. In sepsis, under conditions of severe dysbiosis, some microbial metabolites enter the bloodstream abundantly, which can influence the development and progression of systemic inflammation. In the present study, we evaluated the influence of some selected compounds from various groups of microbiota-related metabolites on the activity of COX. Among these are the microbial derivatives of phenylalanine and tryptophan: phenylpropionic (PPA), 4-hydroxyphenyllactic (HPLA), and indolacetic acids (IAA), the concentrations of which change in the blood during sepsis [17]; itaconic acid (ITA), which is formed in macrophages under the action of bacterial LPS and is detected in the blood in the early stages of infection [18]; and metabiotic acetyl phosphate (AcP) as a strong acetylating agent. Their influence on COX activity was tested using the commercial pure enzyme, cultured healthy monocytes, and the human acute monocytic leukemia cell line THP1.

2. Results

Firstly, we tested the influence of microbial metabolites on the activity of the pure enzyme to identify the most active among them that are involved in various ways in the inflammatory process. Figure 1 shows the effect of microbial metabolites on COX activity, determined by the oxidation of TMPD, which acts as an electron donor, the intensity of color and adsorption of which increase as it oxidizes. All metabolites tested decreased enzyme activity, with the degree of inhibition ranging from 10 to 45% (Figure 1a). As shown in Figure 1b, at concentrations of 100 µM and 500 µM and after a 10 min of incubation of the enzyme with arachidonic acid and cofactors, phenylpropionic acid (PPA) decreased the optical density of TMPD by 10% and 15%, that of 4-hydroxyphenyllactic (HPLA) by 17–25%, and that of indolacetic acids (IAA) by 20–25%. The effect of itaconic acid (ItA) was close to these values (20–25% inhibition), while the influence of acetyl phosphate (AcP) was much more pronounced, reaching 45% inhibition. Figure 1c shows COX activity measured via the rate of TMPD oxidation per minute in the control and in the presence of 100 µM microbiota-associated metabolites.

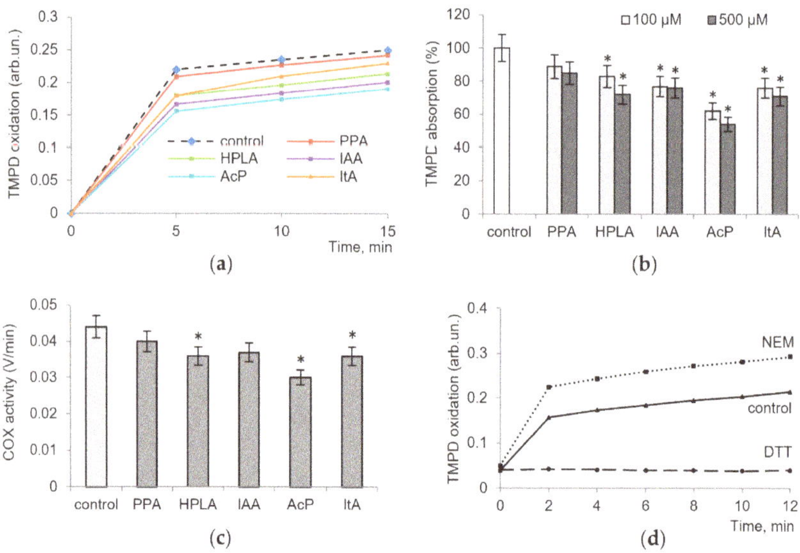

Figure 1. Influence of microbiota-associated metabolites and thiol reagents on commercially pure enzyme activity, as measured via TMPD oxidation. TMPD oxidation in the control and in the presence of 100 µM microbiota-associated metabolites (**a**), TMPD absorption in the control and in the presence of 100 µM and 500 µM microbiota-associated metabolites after 10 min of incubation (**b**), COX activity measured via the rate of TMPD oxidation per minute in the control and in the presence of 100 µM microbiota-associated metabolites (**c**), and the influence of NEM (25 µM) and DTT (250 µM) on TMPD oxidation (**d**). All measurements were performed in the presence of equal amounts of the commercially pure enzyme and arachidonic acid. An asterisk (*) indicates values that differ significantly from the control values ($p < 0.05$).

In the next experiments, we evaluated the role of the redox-dependent reactions in changes in enzyme activity. For this purpose, we used the thiol antioxidant dithiothreitol (DTT) and the thiol-blocking agent N-ethylmaleimide (NEM), the redox effects of which were similar to those shown by some microbial phenolic acids on neutrophils and mitochondria [17]. As seen in Figure 1d, NEM activated and DTT inhibited the activity of the enzyme, which indicates the possible influence of the tested microbial metabolites through thiol-dependent pathways.

The influence of microbiota-associated metabolites on COX activity was studied on human healthy monocytes and the human acute monocytic leukemia cell line THP1, activated by phorbol myristate acetate (PMA) or LPS. The lysates responded to the addition of the COX substrate's arachidonic acid with an increase in TMPD adsorption and to the addition of the COX inhibitor SC-560 with a decrease in TMPD adsorption (Figure 2a). However, these cells differed greatly in their sensitivity to TMPD-based detection. To achieve approximately equal values of TMPD adsorption, one order of magnitude more THP1 cells were required compared to healthy monocytes so that a lysate from 1 million healthy cells and a lysate from 15–20 million THP1 cells were used in each sample (Figure 2b). This ratio remained the same, although we used various methods to obtain cell lysates, including hypotonic lysis and ultra-sonication to disrupt cells.

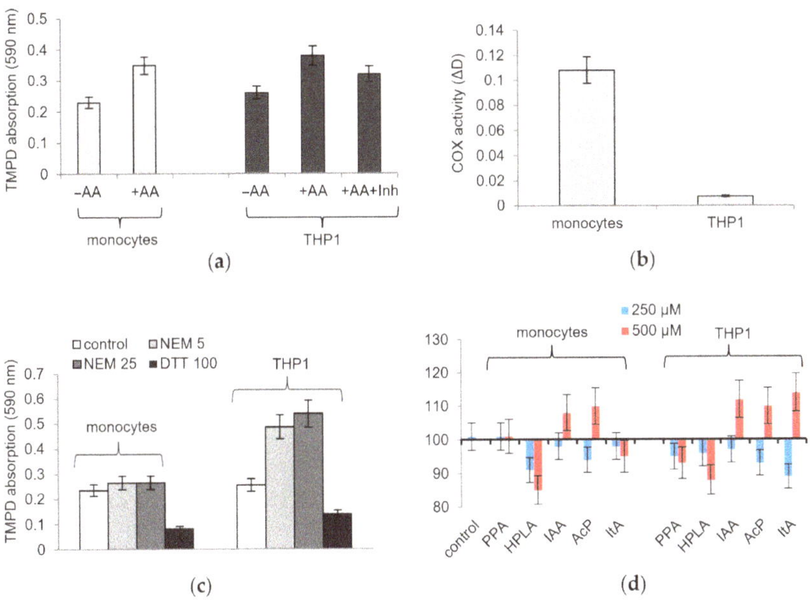

Figure 2. Influence of microbiota-associated metabolites and thiol reagents on COX activity in healthy and leukemic monocytes. An increase in TMPD absorption in response to the addition of arachidonic acid (AA) and the influence of the inhibitor SC-560 (**a**), COX activity as estimated via the difference in TMPD adsorption after 10 min of incubation of cell lysates with or without AA, per million cells (**b**), the influence of NEM and DTT on COX activity measured via TPMD absorption after 10 min of incubation with cell lysates (**c**), and the influence of microbiota-associated metabolites on COX activity (%) where the value of TMPD absorption after 10 min of incubation of cell lysates in the control is taken as 100% (**d**).

As shown in Figure 2d, the effect of microbial metabolites on COX activity in both lysates was weaker than that on the pure enzyme; it slightly increased with an increasing concentration of them and did not exceed a 10–20% deviation from the control. Some of the metabolites reversed their effect as the concentration increased; thus, instead of inhibition, an increase in activity was observed. This refers to IAA, AcP, and ItA, which at a concentration of 500 μM caused an increase in TMPD oxidation.

Thiol reagents had the same effect on cell lysates as that on the pure enzyme. Both compounds had a strong opposite effect on the activity of the enzyme in cell lysates; namely, NEM activated and DTT inhibited TMPD oxidation (Figure 2c). However, healthy monocyte lysates and THP1 monocyte lysates differed in sensitivity to NEM. As shown in Figure 2c, DTT almost equally inhibited TMPD oxidation in both types of lysates, while NEM activated TMPD oxidation by 15% in healthy monocytes and by 70% in THP1 monocyte lysates. It follows from these data that the redox state of thiol groups in these lysates and, accordingly, cells, varies significantly, which can affect COX activity. Since the most redox-dependent COX reaction is the glutathione- and heme-dependent peroxidase step, the influence of microbial metabolites in the following experiments was evaluated under model conditions imitating the peroxidase reaction.

We suggested that *tert*-butyl hydroperoxide (TBH) can serve as an analog of the peroxy intermediate PGG2 formed during the cyclooxygenase step and can replace PGG2 in the peroxidase step. As shown in Figure 3, TBH indeed replaced the substrate in the assay with TMPD. TBH induced TMPD oxidation, which depended on TBH concentration, required the presence of a heme, and occurred in the absence of the enzyme, i.e., in a nonenzymatic manner. The reaction was initiated by TBH addition. Thus, these model conditions reproduced the peroxidase step of the enzyme and could be applied in our experiments to evaluate the role of the redox-dependent influence of microbial metabolites on COX activity. It can be seen that none of the components by itself causes the oxidation of TMPD. The same applies to microbial metabolites, each of which was tested separately (Figure 3b). As shown in Figure 3c,d,e, microbial metabolites affected TMPD oxidation to different degrees, depending on their concentrations and the concentration of TBH. At high concentration of TBH (500 μM), all metabolites increased the rate of TMPD oxidation (Figure 3c). At a moderate concentration of TBH (250 μM), they decreased the rate of TMPD oxidation (Figure 3d). At lower concentrations of both components (TBH and a metabolite), the rate of TMPD oxidation varied in the range of 15–20% both towards activation and inhibition (Figure 3e). Furthermore, the ratio of concentrations of metabolites and hydroperoxide was of importance. Despite the variations, AcP, ItA, and HPLA can be considered the most active metabolites (Figure 3f).

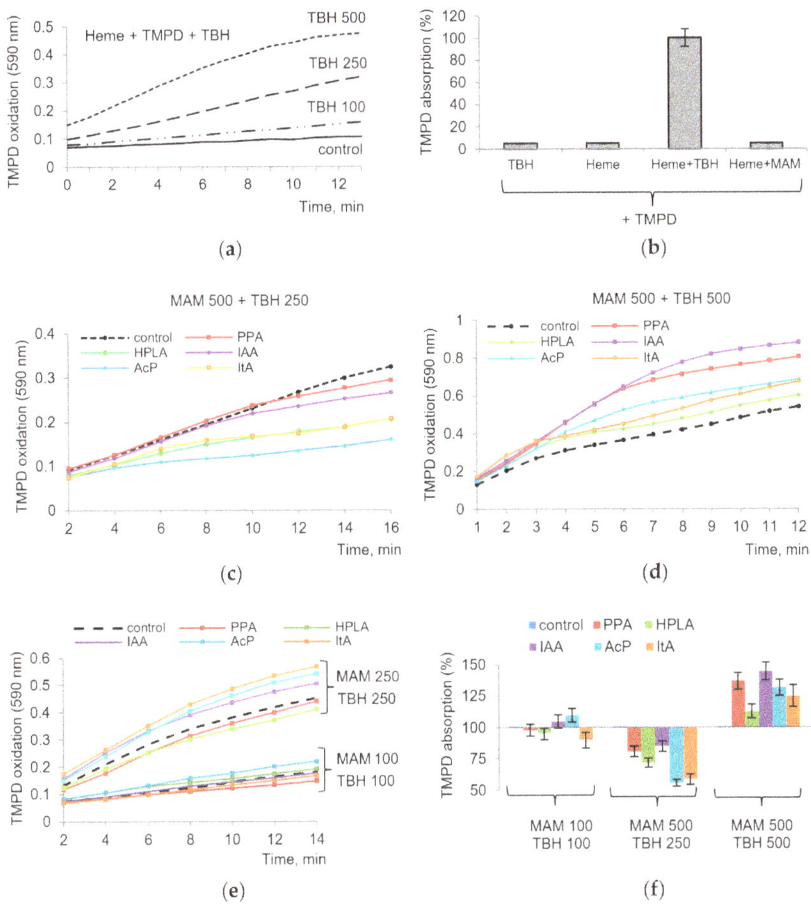

Figure 3. Influence of microbiota-associated metabolites on peroxidase activity in model conditions. An enhancement of TMPD oxidation with increasing TBH concentration (**a**), the role of each component in the induction of TMPD oxidation (**b**), TMPD oxidation at varying concentrations of TBH and microbiota-associated metabolites (MAM) (**c**–**e**), and the activation and inhibition of TMPD oxidation depending on the concentrations of TBH and microbiota-associated metabolites (MAM), where the absorption of TMPD after 10 min of incubation with TBH at each indicated concentration is taken as 100% (**f**).

3. Discussion

The development of a systemic inflammatory reaction in the human body can be observed in various nosological forms of disease, including local infections of the skin and soft tissues, pneumonia, osteomyelitis, and others, as well as upon tissue damage (trauma and surgery). An increase in clinical and laboratory signs of inflammation is always regarded as a progression of the pathological process, since it is associated with the development of certain complications; in the most severe cases, it is accompanied by the development of multiple organ dysfunctions and sepsis with the risk of an unfavorable outcome. Given the extensive evidence on the role of the structural components of the bacterial cell wall (LPS, etc.), as well as immunological, genetic, and many other etiopathogenetic factors involved in triggering systemic inflammation and sepsis, today, very little is known about the role of the products of microbial metabolism in the regulation of the inflammatory process.

Our results showed the participation of microbiota-related metabolites in the regulation of the activity of COX as a key enzyme of inflammation. Their action depends on the concentration, the initial activity of the enzyme, and the redox state of cells. As we have shown earlier, phenolic and indolic acids of microbial origin participate in the regulation of ROS production in both the circulation and tissues, and affect the activity of oxidative enzymes and membrane permeability [17,18]. Their effects strongly increase in the presence of iron ions and in acidosis inherent in infectious diseases and sepsis [19,20]. These observations also apply to itaconic acid, which affects redox-dependent enzymes and processes [20,21]. Thus, these data served as the basis for evaluating their influence on the activity of the key inflammatory enzyme COX in monocytes, the activation of which is dependent on bacterial LPS and is accompanied, among other things, by the production of itaconic acid in macrophages.

As our data show, all metabolites tested decreased the activity of the pure enzyme. Their effect diminished on lysates of both types of monocytes, healthy and THP1. Interestingly, a similar difference was observed earlier when studying the influence of some fecal microbial phenolic acids, such as 3-phenylpropionic acid, 3-hydroxyphenylacetic acid, and 3-(4-hydroxyphenyl)-propionic acid, on the activity of the pure enzyme, and the human colon adenocarcinoma cell line HT-29 [16]. Then, it was shown that these phenolic acids inhibited the biosynthesis of prostaglandin by the purified enzyme and did not affect its production in cancer cells. It was suggested that phenolic acids affect the peroxidase activity of COX. Our data obtained under model conditions are consistent with this assumption and also indicate the reasons for the observed variations.

The model system, in which the intrinsic intermediate of the cyclooxygenase step, hydroperoxy-endoperoxide PGG2, was replaced by *t*-butyl-hydroperoxide, made it possible to identify the conditions that promote either the inhibition or activation of the enzyme by these metabolites. The strong dependence of their effect on the TBH concentration suggests that the influence of metabolites is directed to the peroxidase step of the enzyme and, therefore, depends on the amount of hydroperoxide formed in the cyclooxygenase step. Also, the model conditions show a strict dependence on the presence of the heme. As known, the heme is a key participant of both steps of the cyclooxygenase reaction [22]. In the model system, the heme iron is required for an interaction with hydroperoxide by the type of Fenton reaction, while TMPD donates electrons to this system.

There are several types of reactions of phenolic compounds with iron. These are redox reactions with the participation of iron ions, where phenol-containing acids can act as reducing agents [23]. Phenolic acids containing hydroxyl groups in the phenolic ring have been shown to bind iron ions to form complexes in a 3:1 ratio, which suggests that they can act as iron chelators [24]. In addition, there is evidence indicating that the indole derivatives indole-3-propionic acid and 5-hydroxy-indole-3-acetic acid in the concentration range of 0.25–4.0 mM can act as the inhibitors of lipid peroxidation induced by iron ions [25]. Therefore, phenolic and indolic acids, being iron chelators and antioxidants, can act as the inhibitors of COX activity. Probably, this effect shows up in our experiments on the pure enzyme and at moderate concentrations of TBH in model conditions. On the contrary, the activation of the peroxidase reaction by metabolites, observed at high concentrations of TBH, may be associated with enhanced interactions of all three redox-active components, namely iron, hydroperoxide, and the metabolite. Then, the inhibition or the activation of the enzyme depends on the ratio of their concentrations in each individual case, as indicated by the variations observed at different concentrations of TBH and metabolites.

In the context of the key role of heme iron in this reaction, it can be noted that itaconate also alters iron metabolism in macrophages by degrading Fe–S clusters in both mitochondrial and cytosol aconitase, as has recently been shown [26]. Therefore, itaconic acid can affect COX activity not only as a redox-active molecule, but also as a modulator of free iron. Moreover, itaconate could directly modify cysteine sites on functional proteins involved in inflammation [27].

Until recently, there was no evidence on the presence of AcP in the human blood. In bacteria, AcP is present in millimolar concentrations [28]. It was also identified in mammalian mitochondria as an intermediate of the pyruvate dehydrogenase complex [29]. Recently, an elevated level of AcP in the blood was found in pathologies associated with spinal cord injury and interpreted as an indicator of myopathy and mitochondrial breakdown [30]. In our experiments, AcP inhibited the activity of the pure enzyme, which can be explained via the acetylation of COX in a manner acetylsalicylic acid does. However, AcP affected TMPD oxidation under model conditions in the absence of the enzyme, and the direction of its effect depended on the concentration of TBH. These data show the possible involvement of AcP in iron-dependent reactions, which is consistent with our previous data on the opposite effects of AcP on membrane permeability in normal conditions and in the presence of iron ions [19].

These data support the assumption that bacterial-derived metabolites are involved in the control of inflammatory responses via the regulation of the cell redox balance. As supposed, a balanced microbiome provides a stable redox balance, whereas dysbiosis disturbs the redox status, thereby influencing the immune system and inflammatory responses [31]. This concept is also confirmed by our data on differences in the redox states of healthy and leukemic cells. Our experiments revealed significant differences between healthy and leukemic monocytes in their sensitivity to redox-active compounds including NEM, DTT, and TMPD. The low sensitivity of THP-1 cells to TMPD and the high sensitivity to NEM may indicate an elevated redox status of THP-1 cells. According to current views, the redox state determines the response of tumor cells, including the immune response, to various drugs and treatments. It is assumed that the level of antioxidant defense, including the protection of thiol groups, allows them to obviate the deleterious effects of inductors of oxidative stress [32]. To what extent these indicators are applicable to THP-1 cells as a whole and to COX activity in particular requires further research. However, our results obtained under model conditions show a strong dependence of the direction of the reaction on the concentration of redox active participants. As was recently shown, all isoforms of glutathione peroxidase can reduce not only fatty acid-derived hydroperoxides, but also small hydroperoxides such as TBH [33]. If the intrinsic glutathione is considered instead of the artificial electron donor TMPD, then it can be assumed that the oxidized state of cells would decrease glutathione-dependent prostaglandin synthesis, and, vice versa, the reduced state would promote it, which also needs clarification.

The results of clinical studies in different groups of patients confirm the above. It has been shown that normally, with a preserved microbiome, the intestinal contents of healthy people are dominated by the end products of microbial metabolism of aromatic amino acids (phenylpropionic acid, phenylacetic acid, and indolacetic acid) and, accordingly, they can be detected in the systemic bloodstream. They are formed from aromatic amino acids as a result of biotransformation by strictly anaerobic intestinal bacteria. Conversely, intermediate products (4-hydroxyphenyllactic acid and 4-hydroxyphenylacetic acid) are predominantly the metabolites of facultative anaerobes (e.g., Gram-negative enterobacteria (*Klebsiella pneumonia*, *Escherichia coli*, and *Staphylococci* spp.), which multiply well in the presence of oxygen and actively participate in the development of postoperative complications (pneumonia, bacteriemia, sepsis, etc.) [17]. These metabolites are absent in the intestines; they do not enter the bloodstream and have no effect on the immune system. On the contrary, in patients with documented bacterial inflammation, hydroxylated phenolic metabolites are always detected in the blood [34,35]. In the case of local forms of infection, the level of hydroxyphenyllactic acid in the serum is within 2–4 μM, and with sepsis, its concentration in the circulation increases and reaches the highest values (up to 10–100 μM or more); that is, it occurs in parallel with the imbalance of various processes in the body of a septic patient [34,35]. Thus, the results obtained shed light on the molecular mechanisms of the regulation of the inflammatory process involving the microbiota and its metabolites.

Thus, our research confirms the assumption that the regulation of a systemic inflammatory process is associated with the state of the microbiota and its metabolites. Our data show that the involvement of microbial metabolites in the regulation of the activity of COX is mainly due to their influence on the peroxidase activity of the enzyme. As a consequence, the redox state of cells, including the protection of thiol groups, allows them to obviate the deleterious effects of the inductors of oxidative stress and inflammation. Our results obtained under model conditions show a strong dependence of the direction of the action of microbial metabolites on the concentration of redox-active participants. The model system in which the intrinsic intermediate of the cyclooxygenase step, endoperoxide, was replaced by *t*-butyl-hydroperoxide made it possible to identify the factors that promote either the inhibition or activation of the enzyme.

4. Materials and Methods

4.1. Reagents and Chemicals

The colorimetric COX Activity Assay Kit (Cayman Chemical Company, Ann Arbor, MI, USA) was used in experiments. All other reagents were obtained from the Sigma–Aldrich Corporation (St. Louis, MO, USA).

4.2. Cell Cultures

Human acute myeloid leukemia cell line THP-1 was obtained from the ATCC (Manassas, VA, USA). Cells were cultured in RPMI 1640 (Sigma-Aldrich, St. Louis, MO, USA) medium supplemented with 10% fetal bovine serum (Thermo Scientific, Waltham, MA, USA), 40 μg/mL gentamicin sulfate (Sigma-Aldrich, St. Louis, MO, USA), and 0.05 mM 2-mercaptoethanol (Sigma-Aldrich, St. Louis, MO, USA) at 37 °C in a humidified atmosphere of 5% CO_2, as described previously [36,37].

Human peripheral blood CD14+ monocytes (HPBMs) were obtained from the CLS Cell Lines Service GmbH (Eppelheim, Deutschland). HPBMs were cultured in Human Blood Cell Culture Medium (CLS Cell Lines Service GmbH, Eppelheim, Deutschland) with 40 μg/mL of gentamicin sulfate (Sigma-Aldrich, St. Louis, MO, USA) at 37 °C in a humidified atmosphere of 5% CO_2.

Monocyte-like THP-1ATRA cells were obtained via the treatment of THP-1 cells with ATRA. THP-1 cells were cultured in RPMI 1640 (Sigma-Aldrich, St. Louis, MO, USA) supplemented with 10% FBS (Thermo Scientific, Waltham, MA, USA) and 2 μM ATRA (Sigma-Aldrich, St. Louis, MO, USA) for 6 days. Then, cells were washed three times with RPMI 1640 and used for experiments.

The testing of cell cultures for mycoplasma infection was performed using the MycoFluor™ mycoplasma detection kit (Thermo Scientific, Waltham, MA, USA). The infection of cell cultures with mycoplasma was not detected.

4.3. Cell Lysate Preparation

Cells in cultures were activated by adding PMA (100 nM) one day before the experiment. Then, cells were washed with RPMI 1640 medium and used for experiments. Prior to obtaining cell lysates, the number of cells in culture in each experiment was determined using a TC20 Cell Counter (Bio-Rad, Hercules, CA, USA). This indicator was used for the normalization of the enzyme activity in lysates per million cells of the tested cell cultures. For cell lysate preparation, cells were collected via centrifugation (2000× *g* for 10 min at 4 °C). The cell pellet was homogenized in a cold buffer (0.1 M Tris-HCL, pH 7.8 with 1 mM EDTA) by a Dounce homogenizer (Sigma-Aldrich, St. Louis, MO, USA). Then, the cell lysate was centrifuged (10,000× *g* for 15 min at 4 °C) and the supernatant for the assay was stored on ice. Cell lysates were also obtained using the second method using ultrasonication to disrupt cells. For this, the cells were subjected to the ultrasonic treatment at certain parameters (amplitude: 30; process time: 20 s; pulse on time: 5 s; pulse off time: 20 s) using a Q700 sonicator (Qsonica, Newtown, CT, USA). Destroyed cells were centrifuged (10,000× *g* for 15 min at 4 °C), and the supernatant was collected and stored on ice.

4.4. Determination of the Cyclooxygenase Activity and Influence of Microbiota-Related Metabolites on the Enzyme Activity in Cell Lysates

Cyclooxygenase activity was assayed using the colorimetric method by monitoring the oxidation of N,N,N′,N′-tetramethyl-p-phenylenediamine (TMPD) at 590 nm. Measurements were carried out in accordance with the instructions of the assay kit and the description of the method [38]. The influence of microbiota-related metabolites on COX activity was determined in a 96-well plate with the following order of additions of reaction components; the enzyme was incubated with the microbial metabolite for 5 min, followed by the addition of heme, arachidonic acid, and TMPD. The influence of microbiota-related metabolites on COX activity in cell lysates was determined in a similar mode: their incubation with the lysate preceded the addition of the substrate, the cofactor, and TMPD. Changes in the optical density of TMPD during the reaction were recorded using a microplate reader iMark (Bio-Rad, Hercules, CA, USA).

4.5. Statistical Analysis

The data given represent the means ± standard error of means (SEMs) from five experiments or are the typical traces of three to five identical experiments. Statistical significance was estimated via the Student's t-test with $p < 0.05$ as the criterion of significance.

Author Contributions: N.B. and N.F. formed the concept; N.F. and N.B. designed the study; N.F. and R.F. tested the activity of the enzyme; R.F. worked with cell cultures; N.F. prepared the original project and contributed a visualization of the results; all authors wrote and edited the manuscript. All authors have read and agreed to the published version of the manuscript.

Funding: The work is supported by Russian government contract FGWS-2022-0005 (2022–2024) of the Federal Research and Clinical Center of Intensive Care Medicine and Rehabilitology and the Russian government contract 075-00381-21-00 (2021–2023) of the Institute of Theoretical and Experimental Biophysics, Russian Academy of Sciences.

Institutional Review Board Statement: Not required.

Informed Consent Statement: Not applicable.

Data Availability Statement: The datasets generated during and/or analyzed during the current study are available from the corresponding author on reasonable request.

Conflicts of Interest: The authors declare no conflict of interest.

References

1. Needham, B.D.; Kaddurah-Daouk, R.; Mazmanian, S.K. Gut microbial molecules in behavioural and neurodegenerative conditions. *Nat. Rev. Neurosci.* **2020**, *21*, 717–731. [CrossRef] [PubMed]
2. Caspani, G.; Swann, J. Small talk: Microbial metabolites involved in the signaling from microbiota to brain. *Curr. Opin. Pharmacol.* **2019**, *48*, 99–106. [CrossRef] [PubMed]
3. Visconti, A.; Le Roy, C.I.; Rosa, F.; Rossi, N.; Martin, T.C.; Mohney, R.P.; Li, W.; de Rinaldis, E.; Bell, J.T.; Venter, J.C.; et al. Interplay between the human gut microbiome and host metabolism. *Nat. Commun.* **2019**, *10*, 4505. [CrossRef]
4. Bhat, M.I.; Kapila, R. Dietary metabolites derived from gut microbiota: Critical modulators of epigenetic changes in mammals. *Nutr. Rev.* **2017**, *75*, 374–389. [CrossRef] [PubMed]
5. Kim, C.H. Immune regulation by microbiome metabolites. *Immunology* **2018**, *154*, 220–229. [CrossRef] [PubMed]
6. Nesci, A.; Carnuccio, C.; Ruggieri, V.; D'Alessandro, A.; Di Giorgio, A.; Santoro, L.; Gasbarrini, A.; Santoliquido, A.; Ponziani, F.R. Gut Microbiota and Cardiovascular Disease: Evidence on the Metabolic and Inflammatory Background of a Complex Relationship. *Int. J. Mol. Sci.* **2023**, *24*, 9087. [CrossRef]
7. Peace, C.G.; O'neill, L.A. The role of itaconate in host defense and inflammation. *J. Clin. Investig.* **2022**, *132*, e148548. [CrossRef]
8. Maassen, S.; Coenen, B.; Ioannidis, M.; Harber, K.; Grijpstra, P.; Van den Bossche, J.; van den Bogaart, G. Itaconate promotes a wound resolving phenotype in pro-inflammatory macrophages. *Redox Biol.* **2023**, *59*, 102591. [CrossRef]
9. Aso, K.; Kono, M.; Kanda, M.; Kudo, Y.; Sakiyama, K.; Hisada, R.; Karino, K.; Ueda, Y.; Nakazawa, D.; Fujieda, Y.; et al. Itaconate ameliorates autoimmunity by modulating T cell imbalance via metabolic and epigenetic reprogramming. *Nat. Commun.* **2023**, *14*, 984. [CrossRef]
10. Aoki, T.; Narumiya, S. Prostaglandins and chronic inflammation. *Trends Pharmacol. Sci.* **2012**, *33*, 304–311. [CrossRef]
11. Yamaguchi, A.; Botta, E.; Holinstat, M. Eicosanoids in inflammation in the blood and the vessel. *Front. Pharmacol.* **2022**, *13*, 997403. [CrossRef] [PubMed]

12. Cox, M.A.; Jackson, J.; Stanton, M.; Rojas-Triana, A.; Bober, L.; Laverty, M.; Yang, X.; Zhu, F.; Liu, J.; Wang, S.; et al. Short-chain fatty acids act as antiinflammatory mediators by regulating prostaglandin E_2 and cytokines. *World J. Gastroenterol.* **2009**, *15*, 5549–5557. [CrossRef] [PubMed]
13. Soliman, M.L.; Smith, M.D.; Houdek, H.M.; A Rosenberger, T. Acetate supplementation modulates brain histone acetylation and decreases interleukin-1β expression in a rat model of neuroinflammation. *J. Neuroinflamm.* **2012**, *9*, 51. [CrossRef] [PubMed]
14. Moon, H.-R.; Yun, J.-M. Sodium butyrate inhibits high glucose-induced inflammation by controlling the acetylation of NF-κB p65 in human monocytes. *Nutr. Res. Pract.* **2023**, *17*, 164–173. [CrossRef]
15. Waldecker, M.; Kautenburger, T.; Daumann, H.; Busch, C.; Schrenk, D. Inhibition of histone-deacetylase activity by short-chain fatty acids and some polyphenol metabolites formed in the colon. *J. Nutr. Biochem.* **2008**, *19*, 587–593. [CrossRef]
16. Karlsson, P.C.; Huss, U.; Jenner, A.; Halliwell, B.; Bohlin, L.; Rafter, J.J. Human Fecal Water Inhibits COX-2 in Colonic HT-29 Cells: Role of Phenolic Compounds. *J. Nutr.* **2005**, *135*, 2343–2349. [CrossRef]
17. Beloborodova, N.; Bairamov, I.; Olenin, A.; Shubina, V.; Teplova, V.; Fedotcheva, N. Effect of phenolic acids of microbial origin on production of reactive oxygen species in mitochondria and neutrophils. *J. Biomed. Sci.* **2012**, *19*, 89. [CrossRef]
18. Beloborodova, N.; Pautova, A.; Sergeev, A.; Fedotcheva, N. Serum Levels of Mitochondrial and Microbial Metabolites Reflect Mitochondrial Dysfunction in Different Stages of Sepsis. *Metabolites* **2019**, *9*, 196. [CrossRef]
19. Fedotcheva, N.; Olenin, A.; Beloborodova, N. Influence of Microbial Metabolites on the Nonspecific Permeability of Mitochondrial Membranes under Conditions of Acidosis and Loading with Calcium and Iron Ions. *Biomedicines* **2021**, *9*, 558. [CrossRef]
20. Fedotcheva, N.; Beloborodova, N. Influence of Microbial Metabolites and Itaconic Acid Involved in Bacterial Inflammation on the Activity of Mitochondrial Enzymes and the Protective Role of Alkalization. *Int. J. Mol. Sci.* **2022**, *23*, 9069. [CrossRef]
21. Belosludtsev, K.N.; Belosludtseva, N.V.; Kosareva, E.A.; Talanov, E.Y.; Gudkov, S.V.; Dubinin, M.V. Itaconic acid impairs the mitochondrial function by the inhibition of complexes II and IV and induction of the permeability transition pore opening in rat liver mitochondria. *Biochimie* **2020**, *176*, 150–157. [CrossRef] [PubMed]
22. Rouzer, C.A.; Marnett, L.J. Cyclooxygenases: Structural and functional insights. *J. Lipid Res.* **2009**, *50*, S29–S34. [CrossRef] [PubMed]
23. Bai, Y.; Sun, T.; Angenent, L.T.; Haderlein, S.B.; Kappler, A. Electron hopping enables rapid electron transfer between quinone-/hydroquinone-containing organic molecules in microbial iron(III) mineral reduction. *Environ. Sci. Technol.* **2020**, *54*, 10646–10653. [CrossRef] [PubMed]
24. Yang, S.; Bai, G.; Chen, L.; Shen, Q.; Diao, X.; Zhao, G. The interaction of phenolic acids with Fe(III) in the presence of citrate as studied by isothermal titration calorimetry. *Food Chem.* **2014**, *157*, 302–309. [CrossRef]
25. Karbownik, M.; Gitto, E.; Lewiñski, A.; Reiter, R.J. Relative efficacies of indole anti-oxidants in reducing autoxidation and iron-induced lipid peroxidation in hamster testes. *J. Cell. Biochem.* **2001**, *81*, 693–699. [CrossRef]
26. Liu, X.; Shi, B.; Suo, R.; Xiong, S.; Wang, X.; Liang, X.; Li, X.; Li, G. Itaconate regulates macrophage function through stressful iron-sulfur cluster disrupting and iron metabolism rebalancing. *FASEB J.* **2021**, *35*, e21936. [CrossRef]
27. Lin, J.; Ren, J.; Gao, D.S.; Dai, Y.; Yu, L. The Emerging Application of Itaconate: Promising Molecular Targets and Therapeutic Opportunities. *Front. Chem.* **2021**, *9*, 669308. [CrossRef]
28. Klein, A.H.; Shulla, A.; Reimann, S.A.; Keating, D.H.; Wolfe, A.J. The Intracellular Concentration of Acetyl Phosphate in *Escherichia coli* Is Sufficient for Direct Phosphorylation of Two-Component Response Regulators. *J. Bacteriol.* **2007**, *189*, 5574–5581. [CrossRef]
29. Xu, W.J.; Wen, H.; Kim, H.S.; Ko, Y.-J.; Dong, S.-M.; Park, I.-S.; Yook, J.I.; Park, S. Observation of acetyl phosphate formation in mammalian mitochondria using real-time in-organelle NMR metabolomics. *Proc. Natl. Acad. Sci. USA* **2018**, *115*, 4152–4157. [CrossRef]
30. Bykowski, E.A.; Petersson, J.N.; Dukelow, S.; Ho, C.; Debert, C.T.; Montina, T.; Metz, G.A.S. Identification of Serum Metabolites as Prognostic Biomarkers Following Spinal Cord Injury: A Pilot Study. *Metabolites* **2023**, *13*, 605. [CrossRef]
31. Kunst, C.; Schmid, S.; Michalski, M.; Tümen, D.; Buttenschön, J.; Müller, M.; Gülow, K. The Influence of Gut Microbiota on Oxidative Stress and the Immune System. *Biomedicines* **2023**, *11*, 1388. [CrossRef] [PubMed]
32. Cheung, E.C.; Vousden, K.H. The role of ROS in tumour development and progression. *Nat. Rev. Cancer* **2022**, *22*, 280–297. [CrossRef] [PubMed]
33. Schwarz, M.; Löser, A.; Cheng, Q.; Wichmann-Costaganna, M.; Schädel, P.; Werz, O.; Arnér, E.S.; Kipp, A.P. Side-by-side comparison of recombinant human glutathione peroxidases identifies overlapping substrate specificities for soluble hydroperoxides. *Redox Biol.* **2023**, *59*, 102593. [CrossRef] [PubMed]
34. Beloborodova, N.V. Metabolism of Microbiota in Critical Illness (Review and Postulates). *Gen. Reanimatol.* **2019**, *15*, 62–79. [CrossRef]
35. Beloborodova, N.V. Serum Aromatic Microbial Metabolites as Biological Markers in Intensive Care. In *Biomarkers in Trauma, Injury and Critical Care. Biomarkers in Disease: Methods, Discoveries and Applications*; Rajendram, R., Preedy, V.R., Patel, V.B., Eds.; Springer: Cham, Switzerland, 2023; Chapter 13, pp. 245–268.
36. Kobyakova, M.; Lomovskaya, Y.; Senotov, A.; Lomovsky, A.; Minaychev, V.; Fadeeva, I.; Shtatnova, D.; Krasnov, K.; Zvyagina, A.; Odinokova, I.; et al. The Increase in the Drug Resistance of Acute Myeloid Leukemia THP-1 Cells in High-Density Cell Culture Is Associated with Inflammatory-like Activation and Anti-Apoptotic Bcl-2 Proteins. *Int. J. Mol. Sci.* **2022**, *23*, 7881. [CrossRef]

37. Lomovskaya, Y.V.; Kobyakova, M.I.; Senotov, A.S.; Lomovsky, A.I.; Minaychev, V.V.; Fadeeva, I.S.; Shtatnova, D.Y.; Krasnov, K.S.; Zvyagina, A.I.; Akatov, V.S.; et al. Macrophage-like THP-1 Cells Derived from High-Density Cell Culture Are Resistant to TRAIL-Induced Cell Death via Down-Regulation of Death-Receptors DR4 and DR5. *Biomolecules* **2022**, *12*, 150. [CrossRef]
38. Petrovic, N.; Murray, M. Using *N,N,N′,N′*-tetramethyl-*p*-phenylenediamine (TMPD) to Assay Cyclooxygenase Activity In Vitro. In *Advanced Protocols in Oxidative Stress II, Methods in Molecular Biology*; Armstrong, D., Ed.; Humana Press: Totowa, NJ, USA, 2010; Volume 594, Chapter 9. [CrossRef]

Disclaimer/Publisher's Note: The statements, opinions and data contained in all publications are solely those of the individual author(s) and contributor(s) and not of MDPI and/or the editor(s). MDPI and/or the editor(s) disclaim responsibility for any injury to people or property resulting from any ideas, methods, instructions or products referred to in the content.

International Journal of *Molecular Sciences*

MDPI

Article

Impact of Hypoxia–Hyperoxia Exposures on Cardiometabolic Risk Factors and TMAO Levels in Patients with Metabolic Syndrome

Afina Bestavashvili [1,*], Oleg Glazachev [2], Shabnam Ibragimova [3], Alexander Suvorov [1], Alexandros Bestavasvili [4], Shevket Ibraimov [5], Xinliang Zhang [2], Yong Zhang [6], Chavdar Pavlov [3,7], Elena Syrkina [5], Abram Syrkin [5] and Philipp Kopylov [1]

1 World-Class Research Center "Digital Biodesign and Personalized Healthcare", I. M. Sechenov First Moscow State Medical University, 119991 Moscow, Russia; suvorov_a_yu_1@staff.sechenov.ru (A.S.); kopylov_f_yu@staff.sechenov.ru (P.K.)
2 Department of Normal Physiology, N.V. Sklifosovsky Institute of Clinical Medicine, I. M. Sechenov First Moscow State Medical University, 119991 Moscow, Russia; glazachev_o_s@staff.sechenov.ru (O.G.); zxl0620@yandex.ru (X.Z.)
3 Department of Therapy of the Institute of Professional Education, I. M. Sechenov First Moscow State Medical University, 119991 Moscow, Russia; shabnam.olimp@mail.ru (S.I.); pavlov_ch_s@staff.sechenov.ru (C.P.)
4 Department of Medicine, Salem Hospital, Mass General Brigham, Salem, MA 01970, USA; abestavasvili@mgb.org
5 Department of Cardiology, Functional and Ultrasound Diagnostics, N.V. Sklifosovsky Institute of Clinical Medicine, I. M. Sechenov First Moscow State Medical University, 119991 Moscow, Russia; sheva7864@gmail.com (S.I.); syrkina_e_a@staff.sechenov.ru (E.S.); previntenscardiology@staff.sechenov.ru (A.S.)
6 The State-Province Key Laboratories of Biomedicine-Pharmaceutics of China, Key Laboratory of Cardiovascular Research, Ministry of Education, Department of Pharmacology, Harbin Medical University, Harbin 150081, China; hmzhangyong@hotmail.com
7 Department of Gastroenterology, Botkin Hospital, 125284 Moscow, Russia
* Correspondence: bestavashvili_a_a@staff.sechenov.ru; Tel.: +7-916-338-35-95

Abstract: Along with the known risk factors of cardiovascular diseases (CVDs) constituting metabolic syndrome (MS), the gut microbiome and some of its metabolites, in particular trimethylamine-N-oxide (TMAO), are actively discussed. A prolonged stay under natural hypoxic conditions significantly and multi-directionally changes the ratio of gut microbiome strains and their metabolites in feces and blood, which is the basis for using hypoxia preconditioning for targeted effects on potential risk factors of CVD. A prospective randomized study included 65 patients (32 females) with MS and optimal medical therapy. Thirty-three patients underwent a course of 15 intermittent hypoxic–hyperoxic exposures (IHHE group). The other 32 patients underwent sham procedures (placebo group). Before and after the IHHE course, patients underwent liver elastometry, biochemical blood tests, and blood and fecal sampling for TMAO analysis (tandem mass spectrometry). No significant dynamics of TMAO were detected in both the IHHE and sham groups. In the subgroup of IHHE patients with baseline TMAO values above the reference (TMAO $\geq$ 5 µmol/l), there was a significant reduction in TMAO plasma levels. But the degree of reduction in total cholesterol (TCh), low-density lipoprotein (LDL), and regression of liver steatosis index was more pronounced in patients with initially normal TMAO values. Despite significant interindividual variations, in the subgroup of IHHE patients with MS and high baseline TMAO values, there were more significant reductions in cardiometabolic and hepatic indicators of MS than in controls. More research is needed to objectify the prognostic role of TMAO and the possibilities of its correction using hypoxia adaptation techniques.

Keywords: gut microbiome; intermittent hypoxic training; inflammatory response; lipid oxidation; metabolic syndrome; reactive oxygen species; trimethylamine-N-oxide

1. Introduction

The effect of the gut microbiome, which has been described as a separate organ in some studies, on cardiometabolic pathology (atherosclerosis, hypertension, obesity, insulin resistance (IR)) is now well established [1–5]. A number of studies have shown that certain metabolites of the gut microbiome, such as increased trimethylamine-N-oxide (TMAO), which is formed in the liver after the splitting of food with a high content of choline, phosphatidylcholine, L-carnitine, and other trimethylamine (TMA)-containing nutrients, are associated with a high risk of the initiation and progression of coronary heart disease (CHD), heart failure, hypertension, type 2 diabetes mellitus (T2DM), obesity, and peripheral artery disease [6–9].

In particular, elevated plasma levels of TMAO have been shown to be associated with diastolic dysfunction, contributing to the development of early vascular atherosclerosis through accelerating endothelial injury and dysfunction [10,11]. TMAO initiates pro-atherosclerotic mechanisms: inhibition of reverse cholesterol transport by macrophages, increased platelet responsiveness and risk of thrombosis, and induction of inflammatory vascular endothelial damage [12–15]. Thus, several studies have shown a direct relationship between TMAO levels, inflammatory processes, atherosclerosis, T2DM, and other cardiovascular diseases (CVD) [12,13,16–18]. A systematic review and meta-analysis by Schiattarella et al. demonstrated that there is a positive dose-dependent correlation between blood TMAO levels and increased cardiovascular risk as well as overall mortality in humans [17,19]. According to Kanitsoraphan et al., high circulating blood TMAO levels were associated with a 2.07–2.7-fold higher overall mortality in patients with T2DM, including after adjustment for body mass index (BMI) [20].

Recent data demonstrate the potential role of the gut microbiota as a pathogenic factor associated with metabolic syndrome (MS). Qualitative and quantitative changes in the gut microbiota are associated with various clinical conditions, such as obesity, insulin resistance and T2DM, and steatohepatitis [21–23]. Barrea L. et al. have shown that circulating blood levels of TMAO correlate positively with the degree of obesity, as well as with other MS components, including a surrogate index of IR-HoMA-IR. When the markers of cardiometabolic risks and early predictors of nonalcoholic fatty liver disease (NAFLD), the Visceral Adiposity Index (VAI) and the Fatty Liver Index (FLI), were analyzed, their association with the gut microflora metabolite TMAO was also found [24,25]. In addition, Chen et al., in a study conducted in China and including a sample of 1688 patients diagnosed with NAFLD, a hepatic manifestation of MS, found an association between the presence and severity of NAFLD and TMAO levels, which they positioned as an independent marker and possible risk factor for NAFLD [25,26].

Given the significant potential involvement of TMAO in the development of cardiovascular and metabolic diseases, there is an active search for TMAO inhibitors. No direct inhibitors of TMAO have yet been found, but the compounds 3,3-dimethyl-1-butanol (DMB) and resveratrol are known to inhibit TMAO formation and its negative effects. Through inhibiting the formation of TMA by gut microbes, DMB thereby reduces plasma TMAO levels and also attenuates the development of atherosclerosis caused by a diet high in choline [27]. In addition, DMB treatment significantly reduced plasma TMAO levels and prevented the development of cardiac dysfunction in mice fed with a "Western" diet [28]. Resveratrol isolated from red wine attenuates TMAO-induced atherosclerosis via regulating TMAO synthesis and bile acid metabolism through gut microbiota remodeling [29].

In addition, it was found that episodes of "acute" hypoxia as well as a prolonged stay under conditions of medium-high altitude significantly and multi-directionally change the ratio of intestinal microbiome strains and their metabolites in feces and blood [30]. At the same time, a stay under moderate hypoxia conditions is accompanied by the inhibition of certain gut strains most actively producing TMA, which is considered a definite basis for the application of procedures for adaptation to hypoxia for targeted influence on TMAO production and its associated potential risk factors, CVD and MS.

Intermittent normobaric hypoxic–hyperoxic exposures (IHHEs), as one of the variants of the technology of adaptation to periodic hypoxia in recent years, are actively used in the treatment and complex rehabilitation of patients with multiple pathologies and comorbidities [31]. For example, it was found that a 3-week course of IHHE in CHD patients leads to a significant decrease in blood pressure and resting heart rate, and an increase in exercise tolerance [32]. Another study has shown that adding an IHHE course to the physical rehabilitation program in geriatric patients with initial dementia is accompanied by improved cognitive function and physical mobility [33].

Interval hypoxic–hyperoxic training procedures are a new approach in which the patient's breathing through an oronasal mask with a hypoxic gas mixture is alternated with episodes of breathing with a hyperoxic mixture (FiO_2 35%), which shortens the reoxygenation period and stimulates the redox regulation system. It is known that hypoxia, short-term or long-term, triggers an HIF-induced response with the expression of genes and proteins (EPO, NOS, VEGF, glycolytic enzymes, etc.) providing vasculogenesis, mitochondrial biogenesis, and an increase in the adaptive capabilities of cells and the whole organism, while episodes of moderate hyperoxia are accompanied by the activation of ROS-induced mechanisms activating the antioxidant defense systems of cells, preventing the development of potentially harmful oxidative stress, as well as upregulating HIF-dependent adaptive mechanisms due to other molecular cascades and ROS sensors [34,35].

Intermittent hypoxic conditioning and IHHE, as one of effective hypoxic conditioning protocols, have been used in rehabilitation courses for a wide number of different diseases, but no studies have shown any significant adverse side effects [36]. A review with one of our co-authors on the use of IHHE in cardiological patients also showed no adverse side effects [37]. Data regarding the effects of passive intermittent hypoxic exposures on the metabolism of lipids, carbohydrates, and oxidative status in patients remain rather contradictory [36,38,39]. We have not found any data regarding the possible effects of hypoxic conditioning sessions in different modes on the state of the intestinal microbiome and the level of individual indicators of the gut microflora metabolism.

The aim of this pilot study was to evaluate the possible effects of IHHE on the metabolic product of the intestinal microbiome (TMAO) and the relationship of its dynamics with cardiometabolic risk factors and components of MS.

2. Results

At baseline, the compared groups did not differ in TMAO and most of the controlled variables, except for systolic blood pressure (SBP), total cholesterol (TCh), and low-density lipoprotein (LDL), which were significantly higher in the IHHE group (Table 1). After three weeks of IHHE interventions, there was a decrease in the levels of TMAO in the IHHE group, but the values were not significantly different from the controls.

Despite not having found changes in the levels of TMAO in both groups after IHHE interventions, the degree of reduction in most of the MS and inflammation indicators (Z-value) in the IHHE group was significantly higher than in the controls. And the values of SBP, diastolic blood pressure (DBP), fibrosis degree, as well as N-terminal pro-brain natriuretic peptide (Nt-proBNP) became significantly lower compared to the control in the post-intervention examination.

The majority of patients had no TMAO detected in the stool. In a small number of cases (9 patients out of 65), the presence of TMAO in stool samples was probably related to the consumption of products containing large amounts of TMAO or its precursors.

Table 2 shows the dynamics of TMAO and the analyzed cardiometabolic indicators of MS in IHHE patients divided into two subgroups: subgroup 0, with normal baseline TMAO values (<5.0 μM/L), and subgroup 1, or patients with elevated TMAO values (>5 μM/L). In the pre–post dynamics, there was a significant reduction in TMAO in subgroup 1 and reductions in all analyzed indicators in both subgroups. At the same time, the degree of reduction in TCh, LDL, and regression of liver steatosis index (Pre–Post Δ) was more pronounced in patients in subgroup 0, with initially normal TMAO values.

Table 1. Pre- and post-intervention data for cardio-metabolic parameters with the main analysis of covariance results and normalized deltas.

Variables	Groups	Pre-Test	Post-Test	*p*-Value **	Pre–Post Δ, z-Value	*p* Value (Mann–Whitney U) ***
SBP, mm Hg	IHHE	148 [136; 164] * *p* = 0.024	131.5 [121.7;144.0] * *p* = 0.012	<0.001	−0.61 [−1.21; −0.45]	<0.001
	Sham	142.5 [127.5; 152.2]	141.0 [130.0; 151.2]	0.161	0.08 [−0.14; 0.32]	
DBP, mm Hg	IHHE	92.0 [85.7; 102.0]	84.5 [79.7; 93.2] * *p* = 0.043	<0.001	−0.48 [−0.87; −0.29]	<0.001
	Sham	88.0 [81.5; 96.2]	92.0 [83.0; 98.2]	0.463	0.14 [−0.02; 0.48]	
Heart Rate, bpm	IHHE	71.0 [61.0; 77.2]	64.0 [58.0; 70.2]	<0.001	−0.39 [−0.54; −0.07]	<0.001
	Sham	66.5 [61.0; 75.7]	68.5 [60.7; 77.2]	0.561	0.00 [−0.15; 0.22]	
Total Cholesterol, Mmol/L	IHHE	5.86 [5.14; 6.85] * *p* < 0.001	5.01 [4.31; 5.86]	<0.001	−0.41 [−0.84; −0.25]	<0.001
	Sham	4.62 [3.72; 5.69]	4.93 [4.40; 5.73]	0.722	0.03 [−0.00; 0.26]	
LDL-Cholesterol, Mmol/L	IHHE	3.83 [2.77; 4.57] * *p* < 0.001	2.86 [1.95; 3.82]	<0.01	−0.51 [−0.75; −0.29]	<0.001
	Sham	2.65 [1.75; 3.56]	2.98 [2.29; 3.74]	0.763	0.04 [0.00; 0.28]	
TMAO, μM/L	IHHE	3.25 [2.32; 5.16]	2.76 [2.07; 3.46]	0.011	−0.16 [−0.57; 0.03]	0.320
	Sham	3.09 [2.15; 3.67]	2.73 [2.54; 3.74]	0.724	−0.14 [−0.32; 0.22]	
Liver Steatosis, stage	IHHE	7.50 [5.05; 11.10]	5.20 [3.45; 6.67]	<0.001	−0.14 [−0.29; −0.05]	<0.001
	Sham	6.05 [4.07; 9.90]	6.45 [4.40; 9.45]	0.651	0.00 [0.00; 0.02]	
Liver Fibrosis (LSM), kPa	IHHE	1.00 [0.00; 2.25]	0.00 [0.00; 1.00], *p* * = 0.025	<0.001	−0.24 [−0.37; −0.08]	0.420
	Sham	0.50 [0.00; 2.00]	1.00 [0.00; 2.00]	0.701	0.15 [0.05; 0.45]	
CRP-hs, mg/L	IHHE	2.33 [1.50; 4.30]	1.97 [1.15; 2.69]	<0.001	−0.07 [−0.45; 0.00]	0.031
	Sham	1.88 [1.51; 3.96]	2.36 [1.62; 4.52]	0.633	0.00 [−0.05; 0.07]	
NTproBNP, pmol/L	IHHE	5.83 [3.00; 17.2]	3.33 [3.00; 13.4] * *p* = 0.012	0.124	−0.03 [−0.12; 0.00]	<0.001
	Sham	11.2 [3.09; 32.1]	15.9 [5.87; 40.6]	<0.01	0.05 [0.00; 0.19]	

Values are expressed as median and first–fourth quartiles. *—significant difference between the groups at the same study time (Mann–Whitney U); **—significantly different pre- vs. post-treatment; ***—significantly different in Z-values between the groups. SBP—systolic blood pressure; DBP—diastolic blood pressure; TMAO—Trimethylamin-N-oxide; LSM—liver stiffness measurement; LDL—cholesterol, low-density lipoprotein; CRP-hs—high-sensitivity C-reactive protein; NTproBNP—N-terminal prohormone of brain natriuretic peptide.

Table 2. Pre- and post-intervention data for cardio-metabolic parameters with the main analysis of covariance results and normalized deltas for IHHE subgroups (Subgroup 0—TMAO < 5 μM/L, n = 24; subgroup 1—TMAO > 5 μM/L, n = 8).

Variables	Groups	Pre-Test	Post-Test	*p*-Value **	Pre–Post Δ, z-Value	*p* Value (Mann–Whitney U) ***
SBP, mm Hg	Subgroup 0	150.0 [141.0; 161.0]	130.5 [121.5; 141.0]	<0.001	−0.631 [−1.441; −0.439]	0.317
	Subgroup 1	137.0 [132.7; 167.2]	133.0 [125.5; 148.0]	<0.001	−0.467 [−0.864; −0.261]	
DBP, mm Hg	Subgroup 0	93.0 [87.5; 102.0]	84.5 [79.7; 92.2]	<0.001	−0.601 [−1.155; −0.346]	0.074
	Subgroup 1	89.500 [83.0; 100.2]	85.0 [79.5; 95.7]	0.741	−0.277 [−0.531; −0.185]	
Heart Rate, bpm	Subgroup 0	71.5 [64.0; 78.2]	64.5 [58.0; 69.7]	<0.001	−0.350 [−0.594; −0.192]	0.043
	Subgroup 1	68.5 [60.2; 73.2]	62.5 [60.0; 70.2]	0.053	−0.070 [−0.227; 0.000]	

Table 2. *Cont.*

Variables	Groups	Pre-Test	Post-Test	p-Value **	Pre–Post Δ, z-Value	p Value (Mann–Whitney U) ***
Total Cholesterol, Mmol/L	Subgroup 0	5.54 [4.82; 6.60]	5.32 [4.30; 5.86]	<0.001	−0.359 [−0.674; −0.190]	0.018
	Subgroup 1	6.06 [5.75; 7.57]	4.72 [4.44; 5.65]	<0.001	−0.934 [−1.479; −0.719]	
LDL-Cholesterol, Mmol/L	Subgroup 0	3.77 [2.71; 4.55]	3.16 [1.93; 3.82]	<0.001	−0.410 [−0.588; −0.263]	0.010
	Subgroup 1	3.83 [3.54; 5.41]	2.71 [2.49; 3.12]	0.017	−0.827 [−1.677; −0.714]	
TMAO, μM/L	Subgroup 0	2.70 [2.24; 3.38], p * < 0.001	2.72 [1.86; 3.28]	0.495	−0.040 [−0.171; 0.047]	<0.001
	Subgroup 1	10.25 [9.25; 11.54]	2.81 [2.48; 4.78]	0.008	−1.808 [−2.031; −1.163]	
Liver Steatosis (LSM), stage	Subgroup 0	9.15 [5.17; 14.12]	5.20 [3.80; 6.72]	<0.001	−0.235 [−0.386; −0.123]	0.007
	Subgroup 1	6.20 [4.85; 7.30]	5.00 [3.30; 6.30]	0.008	−0.087 [−0.136; −0.047]	

The study did not reveal any correlations of baseline blood TMAO levels and pre–post dynamics with the values of the main indicators of MS, parameters of chronic inflammation, and liver steatosis. A weak correlation was revealed only between levels of TMAO at baseline and acetyl-carnitine ($r = 0.30$, $p = 0.01$), which can be explained by their formation from a common carnitine precursor [15]. *—significant difference between the groups at the same study time (Mann–Whitney U); **—significantly different pre- vs. post-treatment; ***—significantly different in Z-values between the groups.

3. Discussion

The results of the study show that a three-week course of hypoxic conditioning in the IHHE mode against the background of a balanced medical treatment leads to a more pronounced regression in the values of known components of MS, such as TCh, LDL, BP, and an improvement of the liver fibrotic index than in the placebo group, and can also potentially have beneficial effects on a new marker of cardiovascular health—TMAO in patients with metabolic syndrome.

TMAO, which belongs to the class of aminoxides, is known as a product of the metabolism of the intestinal microbiome. Many studies have established a link between the metabolism of certain microbes and the development of cardiovascular disease. As early as 1988, Saikku et al. noted higher antibody titers to C. pneumoniae in patients who suffered from myocardial infarction [40]. Subsequently, a large cohort study of patients with coronary heart disease identified several metabolites, including trimethylamine-N-oxide, that significantly increase cardiovascular risk [17].

The adverse effects of TMAO are due to several metabolic pathways. TMAO contributes to the early pathological process of atherosclerosis through accelerating endothelial activation and dysfunction, including reduced endothelial self-renewal and increased monocyte adhesion through activation of PKC/NF-κB/vascular cell adhesion molecule-1 (VCAM-1) [41]. The proatherogenic effect of TMAO is also promoted by its adverse effect on cholesterol transport. TMAO inhibits a key enzyme of bile acid synthesis, 7-α-hydroxylase, thus reducing the formation of bile acids from cholesterol [42]. Another point of application of TMAO is the enterocytes, where trimethylamine (TMA), affecting the expression of C1-like protein Newman-Pick, ATP-linked cassette transporters class G1, G5/G8, inhibits cholesterol transport into the intestinal lumen and helps to increase its concentration in the blood [43]. Thus, the level of TMAO has a significant effect on the progression of the atherosclerotic process and is in close interaction with the activity of the intestinal flora.

Dysbiosis, particularly through an unbalanced diet with a high-content carbohydrate intake, contributes to oxidative stress through increased expression of proinflammatory cytokines. This leads to the oxidation of low-density lipoproteins (LDLs), affecting the production of NO and endothelin-1 [11,44]. At the same time, inflammation and oxidative stress are significant in the pathogenesis of the "hepatic component" of MS-NAFLD [45].

One of the mechanisms through which an unbalanced gut microbiota is involved in the pathogenesis and progression of NAFLD is an increase in the permeability of the intestinal wall, development of ischemic necrosis of mesenteric epithelial cells and induction of liver fibrosis development [46–48]. When NAFLD progresses to non-alcoholic steatohepatitis (NASH), changes in the gut microbiome in the form of its overgrowth are also observed [49], which leads to the hyperproduction of reactive oxygen species and increased risk of liver cancer development [50].

Given the role of TMAO in the progression of cardiometabolic disorders, various ways to suppress the production of TMAO are being developed, including antibiotic therapy, but the evidence base for them has not been formed [40].

The fact of different qualitative and structural ratios of the intestinal microbiota under different oxygen regimes seems to be interesting [50]. Thus, an increase in Prevotella species during prolonged systemic or local hypoxia has been noted, which contributes to intestinal dysfunction and inflammation [51,52].

On the other hand, moderate hypoxic stimuli can activate various transcriptional pathways via hypoxia response element (HRE) through activation of HIF-1 providing the production of protective proteins with antioxidant activity, inducible NOs, glucose transporters, glycolytic enzymes, selected growth factors, etc. [30]. Interestingly, short-term exposure to elevated oxygen conditions, interval hyperoxia, also exerts positive effects through other redox-dependent transcriptional cascades, particularly through the Nrf2 pathway and the antioxidant-response element (ARE), triggering the synthesis of antioxidants, anti-inflammatory cytokines, cellular matrix remodeling proteins, mitochondrial biogenesis, etc., providing a state of hormesis [30,53]. Precisely such a regime of hypoxic conditioning—IHHE was applied in our study, in which the patient's breathing with a hypoxic gas mixture was interrupted with short intervals through feeding a hyperoxic gas mixture [32].

Based on the results obtained in this work and the described known effects of hypoxic conditioning, it is shown that the use of an IHHE course has a systemic optimizing effect on various metabolic and molecular mechanisms involved in the pathogenesis of MS. Despite the absence of a significant decrease in TMAO values, it can be reasoned that the normalization of lipid metabolism with a decrease in TCh and LDL, a reduction in CRP values, and a reduction in steatosis and liver fibrosis indicators are also partially associated with the improvement of the intestinal barrier and hepato-biliary functioning. This position corresponds to the work of Serebrovskaya et al. [34,54], where it has been shown that the use of IHHE or a course of IHHT reduces insulin resistance, normalizes carbohydrate metabolism in patients with prediabetes, and reduces the level of circulating proinflammatory molecules in patients with Alzheimer's disease. The work of Timon et al., who noted a positive effect of 24-week hypoxic exposure on the inflammatory marker CRP in elderly people suffering from systemic chronic inflammation, also confirms this finding [55].

The possible involvement of TMAO in the pathogenesis of MS and NAFLD is indirectly confirmed by our data regarding the dynamics of TMAO and cardiometabolic indicators in a 3-week course of IHHE in patients with baseline elevated values of this metabolite (>5 μM/L). In the course of IHHE, their TMAO values decreased significantly, but the degree of reduction in TCh, LDL, and liver steatosis indicators was significantly less pronounced than in the group of patients with baseline normal TMAO values, which may indicate more significant changes in lipid metabolism and hepatic dysfunction and the need for longer IHHE courses to achieve the necessary effects in this group of patients.

It is also important to note the detected effects of IHHE in cardiohemodynamic indicators—a significant decrease in SBP and DBP, resting heart rate, and a significant decrease in the predictor of heart failure, i.e., NT-proBNP, which confirms our previous results, as well as the well-founded conclusion in the systematic review of Behrend T. et al. (2022), that IHHE or continuous sessions of hypoxia at rest or in combination with exercise are generally effective to reduce blood pressure and positively influence vascular health due to the stimulation of vascular adaptions (e.g., increased vascularization and endothelium-dependent vasodilatation) as well as adaptations in the autonomic nervous system disturbed by TMAO's pro-atherogenic and pro-inflammatory activities in patients with MS [31,38].

To understand the positive dynamics of indicators of hepatic steatosis and fibrosis, the fact that hypoxia conditioning is thought to benefit smooth muscles in blood vessels but also elsewhere, including the bladder and gastrointestinal tract, mediated by diverse pathways involving heme oxygenase-1, NO, HIFs, VEGF, ATP-sensitive K+ channels, prostaglandins, bradykinin, different protein kinases, and ROS [55], normalizing the drainage and detoxification functions of the liver [56] and the barrier properties of the intestinal wall. The last is important given the proven direct cardiotoxic effect of molecules of microbial origin (lipopolysaccharides, peptidoglycans, and TMAO), which, interacting with receptors of cardiomyocytes and microenvironmental cells, can cause the development of myocardial remodeling and dysfunction, provoking the progression of heart failure [57]. In our work, on the other hand, we have shown a decrease in postload and NT-proBNP in the IHHE course. Moreover, the hypoxia conditioning-related priming of energy homeostasis molecules, e.g., adenosine, AMPK; the activation of systemic anti-inflammatory signaling, e.g., related to reduced levels of C3, CRP, and MCP-1; and the augmentation of antioxidant molecules, e.g., superoxide dismutase and glutathione [55], are potential contributors to hypoxia conditioning-associated cardiovascular and metabolic benefits.

In addition, changes in TMAO levels may be associated not only with changes in microbiome status, but also with changes in hepatic activity, as demonstrated by our results showing decreased TMAO levels and improved hepatic indicators, which were published earlier in our article [31].

Limitations and Prospects

The sample size in the study was small. A larger study with no different baseline MS parameters should be conducted to obtain more reliable results. This was a single-centered trial using a limited number of patients, which does not allow firm conclusions in regard to rare but important safety events and side effects of IHHE.

Secondly, the patient's dietary regime as well as physical activity were not closely monitored, which may have partly influenced both the dynamics of TMAO and the effectiveness of hypoxic conditioning.

Thirdly, TMAO values above the reference were found at baseline in only a small proportion of patients, which may have influenced the results of the study.

Finally, the study was limited to a survey before–post 3-week interventions; longer monitoring might have revealed the timing of the persistence of the observed effects of IHHE.

4. Materials and Methods

This prospective, single-center, randomized controlled trial enrolled 65 patients (32 women) with MS. Subjects' ages ranged from 29 to 74 years. Patients were randomly allocated into groups: the IHHE group and the control group (sham). Patients in the IHHE group received intermittent hypoxic–hyperoxic treatment for 3 weeks, 5 days a week. The control group was simulated with the same protocol, but instead of intermittent hypoxia–hyperoxia, room air (normoxia, 20.9% O_2) was given for the same training period.

The main endpoints were the evaluation of changes in plasma TMAO levels, the determination and evaluation of changes in TMAO stool samples, and the correlation detection of plasma and stool TMAO levels with other MS parameters. All endpoints were evaluated after three weeks of IHHE training.

We selected a number of parameters as secondary endpoints and compared baseline values with those after 3 weeks of exposure: hemodynamic parameters (systolic (SBP) and diastolic (DBP) blood pressure, heart rate (HR)), liver tissue elasticity and fibrosis stage according to liver elastography, lipid spectrum (total cholesterol—TCh, high- and low-density lipoproteins—HDL and LDL, triglycerides), and inflammatory markers (high sensitive C-reactive protein—CRP-hs, N-terminal prohormone brain natriuretic peptide—Ntpro-BNP).

The lipid profile and AST/ALT were measured using a Siemens Advia 1800 Biochemical Analyzer ("Siemens Healthcare Diagnostics Inc.", Newark, DE, USA) and dedicated test kits "Siemens Healthcare Diagnostics Inc.", USA. High-sensitivity C-reactive protein (CRP-hs, milligram per liter, "Beckman Coulter," Brea, CA, USA) was measured using Siemens Advia 1800 Biochemical Analyzer ("Siemens Healthcare Diagnostics Inc.", Newark, DE, USA) by immunoturbidimetry using latex particles. The levels of brain natriuretic peptide N-terminal prohormone (NTproBNP, "Biomedica", Wien, Austria) were measured using an enzyme-linked immunosorbent assay and a Biochrom Anthos 2020 Jencons Microplate Reader photometer ("Biohrom Ltd.", Cambridge, UK). Weight and height were measured with a Seca gmbh&co.kg scale (Hammer Steindamm 3-25 22089 Hamburg, Germany).

The assessment of the functional liver status and liver fibrosis stage and steatosis severity was performed by a noninvasive Vibration-Controlled Transient Elastography (VCTETM) ultrasound method. The measurements were done using the FibroScan 502 Touch (Echosense, Paris, France) device [31].

4.1. *Methods*

4.1.1. Study Design

Subjects were recruited from March 2019 to March 2020 from the Cardiology Clinic of the First Sechenov Moscow State Medical University, Moscow, Russia. Laboratory and instrumental procedures, as well as the training course, were conducted in the same clinic.

The study was approved by the ethical committee of the I.M. Sechenov First Moscow State Medical University (Local Ethical Protocol No. 05-19 10.04.2019) and was performed according to the ethical standards based on the Declaration of Helsinki—Ethical Principles for Medical Research Involving Human Subjects. Written informed consent was obtained from all trial participants. The study was registered at ClinicalTrials.gov (NCT04791397, protocol identifier A0519).

4.1.2. Participants and Randomization

Eighty-six patients with MS aged 29 to 74 years who had been in a stable clinical condition within the past 3 months were invited to participate in the study. MS was defined according to National Institutes of Health guidelines as the presence of three or more of the following characteristics: waist circumference greater than 102 cm in men and greater than 89 cm in women, blood pressure $\geq$ 130/85 mm Hg for SBP and DBP, dyslipidemia (triglyceride level $\geq$ 150 mg/dL (1.7 mmol/L), high-density lipoprotein cholesterol < 40 mg/dL (1.04 mmol/L) in men or <50 mg/dL (1.3 mmol/L) in women, and elevated fasting blood glucose levels ($\geq$100 mg/dL [5.6 mmol/L]). All subjects had the following exclusion criteria: individual intolerance to hypoxia, cirrhosis, Child–Pugh class C, patients serologically positive for hepatitis B and C, chronic kidney disease (GFR < 30 mL/min/1.73 m^2), pregnancy, serious respiratory distress, acute cardiovascular disease, and neuromuscular disorders.

The study included 65 patients who were randomly assigned to either the IHHE group (32 patients) or the sham group. Withdrawal of consent, intolerance to hypoxia, territorial inconvenience, and inability to perform liver ultrasonography were reasons for excluding 21 patients from the study. We performed a simple blinded prospective randomized controlled study in two parallel groups. Participants were randomized using a 1:1 allocation ratio. Baseline anthropometrics, clinical characteristics, and medications are presented in Table 3.

Table 3. Anthropometric and clinical characteristics of the patients.

Variables	IHHE Group (n = 32)	Sham Group (n = 33)	*p*-Value
Gender, male	14 (43.7%)	19 (57.5%)	NS
Age, years	56.9 ± 11.7	59.8 ± 10.3	NS
Smoking	10 (31.2%)	11 (33.3%)	NS
Hypertension	32 (100%)	33 (100%)	NS
Diabetes Mellitus, type 2	12 (37.5%)	10 (30.3%)	NS
Obesity (BMI > 30)	24 (75.0%)	27 (81.8%)	NS
Weight, kg	92.0 (81.0; 114)	92.5 (82.8; 104)	NS
BMI, kg I m^2	34.3 (30.2; 38.0)	32.4 (30.8; 35.8)	NS
Waist Circumference, cm	114 (108; 124)	110 (105; 120)	NS
Hip Circumference, cm	113.5 (106.8; 119.5)	110 (105;120)	NS
	Regular Medication:		
Aspirin	4 (12%)	8 (24.4%)	NS
ACE inhibitors	16 (50%)	13 (39%)	NS
AT II inhibitors	14 (43.7%)	16 (48.4%)	NS
Calcium channel blockers	12 (37.5%)	12 (36.3%)	NS
Beta-blokers	14 (43.7%)	13 (39.9%)	NS
Diuretics	14 (43.7%)	14 (42.2%)	NS
Statins	14 (43.7%)	17 (51.6%)	NS
Metformine	14 (43.7%)	10 (30%)	NS
Sulfonylureas	6 (18.7%)	4 (12.1%)	NS
Insulin	2 (6.25%)	5 (15%)	NS

Note. The data are expressed as mean ± SD or frequencies (%). IHHE—intermittent hypoxic–hyperoxic training. BMI—body mass index. NS—nonsignificant differences between the groups.

Groups were matched for gender, age, presence of MS components, and comorbidities. Patients were asked to adhere to their daily food intake, physical activity, prescribed medications, and habitual lifestyle throughout the study period.

4.1.3. Program of Interval Hypoxic–Hyperoxic Exposures (IHHE)

Patients included in the study underwent a course of interval hypoxic–hyperoxic therapy using a special ReOxy-Cardio device (AI Mediq S.A., Luxemburg, Luxemburg). At the first visit, after all anthropometric and laboratory-instrumental examinations, the patients of both groups underwent a hypoxic challenge test (HCT) for 10 min. The test was performed in a sitting position in a chair, with hypoxic gas fed through a facial mask with a minimum oxygen content of 11% under the control of arterial blood oxygen saturation (SpO_2) and heart rate, which were measured continuously using a built-in KIT Masimo pulse oximeter (measurement error ± 2%). The device automatically adjusted the individual mode of IHHE procedures for each patient based on the result of the HCT. Data on each procedure, starting from the HCT, were collected on the device. Starting from the second day of the study, the patients in the IHHE group were trained with a gas mixture with interval changes in O_2 level from 10–14% (corresponding to 4000–6500 m above sea level) to 35% and nitrogen (N_2). The patients inhaled on average a mixture of hypoxic gases with 11–12% O_2 content for 4–7 min (depending on individual parameters set during HCT), followed by subsequent exposure to a hyperoxic gas mixture with 30–35% O_2 content for 2–4 min.

The physician-researcher was present at each training session and monitored HR and SpO_2, which were recorded and transmitted to the device monitor. When the individual minimum set SpO_2 level (83–85%) was reached, the device was automatically switched to a hyperoxic gas mixture until full pre-hypoxic SpO_2 recovery occurred (during 1–3 min). After that, the next cycle of hypoxia–hyperoxia was repeated. The training duration ranged from 40 to 45 min in each group and had no differences for anyone other than the researcher who conducted the session. Blood pressure was measured before and at the end of each treatment, using an automatic tonometer AND UA-767 (AND, Tokyo, Japan).

Patients in the control group were trained according to the same scheme as patients in the main group, with the same "exposure time" and a number of workouts, but with a normoxic gas mixture administered throughout the session (room air, with an O_2 concentration of 20.9%).

The total number of training sessions for patients in both groups was 15 hypoxic–hyperoxic or sham treatments, which were performed five times a week with a weekend break (2 days) for three weeks.

At the beginning of treatment, some patients in the IHHE group had brief complaints of light dizziness, which did not require the interruption of the procedure. After the first session, no patient refused to continue participation in the study.

4.2. Procedures

Before IHHE or sham treatment, all patients underwent a routine medical examination, with a collection of medical history, data on the intake of drugs, family history for cardiovascular diseases, carbohydrate metabolism disorders, diabetes mellitus, and other diseases. One day before the experimental sessions, and 2–3 days after the last training, the patients of both groups underwent the same examinations and measurements: an evaluation of hemodynamic parameters (blood pressure (SBP and DBP) at rest, heart rate (HR)), SpO_2, anthropometric data (height, body weight, waist, and hip circumference), measurement of liver tissue elasticity (stiffness), and evaluation of fibrosis stage as described with details in our previous study [31].

TMAO was measured in plasma and fecal samples, as described below. In addition, blood samples were collected to determine serum lipid levels (TCh, HDL, LDL, Mmol/L); liver enzymes—alanine aminotransferase (ALT, units per liter) and aspartate aminotransferase (AST, units per liter); markers of chronic inflammation—CRP-hs (mg/L), heat shock proteins-70 (HSP70, ng/mL); and N-terminal prohormone of brain natriuretic peptide (NTpro-BNP, pmol/L), and were reported in our previous study [39]. Samples (10 mL of venous blood) were collected (into vacuum lithium heparin and EDTA tubes) in the morning between 7 and 10 a.m., after an overnight fast of at least 8 h. In order to minimize platelet count, blood was allowed to clot (BD Vacutainer Plus SST), and serum separated immediately (by centrifugation at 3500 rpm for 15 min after sampling, Eppendorf Centrifuge 5702R, Darmstadt Germany), aliquoted, and stored at −80 °C until being processed. All biochemical analyses were conducted in the University's hospital Blood Analysis Center, a blood biochemistry laboratory certified by the Moscow Department of Health Care.

4.3. Metabolite Profiling

4.3.1. TMAO Plasma Quantification

Plasma aliquots for TMAO analysis were isolated from whole blood, collected in tubes with EDTA, and stored at −80 °C from March 2019 to March 2020. One study indicated that TMAO is stable under these storage conditions for several years [58].

4.3.2. Sample Preparation

Samples were thawed on ice and were vortexed for 1 min at +4 °C. Then, samples were centrifuged for 5 min at 5000× *g* rpm at +4 °C. A 20 µL aliquot was taken from each of the samples, after which the samples were re-frozen. The aliquot was mixed with 20 µL of an ice-cold (−40 °C) methanol solution with an internal standard.

The prepared samples were vortex-mixed for 5 min at +4 °C. Afterward, the samples were centrifuged for 10 min at maximum rpm (18,000× *g* rpm) at +4 °C. We took 25 µL of supernatant from the samples, transferred them to vials with screw-on lids, sealed them, and transferred for analysis using high-performance liquid chromatography-mass spectrometry (HPLC-MS).

A Sciex 4500QTRAP tandem mass spectrometer (AB Sciex Pte. Ltd., Framingham, MA, USA) with a Shimadzu Nexera 30AD chromatograph (Shimadzu, Kyoto, Japan) was used for analysis. Ion source settings were as follows: temperature = 550 °C; GS1 = 55; GS2 = 55; CUR = 25; IS = 5500. Ion detection was performed in the positive ionization mode of the sample.

The chromatographic separation of the components of the test sample was performed via RPLC chromatography using an Agilent ZORBAX Eclipse Plus (Aglient Technologies, Santa Clara, CA, USA) C18 2.1 × 50 mm 1.8 μm chromatographic column: phase A (water; 5 mM ammonium formate); phase B (ACN: MeOH 1:1; 5 mM ammonium formate); sample input volume 1 μL. Chromatographic gradient: 0 min 2% B; 0.25 min 2% B; 1.2 min 98% B; 2.5 min 98% B; 2.6 min 2% B; 3.5 min 2% B; flow 0.45 mL/min, thermostat temperature 45 °C.

The referent values for TMAO were presumed to be 0.5–5.0 μM/L as recommended in certain studies [59,60].

4.3.3. TMAO Fecal Quantification

Fecal material samples for TMAO analysis were stored at −80 °C from March 2019 to March 2020.

Sample Preparation

A representative sample weighing approximately 20 mg was taken from the starting material in the frozen state and transferred into a sterile Eppendorf pipette. Afterward, 200 μL of an aqueous solution of the internal standard in phosphate buffer was added to the sample. The resulting solution was placed in a vortex at +4 °C and centrifuged for 10 min at 1400× *g* rpm. The resulting homogenate was recovered following centrifugation at 4000× *g* rpm for 10 min to maintain the integrity of the microbiota cells. Then, 100 μL of the resulting supernatant was transferred to an Eppendorf pipette with the addition of 20 μL of methanol and 120 μL of chloroform. The resulting solution was vortexed for 10 min at 1400× *g* rpm at +4 °C, after which the sample was centrifuged at a maximum of 18,000× *g* rpm for 10 min. The top layer of liquid was collected, transferred to vials with inserts and screw-on lids, and given for analysis. A Sciex 4500QTRAP tandem mass spectrometer with (AB Sciex Pte. Ltd., Framingham, MA, USA) a Shimadzu Nexera 30AD chromatograph (Shimadzu, Kyoto, Japan) was used for analysis.

4.4. Statistical Methods

All data were analyzed using Python Software Foundation version 3.8 for Windows (DE, USA). The data were reported as median and inter-quartile range 25–75. The assumption of normality and homoscedasticity was verified using Shapiro–Wilk's W-test before parametric tests. To identify the magnitude of the statistical difference between sham and IHHE treatments in each period of the study, Fisher's exact test was used. Because of the non-normal distribution of data, we used Mann–Whitney's U-test or Wilcoxon's test to compare the baseline data between the groups as well as different changes (normalized delta pre–post, z-scores) between the groups. Pearson's or Spearman's correlation analysis was performed to test relationships between the variables. The a-level was set at 0.05 for all statistical analyses.

5. Conclusions

Despite significant interindividual variation and a small number of patients with MS and baseline elevated TMAO in the study, it was noted that a hypoxic conditioning course in the mode of intermittent hypoxic–hyperoxic exposures at rest leads to a downward trend in blood TMAO, which was accompanied by a more significant reduction in cardiometabolic and hepatic indicators of MS than in the placebo group. Patients with baseline elevated values of this metabolite in the IHHE course showed a significant reduction in TMAO but a smaller degree of reduction of total cholesterol, low-density lipoproteins, and hepatic steatosis than in the subgroup of patients with baseline normal TMAO values.

IHHE can potentially be considered as an effective and safe (well-tolerated) technique to improve cardiometabolic health, presumably through normalization of the gut microbiome, the barrier function of the intestinal wall, and hepatic metabolism.

More research is needed to objectify the role of microbiome dysfunction and TMAO in the development and progression of metabolic syndrome components and hepatic and cardiovascular pathology, as well as to identify the cellular and molecular mechanisms of interval hypoxic exposures to TMAO levels as an indicator of the "microbiome-lipid metabolism-oxidative stress-cardiovascular risk factors" interplay. The aspect of individualization of IHHE regimens with regard to the optimal combination of procedure variables to achieve the best physiological and structural adaptations deserves special attention.

Author Contributions: Conceptualization, P.K., O.G., Y.Z., A.S. (Abram Syrkin), and C.P.; methodology, P.K., O.G., Y.Z., and C.P.; formal analysis, A.B. (Afina Bestavashvili), P.K., and O.G.; investigation, A.B. (Afina Bestavashvili) and C.P.; resources, A.B. (Afina Bestavashvili), P.K., and C.P.; data curation, A.B. (Afina Bestavashvili), A.S. (Alexander Suvorov), S.I. (Shevket Ibraimov), and X.Z.; writing—original draft preparation, A.B. (Afina Bestavashvili) and S.I. (Shabnam Ibragimova); writing—review and editing, A.B. (Afina Bestavashvili), O.G., S.I. (Shabnam Ibragimova), and A.B. (Alexandros Bestavasvili); supervision, P.K., E.S., and O.G.; project administration, P.K., O.G., and C.P.; funding acquisition, P.K. and A.B. (Afina Bestavashvili). All authors have read and agreed to the published version of the manuscript.

Funding: The study was supported and financed by the Russian Foundation for Basic Research, Grant No. 19-515-55025.

Institutional Review Board Statement: The studies involving human participants were reviewed and approved by the Local Ethics Committee of I.M. Sechenov First Moscow State Medical University (Sechenov University), protocol No. 05-19, 10.04.2019.

Informed Consent Statement: Informed consent was obtained from all subjects involved in the study.

Data Availability Statement: Raw data supporting the conclusions of this publication will be made available by the authors without undue reservations.

Acknowledgments: The authors would like to thank all the volunteers who participated in the study and AiMediq S.A. (Luxembourg) for supplying the ReOxy-therapy device used in our experimental session. The company and its representatives did not participate in the design of this study; in the collection, analysis, and interpretation of the study data; or in the preparation of the publication.

Conflicts of Interest: The authors declare no conflict of interest.

References

1. Chistiakov, D.A.; Bobryshev, Y.V.; Kozarov, E.; Sobenin, I.A.; Orekhov, A.N. Role of gut microbiota in the modulation of atherosclerosis-associated immune response. *Front. Microbiol.* **2015**, *6*, 671. [CrossRef]
2. Musso, G.; Gambino, R.; Cassader, M. Obesity, diabetes, and gut microbiota: The hygiene hypothesis expanded? *Diabetes Care* **2010**, *33*, 2277–2284. [CrossRef]
3. Diamant, M.; Blaak, E.E.; de Vos, W.M. Do nutrient-gut-microbiota interactions play a role in human obesity, insulin resistance, and type 2 diabetes? *Obes Rev.* **2011**, *12*, 272–281. [CrossRef] [PubMed]
4. Miele, L.; Giorgio, V.; Alberelli, M.A.; De Candia, E.; Gasbarrini, A.; Grieco, A. Impact of Gut Microbiota on Obesity, Diabetes, and Cardiovascular Disease Risk. *Curr. Cardiol. Rep.* **2015**, *17*, 120. [CrossRef] [PubMed]
5. Tai, N.; Wong, F.S.; Wen, L. The role of gut microbiota in the development of type 1, type 2 diabetes mellitus and obesity. *Rev. Endocr. Metab. Disord.* **2015**, *16*, 55–65. [CrossRef] [PubMed]
6. Trøseid, M.; Andersen, G.Ø.; Broch, K.; Hov, J.R. The gut microbiome in coronary artery disease and heart failure: Current knowledge and future directions. *EBioMedicine* **2020**, *52*, 102649. [CrossRef]
7. Barrington, W.T.; Lusis, A.J. Atherosclerosis: Association between the gut microbiome and atherosclerosis. *Nat. Rev. Cardiol.* **2017**, *14*, 699–700. [CrossRef]

8. Brown, J.M.; Hazen, S.L. Microbial modulation of cardiovascular disease. *Nat. Rev. Microbiol.* **2018**, *16*, 171–181. [CrossRef] [PubMed]
9. Chambers, E.S.; Preston, T.; Frost, G.; Morrison, D.J. Role of Gut Microbiota-Generated Short-Chain Fatty Acids in Metabolic and Cardiovascular Health. *Curr. Nutr. Rep.* **2018**, *7*, 198–206. [CrossRef]
10. Tang, W.H.; Wang, Z.; Shrestha, K. Intestinal microbiota-dependent phosphatidylcholine metabolites, diastolic dysfunction, and adverse clinical outcomes in chronic systolic heart failure. *J. Card. Fail.* **2015**, *21*, 91–96. [CrossRef] [PubMed]
11. Ma, G.; Pan, B.; Chen, Y. Trimethylamine N-oxide in atherogenesis: Impairing endothelial self-repair capacity and enhancing monocyte adhesion. *Biosci. Rep.* **2017**, *37*, BSR20160244. [CrossRef]
12. Koeth, R.A.; Wang, Z.; Levison, B.S. Intestinal microbiota metabolism of L-carnitine, a nutrient in red meat, promotes atherosclerosis. *Nat. Med.* **2013**, *19*, 576–585. [CrossRef] [PubMed]
13. Zhu, W.; Gregory, J.C.; Org, E. Gut Microbial Metabolite TMAO Enhances Platelet Hyperreactivity and Thrombosis Risk. *Cell* **2016**, *165*, 111–124. [CrossRef]
14. Seldin, M.M.; Meng, Y.; Qi, H. Trimethylamine N-Oxide Promotes Vascular Inflammation Through Signaling of Mitogen-Activated Protein Kinase and Nuclear Factor-κB. *J. Am. Heart Assoc.* **2016**, *5*, e002767. [CrossRef] [PubMed]
15. Velasquez, M.T.; Ramezani, A.; Manal, A.; Raj, D.S. Trimethylamine N-Oxide: The Good, the Bad and the Unknown. *Toxins* **2016**, *8*, 326. [CrossRef] [PubMed]
16. Lent-Schochet, D.; Silva, R.; McLaughlin, M.; Huet, B.; Jialal, I. Changes to trimethylamine-N-oxide and its precursors in nascent metabolic syndrome. *Horm. Mol. Biol. Clin. Investig.* **2018**, *35*, 20180015. [CrossRef]
17. Wang, Z.; Klipfell, E.; Bennett, B.J. Gut flora metabolism of phosphatidylcholine promotes cardiovascular disease. *Nature* **2011**, *472*, 57–63. [CrossRef] [PubMed]
18. Schugar, R.C.; Shih, D.M.; Warrier, M. The TMAO-Producing Enzyme Flavin-Containing Monooxygenase 3 Regulates Obesity and the Beiging of White Adipose Tissue. *Cell Rep.* **2017**, *19*, 2451–2461. [CrossRef] [PubMed]
19. Schiattarella, G.G.; Sannino, A.; Toscano, E. Gut microbe-generated metabolite trimethylamine-N-oxide as cardiovascular risk biomarker: A systematic review and dose-response meta-analysis. *Eur. Heart J.* **2017**, *38*, 2948–2956. [CrossRef] [PubMed]
20. Kanitsoraphan, C.; Rattanawong, P.; Charoensri, S.; Senthong, V. Trimethylamine N-Oxide and Risk of Cardiovascular Disease and Mortality. *Curr. Nutr. Rep.* **2018**, *7*, 207–213. [CrossRef]
21. Ley, R.E. Obesity and the human microbiome. *Curr. Opin. Gastroenterol.* **2010**, *26*, 5–11. [CrossRef] [PubMed]
22. Qin, J.; Li, Y.; Cai, Z. A metagenome-wide association study of gut microbiota in type 2 diabetes. *Nature* **2012**, *490*, 55–60. [CrossRef]
23. Everard, A.; Cani, P.D. Diabetes, obesity and gut microbiota. *Best. Pract. Res. Clin. Gastroenterol.* **2013**, *27*, 73–83. [CrossRef]
24. Barrea, L.; Annunziata, G.; Muscogiuri, G. Trimethylamine-N-oxide (TMAO) as Novel Potential Biomarker of Early Predictors of Metabolic Syndrome. *Nutrients* **2018**, *10*, 1971. [CrossRef] [PubMed]
25. Chen, Y.M.; Liu, Y.; Zhou, R.F. Associations of gut-flora-dependent metabolite trimethylamine-N-oxide, betaine and choline with non-alcoholic fatty liver disease in adults. *Sci. Rep.* **2016**, *6*, 19076. [CrossRef]
26. Marchesini, G.; Mara Brizi, G.B.; Sara Tomassetti, E.B.; Marco Lenzi, A.J.M. Nonalcoholic fatty liver disease. *Metab. Med. Surg.* **2014**, *50*, 115–132. [CrossRef] [PubMed]
27. Wang, Z.; Roberts, A.B.; Buffa, J.A. Non-lethal Inhibition of Gut Microbial Trimethylamine Production for the Treatment of Atherosclerosis. *Cell* **2015**, *163*, 1585–1595. [CrossRef]
28. Chen, K.; Zheng, X.; Feng, M.; Li, D.; Zhang, H. Gut Microbiota-Dependent Metabolite Trimethylamine N-Oxide Contributes to Cardiac Dysfunction in Western Diet-Induced Obese Mice. *Front. Physiol.* **2017**, *8*, 139. [CrossRef] [PubMed]
29. Chen, M.L.; Yi, L.; Zhang, Y. Resveratrol Attenuates Trimethylamine-N-Oxide (TMAO)-Induced Atherosclerosis by Regulating TMAO Synthesis and Bile Acid Metabolism via Remodeling of the Gut Microbiota. *mBio* **2016**, *7*, 10–1128. [CrossRef]
30. Han, N.; Pan, Z.; Liu, G.; Yang, R.; Yujing, B. Hypoxia: The "Invisible Pusher" of Gut Microbiota. *Front. Microbiol.* **2021**, *12*, 690600. [CrossRef]
31. Bestavashvili, A.; Glazachev, O.; Bestavashvili, A.; Suvorov, A.; Zhang, Y.; Zhang, X.; Rozhkov, A.; Kuznetsova, N.; Pavlov, C.; Glushenkov, D.; et al. Intermittent Hypoxic-Hyperoxic Exposures Effects in Patients with Metabolic Syndrome: Correction of Cardiovascular and Metabolic Profile. *Biomedicines* **2022**, *10*, 566. [CrossRef] [PubMed]
32. Syrkin, A.L.; Glazachev, O.S.; Kopylov, F.Y.; Dudnik, E.N.; Zagaynaya, E.E.; Tuter, D.S. Daptation to intermittent hypoxia-hyperoxia in the rehabilitation of patients with ischemic heart disease: Exercise tolerance and quality of life. *Kardiologiia* **2017**, *57*, 10–16.
33. Behrendt, T.; Bielitzki, R.; Behrens, M.; Glazachev, O.S.; Schega, L. Effects of Intermittent Hypoxia-Hyperoxia Exposure Prior to Aerobic Cycling Exercise on Physical and Cognitive Performance in Geriatric Patients-A Randomized Controlled Trial. *Front. Physiol.* **2022**, *13*, 899096. [CrossRef]
34. Burtscher, J.; Mallet, R.T.; Pialoux, V.; Millet, G.P.; Burtscher, M. Adaptive Responses to Hypoxia and/or Hyperoxia in Humans. *Antioxid. Redox Signal.* **2022**, *37*, 887–912. [CrossRef] [PubMed]
35. Hadanny, A.; Efrati, S. The Hyperoxic-Hypoxic Paradox. *Biomolecules* **2020**, *10*, 958. [CrossRef] [PubMed]
36. Uzun, A.-B.; Iliescu, M.G.; Stanciu, L.-E.; Ionescu, E.-V.; Ungur, R.A.; Ciortea, V.M.; Irsay, L.; Motoașcă, I.; Popescu, M.N.; Popa, F.L.; et al. Effectiveness of Intermittent Hypoxia–Hyperoxia Therapy in Different Pathologies with Possible Metabolic Implications. *Metabolites* **2023**, *13*, 181. [CrossRef] [PubMed]

37. Glazachev, O.S.; Kryzhanovskaya, S.Y.; Zapara, M.A.; Dudnik, E.N.; Samartseva, V.G.; Susta, D. Safety and Efficacy of Intermittent Hypoxia Conditioning as a New Rehabilitation/ Secondary Prevention Strategy for Patients with Cardiovascular Diseases: A Systematic Review and Meta-analysis. *Curr. Cardiol. Rev.* **2021**, *17*, e051121193317. [CrossRef]
38. Behrendt, T.; Bielitzki, R.; Behrens, M.; Herold, F.; Schega, L. Effects of Intermittent Hypoxia-Hyperoxia on Performance- and Health-Related Outcomes in Humans: A Systematic Review. *Sports Med.-Open* **2022**, *8*, 70. [CrossRef]
39. Bestavashvili, A.A.; Glazachev, O.S.; Bestavashvili, A.B.; Dhif, I.; Suvorov, A.Y.; Vorontsov, N.V.; Tuter, D.S.; Gognieva, D.G.; Yong, Z.; Pavlov, C.S.; et al. The Effects of Intermittent Hypoxic-Hyperoxic Exposures on Lipid Profile and Inflammation in Patients With Metabolic Syndrome. *Front. Cardiovasc. Med.* **2021**, *8*, 700826. [CrossRef]
40. Wang, Z.; Levison, B.S.; Hazen, J.E.; Donahue, L.; Li, X.M.; Hazen, S.L. Measurement of trimethylamine-N-oxide by stable isotope dilution liquid chromatography tandem mass spectrometry. *Anal. Biochem.* **2014**, *455*, 35–40. [CrossRef] [PubMed]
41. Subramaniam, S.; Fletcher, C. Trimethylamine N-oxide: Breathe new life. *Br. J. Pharmacol.* **2018**, *175*, 1344–1353. [CrossRef]
42. Ufnal, M.; Zadlo, A.; Ostaszewski, R. TMAO: A small molecule of great expectations. *Nutrition* **2015**, *31*, 1317–1323. [CrossRef] [PubMed]
43. Andraws, R.; Berger, J.S.; Brown, D.L. Effects of antibiotic therapy on outcomes of patients with coronary artery disease: A meta-analysis of randomized controlled trials. *JAMA* **2005**, *293*, 2641–2647. [CrossRef] [PubMed]
44. Huang, F.; Zhang, F.; Xu, D.; Zhang, Z.; Xu, F.; Tao, X.; Qiu, L.; Wei, H. Enterococcus faecium WEFA23 from infants lessens high-fat-diet-induced hyperlipidemia via cholesterol 7-alpha-hydroxylase gene by altering the composition of gut microbiota in rats. *J. Dairy. Sci.* **2018**, *101*, 7757–7767. [CrossRef] [PubMed]
45. Bu, J.; Wang, Z. Cross-Talk between Gut Microbiota and Heart via the Routes of Metabolite and Immunity. *Gastroenterol. Res. Pract.* **2018**, *2018*, 6458094. [CrossRef] [PubMed]
46. Wilson, A.; Teft, W.A.; Morse, B.L.; Choi, Y.H.; Woolsey, S.; DeGorter, M.K.; Hegele, R.A.; Tirona, R.G.; Kim, R.B. Trimethylamine-N-oxide: A Novel Biomarker for the Identification of Inflammatory Bowel Disease. *Dig. Dis. Sci.* **2015**, *60*, 3620–3630. [CrossRef] [PubMed]
47. Katsiki, N.; Mikhailidis, D.P.; Mantzoros, H.S. Non-alcoholic fatty liver disease and dyslipidemia: An update. *Metabolism* **2016**, *65*, 1109–1123. [CrossRef]
48. Levy, M.; Shapiro, H.; Thaiss, C.A.; Elinav, E. NLRP6: A multifaceted innate immune sensor. *Trends Immunol.* **2017**, *38*, 248–260. [CrossRef]
49. Sepehri, Z.; Kiani, Z.; Kohan, F.; Alavian, S.M.; Ghavami, S. Toll like receptor 4 and hepatocellular carcinoma. A systematic review. *Life Sci.* **2017**, *179*, 80–87. [CrossRef]
50. Wree, A.; McGeough, M.D.; Inzaugarat, M.E.; Eguchi, A.; Schuster, S.; Johnson, C.D.; Peña, C.A.; Geisler, L.J.; Papouchado, B.G.; Hoffman, H.M.; et al. NLRP3 inflammasome driven liver injury and fibrosis: Roles of IL-17 and TNF in mice. *Hepatology* **2018**, *67*, 736–749. [CrossRef]
51. Hartmann, P.; Chen, W.C.; Schnabl, B. The intestinal microbiome and the leaky gut as therapeutic targets in alcoholic liver disease. *Front. Physiol.* **2012**, *3*, 402. [CrossRef] [PubMed]
52. Flass, T.; Tong, S.; Frank, D.N.; Wagner, B.D.; Robertson, C.E.; Kotter, C.V.; Sokol, R.J.; Zemanick, E.; Accurso, F.; Hoffenberg, E.J.; et al. Intestinal lesions are associated with altered intestinal microbiome and are more frequent in children and young adults with cystic fibrosis and cirrhosis. *PLoS ONE* **2015**, *10*, e0116967. [CrossRef]
53. Elinav, E.; Strowig, T.; Kau, A.L.; Henao-Mejia, J.; Thaiss, C.A.; Booth, C.J.; Peaper, D.R.; Bertin, J.; Eisenbarth, S.C.; Gordon, J.I.; et al. NLRP6 inflammasome regulates colonic microbial ecology and risk for colitis. *Cell* **2011**, *145*, 745–757. [CrossRef] [PubMed]
54. Scher, J.U.; Sczesnak, A.; Longman, R.S.; Segata, N.; Ubeda, C.; Bielski, C.; Rostron, T.; Cerundolo, V.; Pamer, E.G.; Abramson, S.B.; et al. Expansion of intestinal Prevotella copri correlates with enhanced susceptibility to arthritis. *eLife.* **2013**, *2*, e01202. [CrossRef]
55. Serebrovska, T.V.; Grib, O.N.; Portnichenko, V.I.; Serebrovska, Z.O.; Egorov, E.; Shatylo, V.B. Intermittent Hypoxia/Hyperoxia Versus Intermittent Hypoxia/Normoxia: Comparative Study in Prediabetes. High Alt. *Med. Biol.* **2019**, *20*, 383–391. [CrossRef] [PubMed]
56. Serebrovska, Z.O.; Serebrovska, T.V.; Kholin, V.A.; Tumanovska, L.V.; Shysh, A.M.; Pashevin, D.A.; Goncharov, S.V.; Stroy, D.; Grib, O.N.; Shatylo, V.B. Intermittent hypoxia-hyperoxia training improves cognitive function and decreases circulating biomarkers of Alzheimer's disease in patients with mild cognitive impairment: A pilot study. *Int. J. Mol. Sci.* **2019**, *20*, 5405. [CrossRef] [PubMed]
57. Timon, R.; González-Custodio, A.; Vasquez-Bonilla, A.; Olcina, G.; Leal, A. Intermittent Hypoxia as a Therapeutic Tool to Improve Health Parameters in Older Adults. *Int. J. Environ. Res. Public Health* **2022**, *19*, 5339. [CrossRef]
58. Almohanna, A.M.; Wray, S. Hypoxic conditioning in blood vessels and smooth muscle tissues: Effects on function, mechanisms, and unknowns. American journal of physiology. *Heart Circ. Physiol.* **2018**, *315*, H756–H770. [CrossRef] [PubMed]
59. Arkhipenko, Y.V.; Sazontova, T.G.; Zhukova, A.G. Adaptation to periodic hypoxia and hyperoxia improves resistance of membrane structures in heart, liver, and brain. *Bull. Exp. Biol. Med.* **2005**, *140*, 278–281. [CrossRef]
60. Vlasov, A.A.; Salikova, S.P.; Golovkin, N.V.; Grinevich, V.B. Intestinal Microbial-tissue Complex and Chronic Heart Failure (part 1): Pathogenesis. *Ration. Pharmacother. Cardiol.* **2021**, *17*, 462–469. [CrossRef]

Disclaimer/Publisher's Note: The statements, opinions and data contained in all publications are solely those of the individual author(s) and contributor(s) and not of MDPI and/or the editor(s). MDPI and/or the editor(s) disclaim responsibility for any injury to people or property resulting from any ideas, methods, instructions or products referred to in the content.

International Journal of *Molecular Sciences*

MDPI

Article

Inhibition of Ferroptosis Enables Safe Rewarming of HEK293 Cells following Cooling in University of Wisconsin Cold Storage Solution

Lucas P. Gartzke [1], Koen D. W. Hendriks [1], Femke Hoogstra-Berends [1], Christian P. Joschko [1], Anne-Lise Strandmoe [1], Pieter C. Vogelaar [1,2], Guido Krenning [1,2] and Robert H. Henning [1,*]

1 Department of Clinical Pharmacy and Pharmacology, University Medical Centre Groningen, University of Groningen, 9713 GZ Groningen, The Netherlands; l.p.gartzke@umcg.nl (L.P.G.)
2 Sulfateq B.V. Admiraal de Ruyterlaan 5, 9726 GN Groningen, The Netherlands
* Correspondence: r.h.henning@umcg.nl

Abstract: The prolonged cooling of cells results in cell death, in which both apoptosis and ferroptosis have been implicated. Preservation solutions such as the University of Wisconsin Cold Storage Solution (UW) encompass approaches addressing both. The use of UW improves survival and thus extends preservation limits, yet it remains unclear how exactly organ preservation solutions exert their cold protection. Thus, we explored cooling effects on lipid peroxidation and adenosine triphosphate (ATP) levels and the actions of blockers of apoptosis and ferroptosis, and of compounds enhancing mitochondrial function. Cooling and rewarming experiments were performed in a cellular transplantation model using Human Embryonic Kidney (HEK) 293 cells. Cell viability was assessed by neutral red assay. Lipid peroxidation levels were measured by Western blot against 4-Hydroxy-Nonenal (4HNE) and the determination of Malondialdehyde (MDA). ATP was measured by luciferase assay. Cooling beyond 5 h in Dulbecco's Modified Eagle Medium (DMEM) induced complete cell death in HEK293, whereas cooling in UW preserved ~60% of the cells, with a gradual decline afterwards. Cooling-induced cell death was not precluded by inhibiting apoptosis. In contrast, the blocking of ferroptosis by Ferrostatin-1 or maintaining of mitochondrial function by the 6-chromanol SUL150 completely inhibited cell death both in DMEM- and UW-cooled cells. Cooling for 24 h in UW followed by rewarming for 15 min induced a ~50% increase in MDA, while concomitantly lowering ATP by >90%. Treatment with SUL150 of cooled and rewarmed HEK293 effectively precluded the increase in MDA and preserved normal ATP in both DMEM- and UW-cooled cells. Likewise, treatment with Ferrostatin-1 blocked the MDA increase and preserved the ATP of rewarmed UW HEK293 cells. Cooling-induced HEK293 cell death from hypothermia and/or rewarming was caused by ferroptosis rather than apoptosis. UW slowed down ferroptosis during hypothermia, but lipid peroxidation and ATP depletion rapidly ensued upon rewarming, ultimately resulting in complete cell death. Treatment throughout UW cooling with small-molecule Ferrostatin-1 or the 6-chromanol SUL150 effectively prevented ferroptosis, maintained ATP, and limited lipid peroxidation in UW-cooled cells. Counteracting ferroptosis during cooling in UW-based preservation solutions may provide a simple method to improve graft survival following cold static cooling.

Keywords: 6-chromanols; SUL150; Ferrostatin-1; transplantation; hypothermia; organ preservation; lipid peroxidation; ferroptosis; HEK293

1. Introduction

In clinical kidney transplant procedures, simple cold static hypothermic (4 °C) preservation remains the most widely employed strategy to avoid ischemia reperfusion injury (IRI) [1] and ensure graft health [2]. During this procedure, organs are cold-flushed with and stored in ice-cooled organ preservation solutions such as the University of Wisconsin

Cold Storage Solution (UW) or an alternative organ preservation solution (Custidiol, HTK, etc.). The traditional perspective is that this forced hypothermia protects the graft from IRI by exogenously lowering metabolism and limiting oxidative stress, thereby maintaining energy reserves and cell viability [2,3]. However, cooling efficacy is limited, as it does not fully prevent the formation of reactive oxygen species (ROS) and graft damage [1–10].

Previously, we showed cooling to decrease mitochondrial respiration according to the Q10 effect, which describes the reaction rate changes of a biochemical reaction per 10 °C temperature change. Specifically, cooling at 4 °C resulted in a ~40% (2 h) and 90% (16 h) inhibition of ATP formation, yet lowered ROS production only by half [11]. Additionally, hypothermia increased the ROS burden by impairing ROS-scavenging enzymes such as superoxide dismutase. In keeping with these findings, we showed that 2 h of cooling doubles the levels of lipid peroxidation [12]. Consequently, cooling slows down, but does not fully preclude, the damaging processes. Indeed, although the recommended limit of kidney hypothermia is 18 h [5], graft survival decreases dramatically from 6 h of preservation time onwards [7].

We and others showed earlier that isolated cells in standard media die from or during hypothermia [11–15], although the exact mechanisms are still debated. One of the suggested pathways is (intrinsic) apoptosis triggered by excess calcium influx in response to cell membrane depolarization because of the failure of the sodium–potassium pump due to a hypothermia-induced loss of ATP [15–19]. This suggestion is supported by the prevention of cold-mediated cell death by calcium chelators and calcium channel blockers [15,17]. While calcium chelators such as EDTA, BAPTA-AM, and dopamine protect against hypothermia, they do not exclusively chelate calcium and show functional overlap with other chelators of cations (such as ferrous iron). Therefore, their protective action cannot easily be attributed to a single ion [17,20], which has challenged the understanding of the role of calcium in hypothermic cell death and implicated the involvement of iron in inducing ferroptosis [21–24].

Ferroptosis is uniquely characterized among cell death pathways by its iron-dependent and ROS-mediated accumulation of lipid peroxidation and subsequent depletion of plasma membrane polyunsaturated fatty acids [25–33] without observable nuclear changes (such as chromatin condensation) [34,35]. Ferroptosis is counteracted by the enzyme family of glutathione peroxidases (GPxs), which reduce glutathione (GSH) to neutralize ROS. Specifically, the isoform GPx4 reduces membrane-bound oxidized lipids and serves as the specific endogenous cellular defense against ferroptosis [35–39]. Lack of GPx4 under oxidative conditions promotes cell death via ferroptosis [34,40]. In keeping with this, the targeted depletion of cellular glutathione (GSH) reservoirs or direct inhibition of GPx4 result in the induction of ferroptosis via the increased formation of lipid peroxides [34,40]. Conversely, the reduction of the available intracellular iron pool with deferoxamine protects cells from ferroptosis during in vitro normothermia and—more importantly—during in vivo hypothermic organ preservation [17,21–24]. Ferrous iron (Fe^{2+}) is central in the generation of lipid peroxides, as it initiates the Fenton reaction in the presence of (increased) mitochondrial hydrogen peroxide (H_2O_2), generating ferric iron (Fe^{3+}), HO• (hydroxide radical), and OH- (hydroxide ion) ($Fe^{2+} + H_2O_2 \rightarrow Fe^{3+} + HO^{\bullet} + OH^{-}$). This conversion of H_2O_2 into the more reactive $HO^{\bullet}$, in turn, causes secondary oxidative reactions, among them the oxidation of membrane polyunsaturated fatty acids (PUFAs).

Ferroptosis can be prevented by direct inhibitors of lipid peroxidation, such as Ferrostatin-1, a synthetic antioxidant that scavenges alkoxyl radicals and is regenerated by ferrous iron and, therefore, is not consumed while executing its scavenging properties [41]. A proposed role of ferroptosis is substantiated by our finding that Ferrostatin-1 attenuates hypothermic tissue damage and cell death [14,42]. ROS overload itself, as well as the damage that ensues from it, such as DNA damage or lipid peroxidation, can trigger (intrinsic) apoptosis and ferroptosis, respectively. Whereas ferroptosis, unlike other forms of regulated cell death, such as apoptosis, is strictly dependent on ROS-mediated lipid peroxidation [25–27,30–33], compounds that limit excessive ROS accumulation may protect cells from cell death via both pathways. Consequently, the limitation of mitochondrial ROS production is another approach to limit ferroptosis and explains the cold tolerance induced by 6-chromanols [13,16,43–45]. These compounds are structurally related to both dopamine and vitamin E, and stabilize mitochondrial function by maintaining complex I and IV functions, resulting in stable ATP production under hypothermic conditions [43,44]. In rats, 6-chromanols maintain structural kidney morphology after prolonged total body hypothermia [43].

In clinical practice, the use of UW has improved outcomes and thus extended safe preservation limits, yet it remains unclear how exactly organ preservation solutions exert their cold protection [11,46,47]. Interestingly, UW encompasses approaches addressing both apoptosis and ferroptosis. UW features sodium and potassium concentrations matching those of the intracellular milieu [48], thus alleviating hypothermia-induced plasma membrane depolarization and subsequent excessive calcium influx and ensuing (intrinsic) apoptosis. Furthermore, UW is rich in GSH [49–51], which provides the antioxidative capacity to sustain GPx4 activity, with the potential to prevent ferroptosis [25–27,36–40]. Here, we aimed to establish the mechanism of cell death in a cellular model for renal cold storage in UW and explored the efficacy and timing of pharmacologic interventions with different antioxidants and 6-chromanols.

2. Results

2.1. Hypothermia Causes Cell Death in HEK293 Cells That Is Partially Ameliorated by UW

As an initial step, we determined cell viability by neutral red assay after cooling for 5, 10, and 24 h in both regular medium (Dulbecco's Modified Eagle Medium, DMEM) and University of Wisconsin Cold Storage Solution (UW). In DMEM, 5 h of cooling did not cause HEK293 cell death (Figure 1b), whereas viable cells were absent after 10 or 24 h of cooling. In contrast, 5 h of cooling in UW lowered cell survival by ~40%, after which cell survival gradually decreased at 10 and 24 h of cooling (Figure 1c).

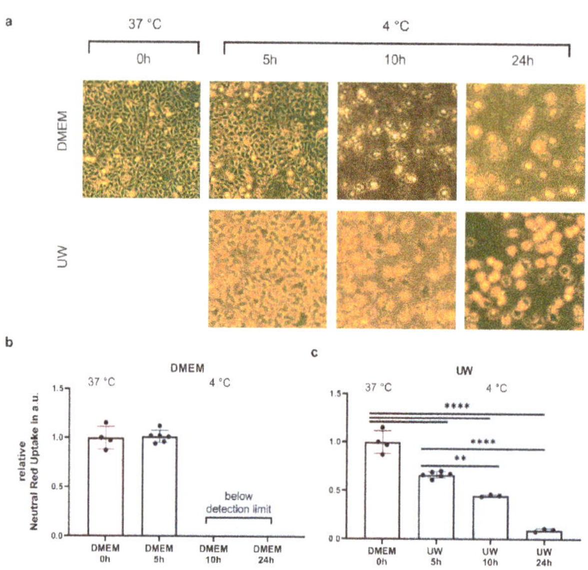

Figure 1. Cell survival in Human Embryonic Kidney (HEK) 293 cells after 5, 10, and 24 h of hypothermia (blue background) in cell medium (DMEM) or University of Wisconsin Cold Storage Solution (UW). (**a**) Typical microscopy pictures of conditions tested (10× magnification). (**b**) Cell survival after cooling in DMEM. (**c**) Cell survival after cooling in UW. All data are expressed relative to standard conditions at 37 °C and presented as means ± SD; ** = $p < 0.005$; **** = $p < 0.0001$; one-way ANOVA with Šidák's post hoc test; n = 3–6 per condition.

2.2. Inhibition of Apoptosis Does Not Increase Cell Survival

To investigate whether cooled HEK293 cells die of apoptosis, we first tested cellular and mitochondrial calcium channel blockades' prevention of cell death during hypothermia and after rewarming. A broad-range blockade of extracellular Ca^{2+} channels by lanthanum chloride (50 mM, Figure 2a,b) and the mitochondrial Ca^{2+} uniporter by Ru360 (5 mM, Figure 2c,d) did not prevent cell death during 24 h cooling in DMEM or UW (Figure 2a–d). Next, we explored apoptosis inhibition by Z-VAD(OH)-FMK, a pan-caspase inhibitor with a specific affinity for the final apoptotic effector caspase 6 and a preserved function in the cold (Figure A3). Like the Ca^{2+} blockade, the inhibition of caspase 6 with Z-VAD(OH)-FMK (50 μM, Figure 2e,f) did not prevent cell death in DMEM-cooled cells, and only very partly rescued UW-cooled cells from cell death following rewarming. Collectively, these results show that hypothermia-induced HEK293 cell death cannot be prevented by inhibiting apoptosis.

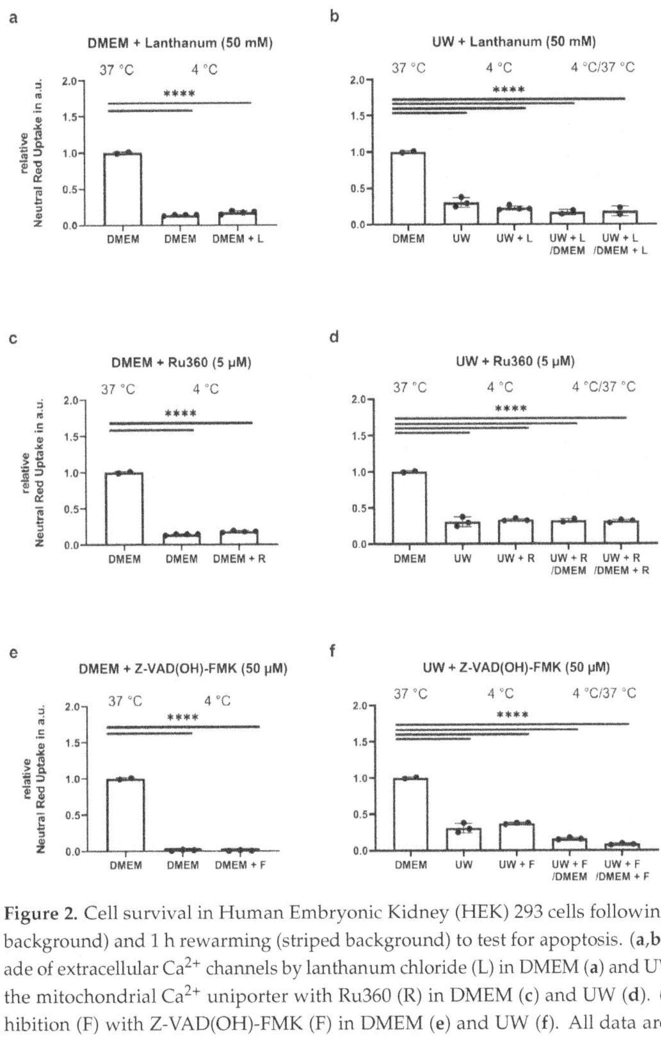

Figure 2. Cell survival in Human Embryonic Kidney (HEK) 293 cells following 24 h cooling (blue background) and 1 h rewarming (striped background) to test for apoptosis. (**a**,**b**) Broad-range blockade of extracellular Ca^{2+} channels by lanthanum chloride (L) in DMEM (**a**) and UW (**b**). (**c**,**d**) Blocking the mitochondrial Ca^{2+} uniporter with Ru360 (R) in DMEM (**c**) and UW (**d**). (**e**,**f**) Pan-caspase inhibition (F) with Z-VAD(OH)-FMK (F) in DMEM (**e**) and UW (**f**). All data are expressed relative to standard conditions at 37 °C and presented as means ± SD; one-way ANOVA with Šidák's post hoc test; **** = $p < 0.0001$; n = 2–4.

2.3. Addition of SUL150 and Ferrostatin-1 Preventing Ferroptosis

Next, we investigated whether the 6-chromanol SUL-150, which preserves mitochondrial function [13,16,43–45], could also attenuate ferroptosis, which it did (Figure A4). We tested the efficacy of SUL150 and Ferrostatin-1, a known inhibitor of lipid peroxidation and ferroptosis, to exert protection from cooling and/or from cooling and rewarming. DMEM or UW were supplemented with SUL150 (10 μM) or Ferrostatin-1 (100 nM) at the start (0 h) of 24 h of cooling. Both SUL150 and Ferrostatin-1 completely inhibited cell death in either DMEM- or UW-incubated cells both at the end of cooling (Figure 3) and following rewarming (Figure 4).

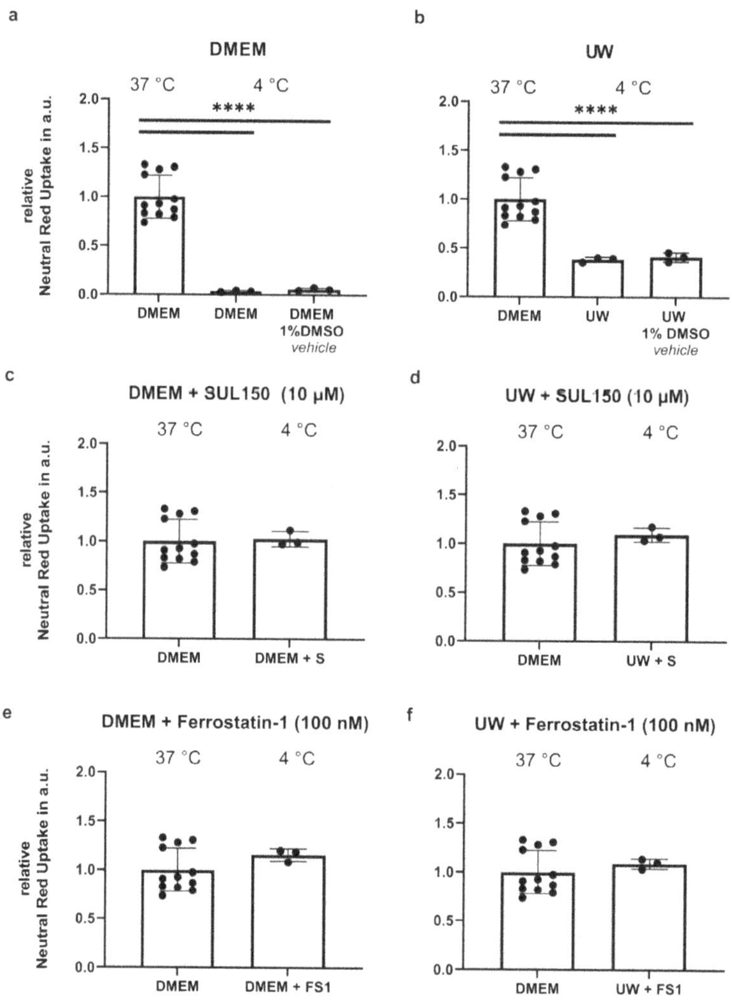

Figure 3. Cell Survival in Human Embryonic Kidney (HEK) 293 cells after 24 h of cooling (blue background) with SUL150 or Ferrostatin-1 supplementation in either DMEM or UW. (**a**,**b**) 24 h cooling with DMSO vehicle in DMEM (**a**) and UW (**b**). (**c**,**d**) 24 h cooling with SUL150 (S) in DMEM (**c**) and UW (**d**). (**e**,**f**) 24 h cooling with Ferrostatin-1 (F) in DMEM (**e**) and UW (**f**). All data are expressed relative to standard conditions at 37 °C and presented as means ± SD; **** = $p < 0.0001$; one-way ANOVA with Dunnett's post hoc test; n = 3–12.

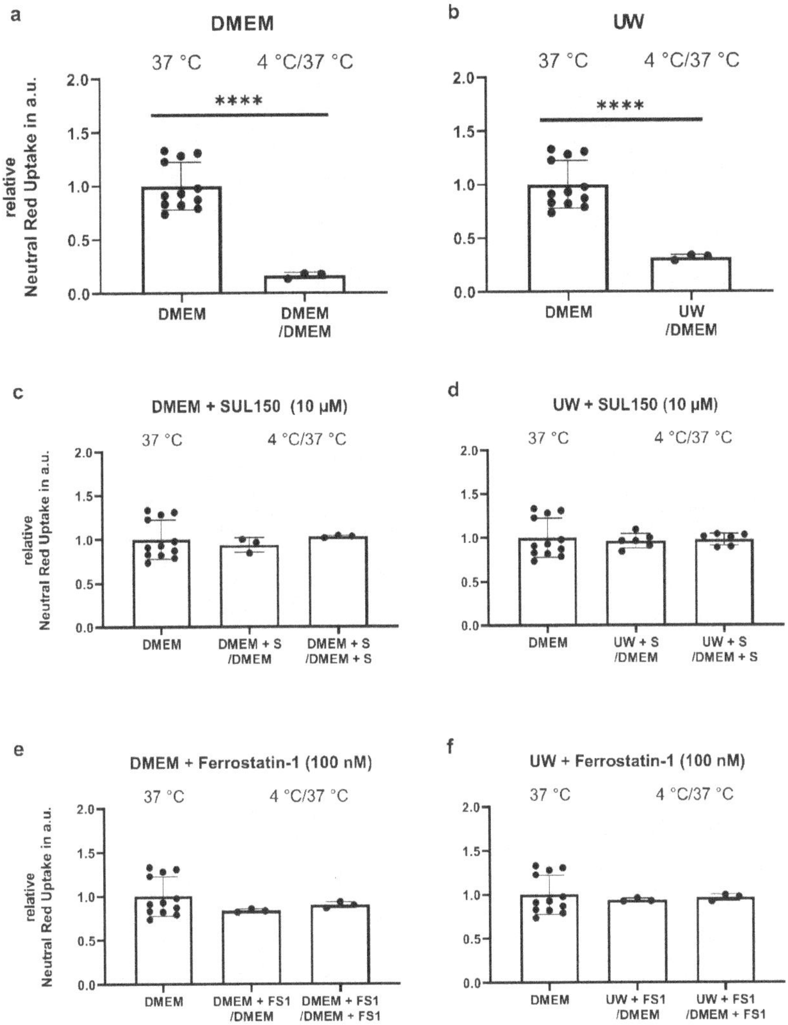

Figure 4. Cell Survival in Human Embryonic Kidney (HEK) 293 cells after 24 h of cooling and 1 h of rewarming (striped background) with SUL150 or Ferrostatin-1 supplementation in either DMEM or UW. (**a**,**b**) Cooling for 24 h and rewarming with DMSO vehicle in DMEM (**a**) and UW (**b**). (**c**,**d**) Cooling for 24 h and rewarming with SUL150 (S) in DMEM (**c**) and UW (**d**). (**d**,**e**) Cooling for 24 h with Ferrostatin-1 (F) in DMEM (**e**) and UW (**f**). All data are expressed relative to standard conditions at 37 °C and presented as means ± SD; **** = $p < 0.0001$; one-way ANOVA with Dunnett's post hoc test; n = 3–12.

2.4. Rewarming Causes ATP Depletion and Lipid Peroxidation in UW Cold-Preserved HEK293 Cells

Paradoxically, whereas preservation in UW conferred robust cell survival during cooling, as shown by light microscopy and neutral red survival assay, the rewarming of cells resulted in imminent cell death. To explore the mechanism of the cell death of cooled UW cells upon rewarming, we measured lipid peroxidation by MDA assay and the ATP content by luciferase assay. The cooling of HEK293 for 24 h in UW followed by rewarming in DMEM for 15 min induced a ~50% increase in MDA (Figure 5b), while concomitantly lowering ATP by >90% (Figure 6b). Treatment with either SUL150 or

Ferrostatin-1 effectively precluded the increase in MDA and decrease in ATP seen in rewarmed UW-cooled cells (Figures 5 and 6). It is of note that Ferrostatin-1 did not fully preserve the ATP content of cooled and rewarmed DMEM cells (Figure 6e). Vehicle-treated DMEM-incubated HEK293 cells could not be explored, as they do not survive without intervention. Thus, we conclude that cooling and rewarming increases lipid peroxidation and depletes ATP in UW-cooled cells upon rewarming, which is precluded by either SUL150 or Ferrostatin-1.

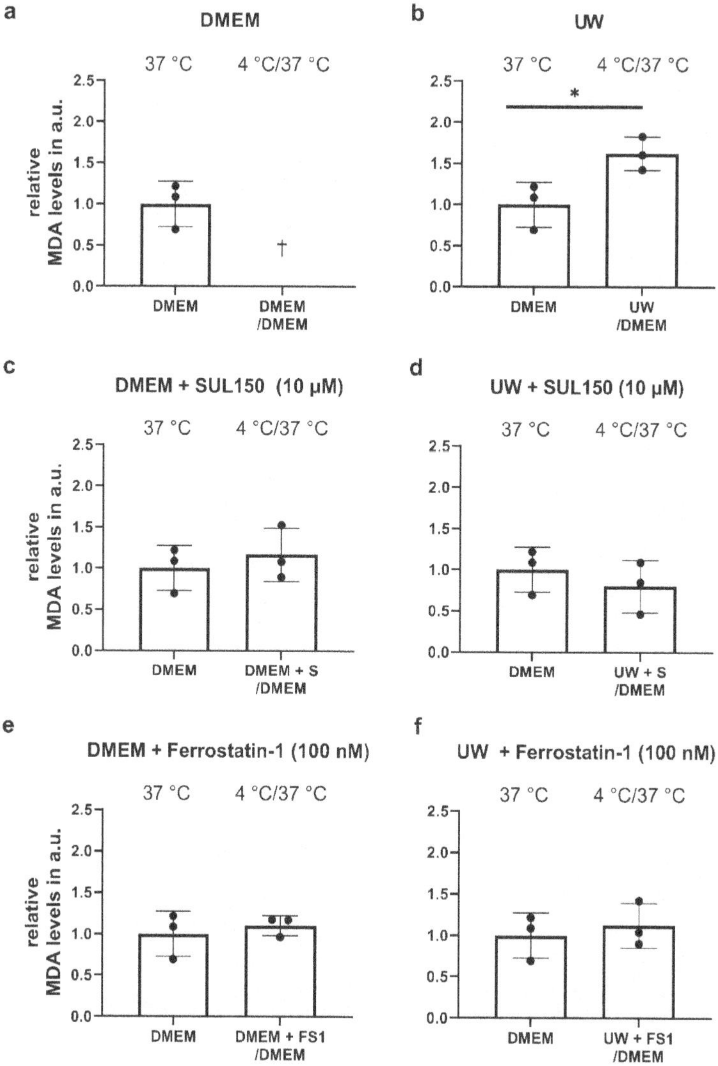

Figure 5. Lipid peroxidation (MDA) in University of Wisconsin Cold Storage Solution in Human Embryonic Kidney (HEK) 293 cells after 24 h cooling and 1 h rewarming (striped background). (**a**,**b**) Lipid peroxidation in DMEM (**a**) and UW (**b**). (**c**,**d**) Lipid peroxidation after SUL150 (S) treatment in DMEM (**c**) and UW (**d**). (**e**,**f**) Lipid peroxidation after Ferrostatin-1 (F) treatment in DMEM (**e**) and UW (**f**). All data are expressed relative to standard conditions at 37 °C and presented as means ± SD; one-way ANOVA with Dunnett's post hoc test; * = $p < 0.05$; $n = 3$.

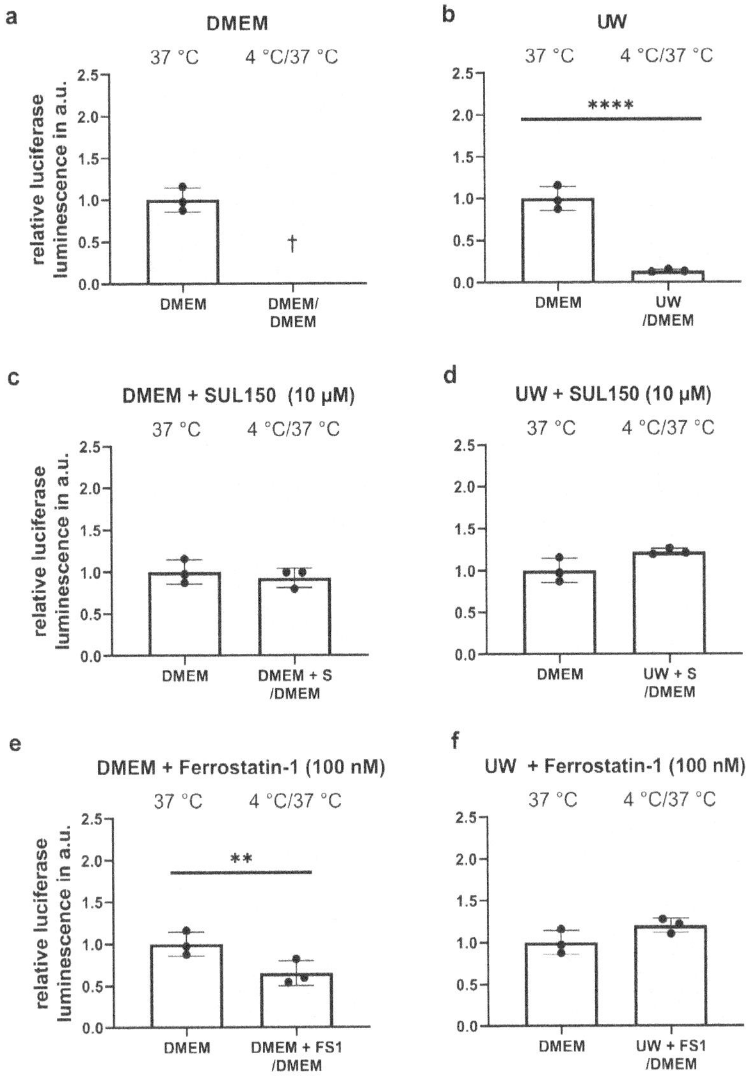

Figure 6. ATP luciferase activity in University of Wisconsin Cold Storage Solution in Human Embryonic Kidney (HEK) 293 cells after 24 h cooling and 1 h rewarming (striped background). (**a**,**b**) ATP in DMEM (**a**) and UW (**b**). (**c**,**d**) ATP after SUL150 (S) treatment in DMEM (**c**) and UW (**d**). (**e**,**f**) ATP peroxidation after Ferrostatin-1 (F) treatment in DMEM (**e**) and UW (**f**). All data are expressed relative to standard conditions at 37 °C and presented as means ± SD; one-way ANOVA with Dunnett's post hoc test; ** = $p < 0.01$; **** = $p < 0.0001$; $n = 3$.

2.5. UW Prevents Oxidative Stress in the Cold

Given the increased lipid peroxidation in UW cooled-and-rewarmed cells, we next asked whether peroxidation is already present during hypothermia or depends on rewarming, and determined this based on an abundance of 4HNE, an obligatory and specific product of lipid peroxidation. 4HNE levels increased after 5 h of cooling of HEK293 cells, both in UW and DMEM (Figure 7b,c). However, in UW-cooled cells, 4HNE abundance returned to baseline at 10 and 24 h of cooling (Figure 7c), whereas DMEM cells did not tolerate cooling beyond 5 h. Treatment with SUL150 (10 μM) or Ferrostatin-1 (100 nM) prevented the increase in lipid peroxidation in cooled UW cells (Figure 7e,g). In DMEM-cooled cells, SUL150 (10 μM) delayed, and Ferrostatin-1 completely prevented, a 4HNE increase (Figure 7d,f). Thus, UW protects cells from increased lipid peroxidation during prolonged cooling, and the initial increase is blocked by SUL150 and Ferrostatin-1.

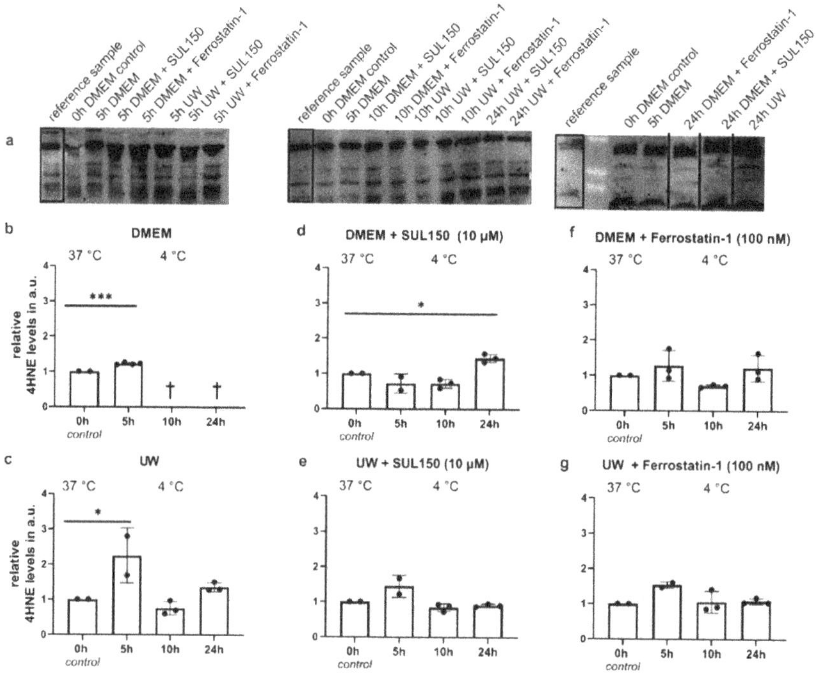

Figure 7. 4HNE expression in Human Embryonic Kidney (HEK) 293 cells in a 5 to 24 h cooling model. (**a**) Typical examples of Western blots. (**b**) Cooling in DMEM. (**c**) Cooling in UW. (**d**) Cooling in DMEM + SUL150 (10 μM). (**e**) Cooling in UW + SUL150 (10 μM). (**f**) Cooling in DMEM + Ferrostatin-1 (100 nM). (**g**) Cooling in UW + Ferrostatin-1 (100 nM). Samples derived from the same experiment and blots were processed in parallel; the same reference sample was used on all blots and is highlighted with a rectangle. Vertical lines indicate the exclusion of lanes on the same blot. All data are expressed relative to standard conditions at 37 °C and presented as means ± SD; unpaired two-tailed t-test (B); one-way ANOVA with Dunnett's post hoc test (C-G); * = $p < 0.05$; *** = $p < 0.001$; n = 2–5.

2.6. Antioxidant Supplementation Shortly before or during Rewarming Does Not Prevent Cell Death

Given the increased lipid peroxidation and, ultimately, cell death in UW-cooled cells following rewarming, we reasoned that UW treatment may provide initial cell protection because of its high glutathione (GSH) content (3 mM), which is lost upon the necessary replacement with DMEM upon rewarming (because of ionic composition). To test this, cell survival was assessed in HEK293 cells supplemented either 30 min before or immediately upon rewarming with oxidized (see Appendix A; Figure A1) and reduced GSH, the lipid-soluble antioxidant vitamin E (Trolox), the 6-chromanol SUL150, and Ferrostatin-1. Compound supplementation either before or during rewarming did not increase cell survival in UW-cooled HEK293 cells (Figures 8 and 9).

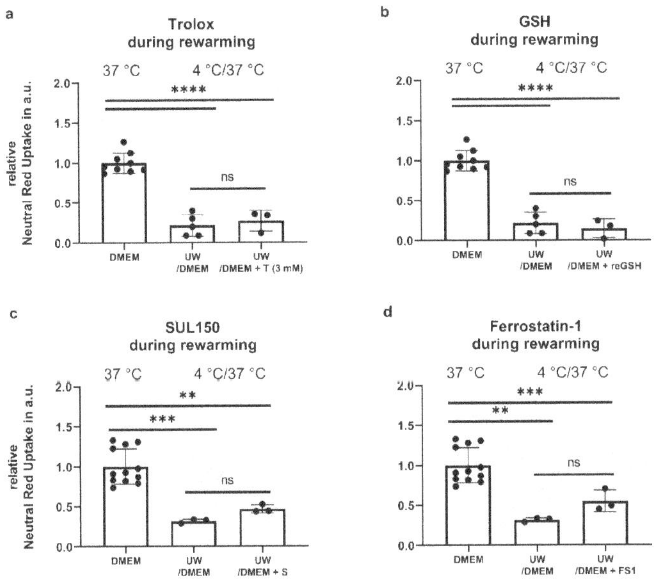

Figure 8. Cell survival in Human Embryonic Kidney (HEK) 293 cells in a 24 h cooling and 1 h rewarming model (striped background) with antioxidants added to the rewarming medium. (**a**) Addition of Trolox (T). (**b**) Addition of reduced GSH (reGSH). (**c**) Addition of SUL150 (S). (**d**) Addition of Ferrostatin-1 (F). All data are expressed relative to standard conditions at 37 °C and presented as means ± SD; one-way ANOVA with Šidák's post hoc test; ** = $p < 0.01$; *** = $p < 0.001$; **** = $p < 0.0001$; n = 3–9.

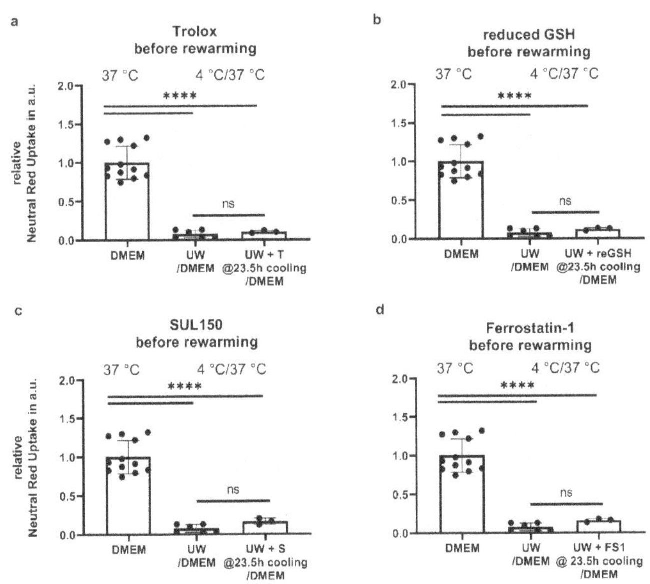

Figure 9. Cell survival in Human Embryonic Kidney (HEK) 293 cells in a 24 h cooling and 1 h rewarming model (striped background) with antioxidants added to the cooling medium 30 min before rewarming. (**a**) Addition of Trolox (T). (**b**) Addition of reduced GSH (reGSH). (**c**) Addition of SUL150 (S). (**d**) Addition of Ferrostatin-1 (F). All data are expressed relative to standard conditions at 37 °C and presented as means ± SD; one-way ANOVA with Šidák's post hoc test; **** = $p < 0.0001$; n = 3–9.

2.7. *UW Does Not Preclude Progressive Loss of GPx4*

Due to the lack of effect of the compounds with administration shortly before or at the time of the initiation of rewarming, we next asked whether cooling affects GPx4 abundance, and measured its protein expression in all conditions and at all time points. The cooling of HEK293 in DMEM for 5 h significantly reduced GPx4 protein levels (Figure 10b), with cooling for 10 or 24 h resulting in complete cell death. In UW-cooled cells, GPx4 levels were unaffected after 5 and 10 h of hypothermia, but significantly decreased at 24 h (Figure 10c).

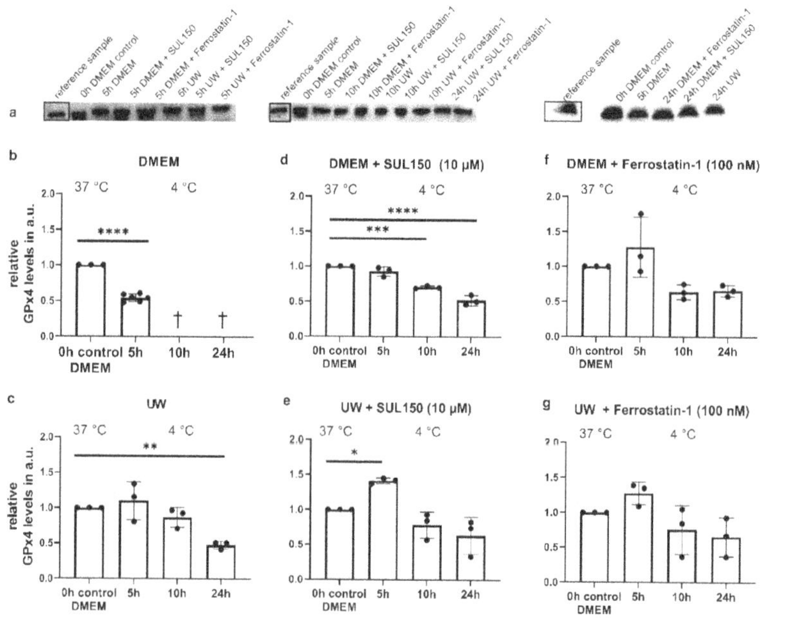

Figure 10. GPx4 expression in Human Embryonic Kidney (HEK) 293 cells in a 5 to 24 h cooling model. (**a**) Typical examples of Western blots. (**b**) Cooling in DMEM. (**c**) Cooling in UW. (**d**) Cooling in DMEM + SUL150 (10 µM). (**e**) Cooling in UW + SUL150 (10 µM). (**f**) Cooling in DMEM + Ferrostatin-1 (100 nM). (**g**) Cooling in UW + Ferrostatin-1 (100 nM). Samples derived from the same experiment and blots were processed in parallel; the same reference sample was used on all blots and is highlighted with a rectangle. Vertical lines indicate the exclusion of lanes on the same blot. All data are expressed relative to standard conditions at 37 °C and presented as means ± SD; unpaired two-tailed *t*-test (**b**); Grubbs outlier test (alpha = 0.05) removed one outlier in DMEM at 5 h; one-way ANOVA with Dunnett's post hoc test (C-G); * = $p < 0.05$; ** = $p < 0.01$; *** = $p < 0.001$; **** = $p < 0.0001$; n = 2–5.

Next, we investigated whether levels of GPx4 were maintained through supplementation of DMEM and UW with SUL150 (10 µM). In DMEM, SUL150 supplementation resulted in stable GPx4 levels following 5 h cooling, but its abundance reduced significantly at 10 and 24 h (Figure 10d). The addition of Ferrostatin-1 (100 nM) to DMEM maintained GPx4 levels throughout the experiment (Figure 10f). In contrast, SUL150 treatment of UW-cooled cells significantly increased GPx4 after 5 h cooling, which returned to baseline level following 10 and 24 h cooling (Figure 10e). Cooling in UW with Ferrostatin-1 did not result in significant changes in GPx4 expression; however, we observed an increase after 5 h of cooling, which did not reach significance (Figure 10f,g). Thus, the time-dependent loss of endogenous antioxidative defenses against ferroptosis can be prevented by SUL150 and Ferrostatin-1 in UW-cooled HEK293 cells.

2.8. Proteasome Inhibition Does Not Prevent GPx4 Loss or Cell Death

Because we observed a reduction of GPx4 levels during hypothermia, both in DMEM and UW, we next investigated whether this was due to proteasomal breakdown. The proteasome inhibitor, MG132 (10 mM), did not increase survival during cooling or rewarming (Figure 11b). It should be noted that MG132 was well tolerated in normothermic cells at 1 h of treatment, yet induced complete cell death after 24 h (Figure 11a).

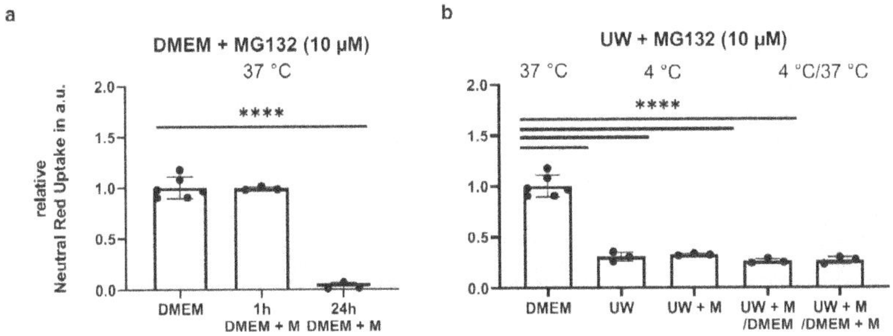

Figure 11. Proteasome blockage in Human Embryonic Kidney (HEK) 293 cells in a cooling (blue background) and rewarming (striped background) model. (**a**) MG132 (M) treatment under normothermic conditions. (**b**) Cell survival measured after 24 h cooling and rewarming using MG132 (proteasome inhibitor). All data are expressed relative to standard conditions at 37 °C and presented as means ± SD; one-way ANOVA with Šidák's post hoc test; **** = $p < 0.0001$; $n = 3$–6.

3. Discussion

The combination of hypothermia and University of Wisconsin Cold Storage Solution (UW) has improved organ preservation outcomes [49–51]. While the mechanism of hypothermia, as well as its benefits and drawbacks, are well documented in the literature [1–11,46,47], the protective mechanism of UW has not been fully elucidated yet. Here, we show that UW indeed protects from hypothermic cell death. However, upon rewarming, cell death is executed through ferroptosis, driven by ATP depletion and lipid peroxidation. Finally, treatments that maintain mitochondrial health or inhibit lipid peroxidation throughout cooling counteract ferroptosis and rescue UW-cooled cells from rewarming-induced cell death.

3.1. UW-Cooled Cell Death Is Executed through Ferroptosis Rather Than Apoptosis

We identified ferroptosis as the pathway conferring cell death following rewarming of UW-cooled cells. Small-molecule Ferrostatin-1, an inhibitor of lipid peroxidation, can prevent ferroptosis. Here and earlier [14], we have demonstrated that Ferrostatin-1 can confer protection from hypothermic cell death. We showed that cooling in DMEM causes rapid and complete cell death by ferroptosis. UW extends cooling time significantly, but rewarming results in massive cell death, likely due to increased lipid peroxidation and ATP depletion. Ferroptosis is caused by lipid peroxidation in the absence of a sufficient antioxidative defense system, i.e., glutathione peroxidase 4 (GPx4), which reduces glutathione (GSH) and thereby counteracts lipid peroxidation [25–27,34–40]. We demonstrated that hypothermia causes a reduction in GPx4 protein abundance in HEK293 cells after 5 h of cooling in DMEM-cooled cells, accompanied by a concomitant increase in lipid peroxidation levels. UW delayed, but did not fully prevent, a cooling-evoked time-dependent decrease in GPx4; i.e., GPx4 levels were maintained at 5 and 10 h of cooling, but were significantly reduced following 24 h of cooling. In addition, UW-cooled HEK293 cells also showed increased lipid peroxidation at 5 h of hypothermia. We conclude that UW alone substantially precludes ferroptosis during hypothermia, most likely due to its high GSH content fueling the activity of Gpx4, thus counteracting the early increase in lipid peroxidation, and effectively extending cell survival.

3.2. ATP Depletion, Calcium, and Hypothermic Cell Death

We and others have shown that during hypothermia, ATP levels are depleted [11,13,14,17,18,44]. Here, we demonstrated ATP depletion in 24 h UW-cooled cells during rewarming. It has been proposed that ATP depletion causes cessation of the sodium–potassium pump, which leads to membrane depolarization, voltage-gated calcium influx, and finally apoptosis [15–18]. While we did show ATP depletion, our data indicate that cells do not die of apoptosis during cooling and rewarming, as solidified both by the ineffectiveness of blocking a calcium influx and by inhibiting the final effector of apoptosis by a pan-caspase inhibitor. Importantly, pharmacologic interventions that allowed for safe cooling and rewarming (SUL150; Ferrostatin-1) maintained ATP levels in UW-cooled cells, but also prevented lipid peroxidation. Conversely, Bayley et al. [15] showed that hypothermic cell death, which they attributed to apoptosis, was prevented by a blockade of voltage-gated calcium channels (lanthanum) in insect muscle tissue, and others have demonstrated the effectiveness of calcium chelation [17]. It is possible that mammalian hypothermic cell death is regulated differently from insect cell death; however, recent literature has shown that ferroptosis is evolutionarily conserved, and also takes place in insects [52]. This corresponds with ferroptosis' independence from transcriptional activity, as it is the effect of a disturbed intracellular microenvironment—the redox balance of the cell [35,41]. In keeping with this, calcium is an important regulator of mitochondrial activity, and an increase in mitochondrial calcium concentrations, i.e., a mitochondrial matrix calcium overload, can cause mitochondrial dysfunction, along with lowered ATP and excessive ROS production [18], the latter of which is central in ferroptosis [25–28,35]. Advanced preservation techniques, such as pressure-controlled oxygenated machine perfusion, create fewer complications, reduced oxidative stress, and increased energy levels [53–55]. An attractive hypothesis is that these techniques maintain or restart mitochondrial respiration through oxygen delivery, and thereby maintain the ATP [53] to support antioxidant transport from the preservation solution, which fails in a static cold storage solution due to progressive ATP depletion. In doing so, oxygenated machine perfusion may prevent mitochondrial dysfunction, maintaining sufficient ATP and antioxidant delivery to prevent ferroptosis by precluding excessive accumulation of lipid peroxides upon rewarming.

3.3. Early Treatment Prevents Ferroptosis

Early treatment with either SUL150 or Ferrostatin-1, i.e., at the start of cooling, proved effective in counteracting increased lipid peroxidation during rewarming, while concomitantly maintaining ATP levels. Importantly, the addition of these compounds and other antioxidants, including glutathione (GSH) and Trolox, at the end of cooling or during rewarming did not increase cell survival. One reason for this could be that GSH is not taken up by the cells, i.e., cannot pass the lipid phase of the cell membrane [56], as it may only be taken up in its ester form [57,58]. However, research in cold-preserved rodent livers has demonstrated the effectiveness of supplying extra GSH during rewarming [59]. Uptake alone, however, does not suffice to explain the failure associated with late administration of these compounds. The water-soluble antioxidant Trolox did not reduce cell death when administered shortly before or during rewarming, although it passes the cellular membrane [60] and prevents ferroptosis [25]. One reason for this could be a failure of antioxidant delivery in the context of ATP depletion. GSH requires active transport via ATPase [56,59], whereas Trolox passes the cell membrane; however, the exact mechanism of action has yet not been elucidated. Alternatively, the loss of GPx4, i.e., the ferroptosis-specific oxidative defense, could explain the ineffectiveness of late intervention. Hypothermia-dependent GPx4 loss is not an active cellular process caused by ubiquitination, as levels of ubiquitin did not differ over time or treatment. Furthermore, cooling with a proteasome inhibitor did not increase cell survival following cooling or rewarming. Oxidative stress itself provides an alternative to the loss of GPx4. Due to its nine cysteines, which can be readily oxidized, GPx4 itself is prone to oxidative damage via disulfide bridge formation, which, in turn, causes confirmation changes [61,62] and may result in protein breakdown. This GPx4 loss may explain why the addition of exogenous GSH during or shortly before rewarming did not prevent cell death, as supplying the substrate GSH in the absence of sufficient GPx4 is ineffective. Alternatively, ATP depletion may inhibit both GSH uptake into the cell and/or the intracellular transport and processing of either GSH or Trolox, and, therefore, fail to exert protection upon rewarming. Finally, the increased effectivity of Ferrostatin-1 compared to other antioxidants may lie in the chain-breaking and iron-scavenging properties of Ferrostatin-1. Trolox is indeed a water-soluble antioxidant that confers its protection through the donation of a hydrogen atom to the lipid peroxyl radicals, thereby preventing their accumulation. Conversely, Ferrostatin-1 scavenges the alkoxyl radicals which are formed from hydroperoxides by ferrous iron (Fe^{2+}), and, additionally, binds ferrous iron from the cytosol which would otherwise initiate lipid peroxidation—a reaction known as the Fenton Reaction. Furthermore, Ferrostatin-1 is self-recycling, as it is regenerated but ferrous iron, which reduces the Ferrostatin-1-radical complex. In lipid peroxidation induced through iron, Ferrostatin-1 has far greater chain-breaking capacity than Trolox [41], which explains Ferrostatin-1's superiority over Trolox, as was also shown by us in earlier work.

3.4. Limitations

While human HEK293 cells are extensively used in research, we should note that the cells are of tumorigenic origin. Although we do not have reason to believe that their response to cold differs from primary kidney cells, we did not prove this.

4. Materials and Methods

4.1. Cellular Cooling and Rewarming Model

Human Embryonic Kidney (HEK293, ATCC CRL-1573) cells were cultured in a humidified incubator with 20% oxygen, 5% carbon dioxide at 37 °C in Dulbecco's Modified Eagle Medium (DMEM) supplemented with 10% fetal bovine serum and 1% penicillin-streptomycin. For cooling and rewarming experiments, six-well plates were coated with 0.03% poly-l-lysine and cells were grown to 90–95% confluency.

To perform hypothermia and rewarming experiments, cells were seeded in six-well plates and after overnight adherence placed at 4 °C in a standard laboratory refrigerator for different time periods, with or without rewarming by reinstitution of standard cell culture conditions (DMEM at 37 °C). Prior to cooling, cell culture plates were placed in open zip-locked bags to maintain the appropriate atmosphere. Subsequently, zip-locks were sealed, and cells were cooled in either DMEM or University of Wisconsin Cold Storage Solution (UW) in the absence or presence of compounds listed in Table 1. To mimic the clinical practices of organ preservation, including rapid temperature changes, a quick wash with 18 °C PBS preceded the final storage in cooling and rewarming media preconditioned at temperatures of 4 °C or 37 °C, respectively. Cells were observed with a Leica DMIL 090-135.002 microscope mounted with a Leica MC120 HD camera at 10× magnification (Leica 506271, 345IH/01). The pictures shown are representative images.

4.2. Neutral Red Cell Survival Assay

After removal of medium, cells were washed once with PBS, neutral red (NR) medium (50% DMEM; 50% HBSS (Hanks' Balanced Salt Solution); 50 mg/mL NR dye) was added, and cells were incubated for 60 min at 37 °C followed by washing with PBS, lysis by NResorb solution (49% demineralized water; 50% ethanol 96% and 1% glacial acid), and incubation in the dark for 10 min on a shaker at room temperature. NR absorbance was measured at 540 nm using a Synergy 2 Multi-Mode plate reader. Data are expressed as relative absorbance with controls set to 1; a one-way ANOVA with Šidák's or Dunnett's post hoc test was used for statistical analysis.

Because non-viable HEK293 cells detach from the surface of the well, trypan blue assay for staining of dead cells could not be reliably performed. To ensure adhesion of viable cells, coating of well surfaces with 0.03% poly-l-lysine was performed prior to seeding.

Poly-l-lysine 0.03% or Gelatin 1–2%, unfortunately, did not enable trypan blue assay for cell death analysis.

Table 1. Used Chemicals.

Compound	Concentration	Manufacturer
SUL150	10 μM	Sulfateq B.V.
Ferrostatin-1	100 nM	
Lanthanum Chloride	50 μM	
Ru360 *	5 μM	
Trolox®	100 μM to 3 mM	Sigma-Aldrich
Reduced GSH	3 mM	
Oxidized GSH	3 mM	
DMSO	n.a.	
Z-VAD(OH)-FMK	50 μM	Selleck Chemicals
MG132	10 μM	
RSL3	1.44 and 2.88 μM	TargetMol
Erastin	5 and 10 μM	

All compounds were dissolved in DMSO, for which vehicle controls were performed. * Compounds were prepared after oxygen was removed through nitrogen displacement for 60 min.

4.3. ATP Measurement

ATP was measured by a luciferase assay (Promega). Cells were collected after the addition of an EDTA buffer and scraping on ice. We employed 6 min boiling (95 °C) to lyse the cells and to denature proteins to stop ATP consumption [63]. Then, samples were centrifugated at 11,000 g for 2 min, and supernatant was collected. ATP was measured at 590 nm using a dark plate in a Synergy 2 Multi-Mode plate reader. Bio-Rad DC protein assay was performed to determine protein concentration, for which all read-outs were corrected. Data are expressed as relative luminescence corrected for sample protein concentration with controls set to 1; a one-way ANOVA with Dunnett's post hoc test was used for statistical analysis.

4.4. MDA Measurement

Cells were harvested with PBS with 1% BHT (Butylhydroxytoluol) and immediately placed on ice. Cells were lysed with SDS for 5 min at room temperature and then boiled for 60 min with TBA reagent. MDA was measured at 590 nm using a dark plate in a Synergy 2 Multi-Mode plate reader. Bio-Rad DC protein assay was performed to determine protein concentration, for which all read-outs were corrected. Data are expressed as relative luminescence corrected for sample protein concentration with controls set to 1; a one-way ANOVA with Dunnett's post hoc test was used for statistical analysis.

4.5. Western Blot

Cell lysates were obtained using RIPA lysis buffer (50 mM Tris-Cl pH = 8.0, 150 mM NaCl, 1% Igepal Ca 630, Sodium Deoxycholate, 1.0% SDS, 0.4% protein inhibitor cocktail, 1 mM sodium orthovanadate, 10 mM NaF). Additionally, cells were mechanically lysed through repetitive aspiration through a 30 G needle. Loading buffer (10% SDS, 50% Glycerol, 0.33 M Tris HCl pH = 6.8, 10% beta-mercaptoethanol, 0.05% bromophenol blue) was added (20% of total volume) to all samples, followed by boiling for 5 min at 95 °C. Samples were loaded on 4–20% SDS precast gels (Bio-Rad TGX gels) and transferred to nitrocellulose membrane (Bio-Rad). Membranes were blocked (50 mL TBST, 2.5 g skim milk powder) for 60 min. Antibodies were diluted 1:1000 in TBST containing 3% BSA (bovine serum albumin). Primary antibodies were incubated overnight at 4 °C. Secondary antibodies were incubated at room temperature for 60 min. Membranes were visualized using SuperSignal (PerkinElmer) and ChemiDoc MP imaging system (Bio-Rad). For incubation with a different primary antibody, membranes were incubated with a stripping buffer (45 mL TBS, 5 mL 20% SDS, 350 μL beta-mercaptoethanol) for 30 min at room temperature. Stripping efficiency was tested with the ChemiDoc MP imaging system and SuperSignal. Antibodies used were anti-4-Hydroxynonenal (Abcam, ab46545), anti-Glutathione-Peroxidase-4 (Abcam, ab125066), anti-Mono- and polyubiquitinylated conjugates (Enzo, ENZ-ABS840-0500), GARPO (Dako), RAMPO (Dako). A master sample derived from pooling all samples was run along with intra- and intermembrane protein normalization and comparison. In ImageLab 6.0 (Bio-Rad), densitometric data was normalized using the stain-free normalization method. Data were expressed as relative protein abundance with normothermic controls set to 1. A one-way ANOVA with Dunnett's post hoc test was used for statistical analysis.

4.6. Caspase 3/7 Glow Assay

To test the efficacy of Z-VAD(OH)-FMK, a pan-caspase inhibitor with specific affinity for the final apoptotic effector caspase 6, rat smooth muscle primary cells (SMAC) were cultured in sterile 96-well dark plates until confluent and incubated for 30 min at 37 °C and subsequently cooled to 4 °C for 30, 60, 90, 120, 150, and 180 min. Caspase activity was assessed using the Caspase-Glo® 3/7 Assay (Promega, Madison, WI, USA) according to manufacturer's protocol, and luminescence was measured using a Synergy 2 Multi-Mode plate reader (BioTek). Data were expressed as total luminescence absorbed. A two-way ANOVA with Tukey's multiple comparisons test was used for statistical analysis.

4.7. Efficacy Assay: SUL150 versus Erastin and RSL3

To test the efficacy of SUL150 to directly counteract ferroptosis, we incubated HEK293 cells under normothermic conditions (37 °C) with the known ferroptosis inducers RSL3 (Ras-selective lethal 3) and Erastin with and without 10 μM of SUL150. RSL3 was incubated for 6 h in concentrations of 1.44 and 2.88 μM. Erastin was incubated for 24 h in concentrations of 5 and 10 μM. To assess cell viability and, by proxy, cell death a neutral red survival assay was performed according to the specifications described above. Data are expressed as relative absorbance with controls set to 1; a one-way ANOVA with Šidák's or Dunnett's post hoc test was used for statistical analysis.

5. Conclusions

In a cellular cooling and rewarming model for renal cold preservation, we showed that cell death from hypothermia and/or rewarming is caused by ferroptosis rather than apoptosis. We showed that University of Wisconsin Cold Storage Solution prevents ferroptosis during hypothermia, but lipid peroxidation and ATP depletion rapidly ensue upon rewarming, ultimately ending in complete cell death. We demonstrated that treatment with small-molecule Ferrostatin-1 or the 6-chromanol SUL150 with UW throughout cooling effectively prevents ferroptosis. Both compounds maintain ATP and limit lipid peroxidation. Because treatments are ineffective when administered shortly before or during rewarming, administration from the initiation of cooling onwards is necessary to maintain adequate ATP levels and counteract lipid peroxidation upon rewarming. Both SUL150 and Ferrostatin-1 might be efficacious, simple, and low-cost additions to the current organ preservation solutions to improve static or machine perfusion cold storage preservation tolerance of allografts and reduce complications such as ischemia reperfusion injury. Future studies should investigate the efficacy of targeting ferroptosis before, or concomitant with, the start of organ preservation protocols.

Author Contributions: L.P.G. participated in designing the research, performing the experiments, analyzing the data, interpreting the results, and writing the manuscript. K.D.W.H. participated in performing and analyzing the data and writing the manuscript. F.H.-B. participated in performing the experiments and analyzing the data. A.-L.S. participated in developing the techniques to assess lipid peroxidation. C.P.J. and P.C.V. participated in designing the experiment. G.K. participated in interpreting the results and revising the manuscript. R.H.H. participated in designing the research, interpreting the results, and revising the manuscript. All authors have read and agreed to the published version of the manuscript.

Funding: Sulfateq B.V. (Groningen, The Netherlands) provided SUL150 free of charge; no research funding was provided.

Data Availability Statement: All data generated or analyzed during this study are included in this published article. The datasets of this article are accessible via the Appendix A.

Acknowledgments: Financial support was received from the University Medical Center Groningen (MD/PhD grant to L.P.G. and C.P.J.) and the Dutch Organization for Scientific Research (NWO; NWA.ID.17.093 to R.H.H. and P.C.V.).

Conflicts of Interest: G.K. is the chief scientific officer at Sulfateq B.V. (Groningen, The Netherlands), a company that owns the patent on the 6-chromanol SUL-150 and produces and markets these and similar compounds. R.H.H. is a member of the Scientific Advisory Board of Sulfateq B.V., a company that holds patents on hydroxychromans, and has received honoraria from his employer. P.C.V. is a PhD student at the UMCG and an employee of Sulfateq B.V., and as such has received personal fees.

Appendix A

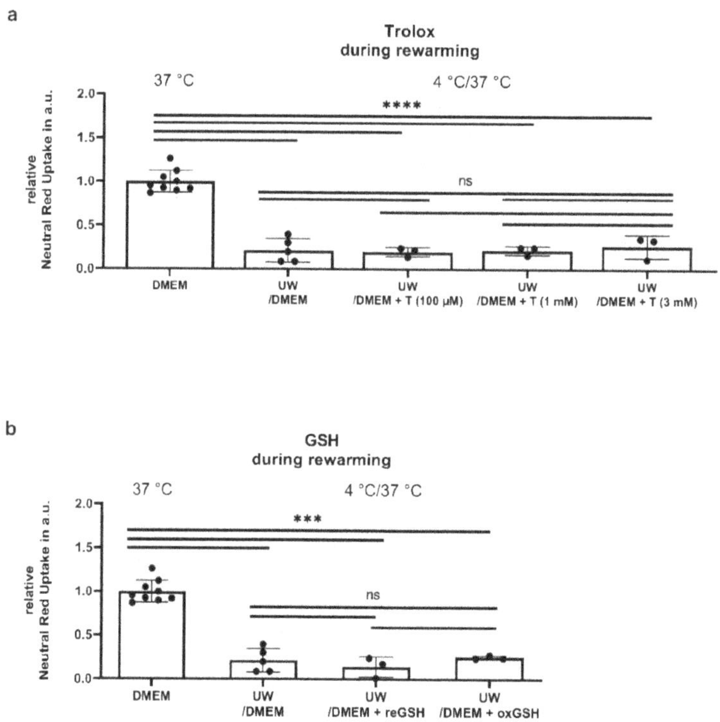

Figure A1. Cell survival in Human Embryonic Kidney (HEK) 293 cells at 24 h cooling and 1 h rewarming (striped background) with antioxidants added to the rewarming medium. (**a**) Addition of Trolox (T) in increasing dosage. (**b**) Addition of reduced GSH (reGSH) or oxidized GSH. All data are expressed relative to standard conditions at 37 °C and presented as means ± SD; one-way ANOVA with Šidák's post hoc test; *** = $p < 0.001$; **** = $p < 0.0001$; n = 3–9.

Supplementation of Trolox does not affect cooling and rewarming outcomes. In keeping with this, there is no difference between oxidized and reduced GSH supplementation; both do not improve cooling and rewarming outcomes.

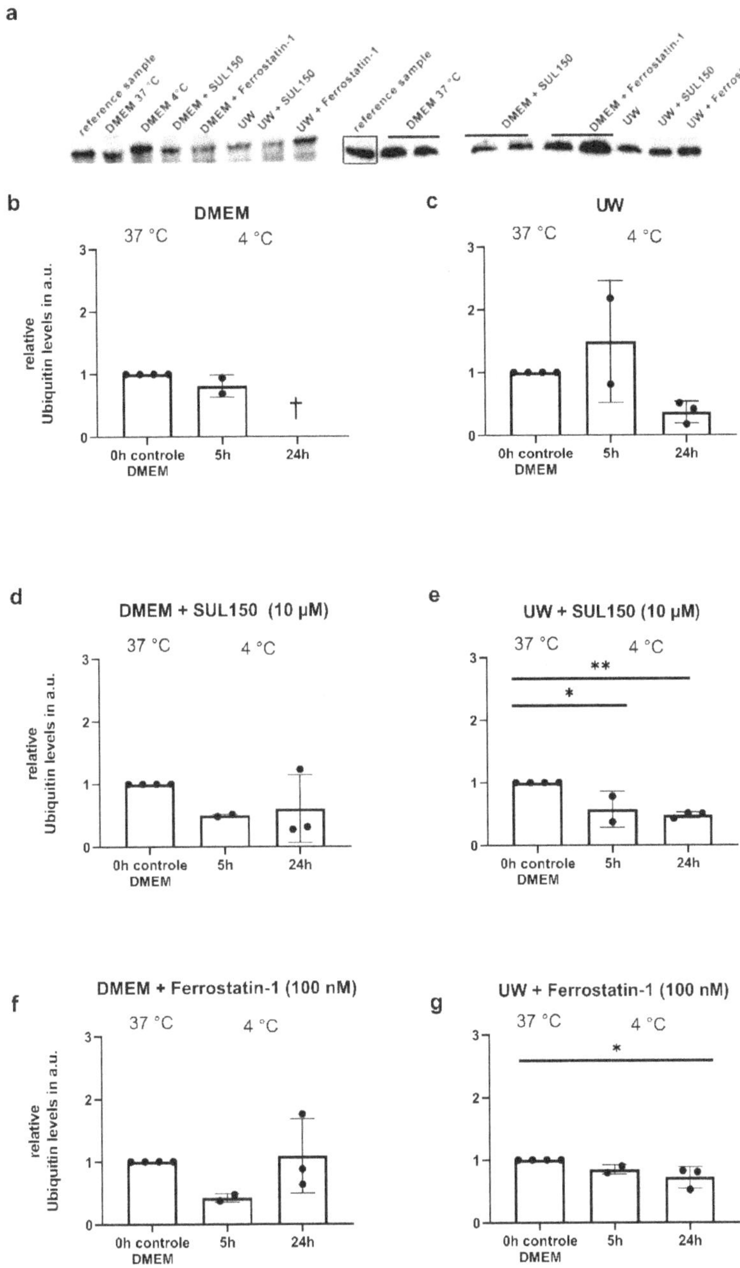

Figure A2. Ubiquitination in Human Embryonic Kidney (HEK) 293 cells at 24 h cooling (blue background). (**a**) In DMEM. (**b**) In UW. (**c**) In DMEM with SUL150 (10 µM). (**d**) In UW with SUL150 (10 µM). (**e**) In DMEM with Ferrostatin-1 (100 nM). (**f**) In DMEM with Ferrostatin-1 (100 nM). (**g**) In UW with Ferrostatin-1 (100 nM). Samples derived from the same experiment and blots were processed in parallel; the same reference sample was used on all blots and is highlighted with a rectangle. All data are expressed relative to standard conditions at 37 °C and presented as means ± SD; one-way ANOVA with Dunnett's (**b**,**c**) or Šidák's (**d**) post hoc test; * = $p < 0.05$; ** = $p < 0.01$; n = 2–6.

In UW-cooled cells, SUL150 (10 µM) significantly decreased protein ubiquitination at 5 and 24 h of cooling. In DMEM, a similar pattern was observed; however, it did not reach significance. Ferrostatin-1 (100 nM) treatment caused a reduction of ubiquitinated proteins after 24 h of cooling. The lower levels of protein ubiquitination during and at the end of cooling may be suggestive of decreased protein damage and breakdown.

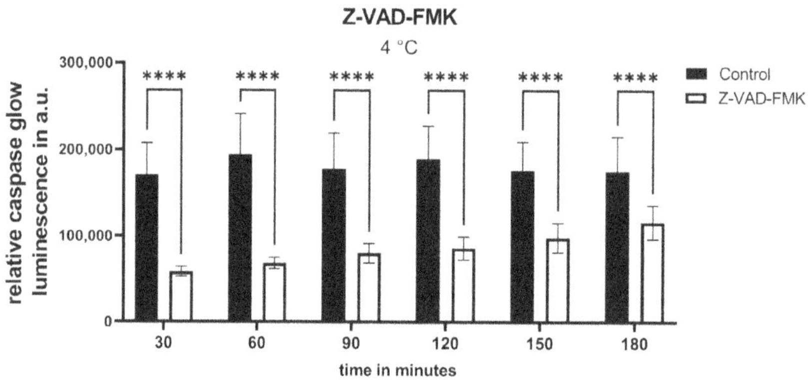

Figure A3. Caspase 3/7 activity in rat smooth muscle primary cells during cooling (blue background) for 30 to 180 min. Z-VAD-FMK concentration was 50 µM. Data are expressed in total caspase glow luminescence in arbitrary units and presented as means ± SD; two-way ANOVA with Tukey's multiple comparisons test; **** = $p < 0.0001$; $n = 4$.

The use of the pan-caspase inhibitor Z-VAD-FMK in a concentration of 50 µM caused effective and prolonged caspase-3 and caspase-7 activity reduction under hypothermic conditions in rat smooth muscle primary cells (SMAC).

SUL150 suppressed cell death evoked by ferroptosis inducers RSL3 and Erastin, which were deliberately administered in low doses to avoid activation of cell death effectors other than ferroptosis. RSL3 triggers ferroptosis by irreversibly inactivating GPx4, leading to an overwhelming accumulation of lipid peroxides. On the other hand, Erastin inhibits the cystine/glutamate antiporter, known as system Xc-, resulting in a reduced level of GSH (the substrate for GPx4), and targets Voltage-Dependent Anion Channels (VDACs). The ability of SUL150 to dampen ferroptosis shows a direct anti-ferroptosis effect of the compound.

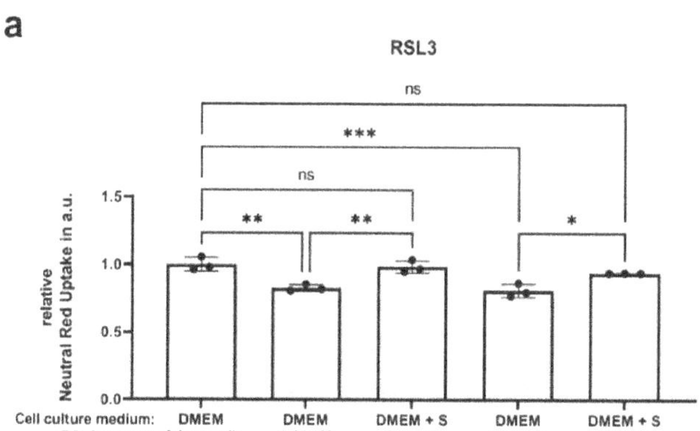

Figure A4. *Cont.*

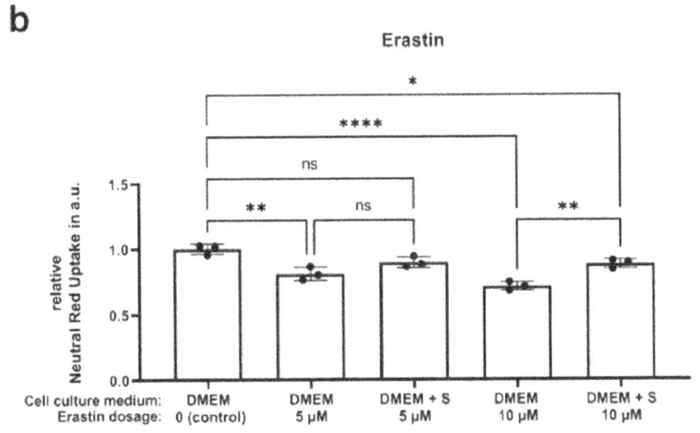

Figure A4. Cell survival in Human Embryonic Kidney (HEK) 293 cells following incubation with Erastin or RSL3 with and without SUL150 under normothermic conditions. (**a**) Incubation for 6 h with RSL3 in concentrations of 1.44 and 2.88 μM. (**b**) Incubation with Erastin in concentrations of 5 and 10 μM. SUL150 was added in a 10 μM concentration. Data are shown as relatives of the control group and presented as means ± SD; one-way ANOVA with Šidák's post hoc analysis; * = $p < 0.5$; ** = $p < 0.01$; *** = $p < 0.001$; **** = $p < 0.0001$; $n = 3$.

References

1. De Rosa, S.; Antonelli, M.; Ronco, C. Hypothermia and kidney: A focus on ischaemia–reperfusion injury. *Nephrol. Dial. Transplant.* **2016**, *32*, 241–247. [CrossRef]
2. Jing, L.; Yao, L.; Zhao, M.; Peng, L.-P.; Liu, M. Organ preservation: From the past to the future. *Acta Pharmacol. Sin.* **2018**, *39*, 845–857. [CrossRef] [PubMed]
3. Steichen, C.; Giraud, S.; Bon, D.; Barrou, B.; Badet, L.; Salamé, E.; Kerforne, T.; Allain, G.; Roumy, J.; Jayle, C.; et al. Barriers and Advances in Kidney Preservation. *BioMed Res. Int.* **2018**, *2018*, 1–15. [CrossRef] [PubMed]
4. van Heurn, L.W.; Talbot, D.; Nicholson, M.L.; Akhtar, M.Z.; Sanchez-Fructuoso, A.I.; Weekers, L.; Barrou, B. Recommendations for donation after circulatory death kidney transplantation in Europe. *Transpl. Int.* **2015**, *29*, 780–789. [CrossRef] [PubMed]
5. Opelz, G.; Döhler, B. Multicenter Analysis of Kidney Preservation. *Transplantation* **2007**, *83*, 247–253. [CrossRef] [PubMed]
6. Dragun, D.; Hoff, U.; Park, J.-K.; Qun, Y.; Schneider, W.; Luft, F.C.; Haller, H. Prolonged cold preservation augments vascular injury independent of renal transplant immunogenicity and function. *Kidney Int.* **2001**, *60*, 1173–1181. [CrossRef]
7. Debout, A.; Foucher, Y.; Trébern-Launay, K.; Legendre, C.; Kreis, H.; Mourad, G.; Garrigue, V.; Morelon, E.; Buron, F.; Rostaing, L.; et al. Each additional hour of cold ischemia time significantly increases the risk of graft failure and mortality following renal transplantation. *Kidney Int.* **2015**, *87*, 343–349. [CrossRef]
8. Peters-Sengers, H.; Houtzager, J.H.; Idu, M.M.; Heemskerk, M.B.; Van Heurn, E.L.; Van Der Heide, J.J.H.; Kers, J.; Berger, S.P.; Van Gulik, T.M.; Bemelman, F.J. Impact of Cold Ischemia Time on Outcomes of Deceased Donor Kidney Transplantation: An Analysis of a National Registry. *Transplant. Direct* **2019**, *5*, e448. [CrossRef]
9. Tennankore, K.K.; Kim, S.J.; Alwayn, I.P.; Kiberd, B.A. Prolonged warm ischemia time is associated with graft failure and mortality after kidney transplantation. *Kidney Int.* **2016**, *89*, 648–658. [CrossRef]
10. Salvadori, M.; Rosso, G.; Bertoni, E. Update on ischemia-reperfusion injury in kidney transplantation: Pathogenesis and treatment. *World J. Transplant.* **2015**, *5*, 52–67. [CrossRef]
11. Hendriks, K.D.W.; Brüggenwirth, I.M.A.; Maassen, H.; Gerding, A.; Bakker, B.; Porte, R.J.; Henning, R.H.; Leuvenink, H.G. Renal temperature reduction progressively favors mitochondrial ROS production over respiration in hypothermic kidney preservation. *J. Transl. Med.* **2019**, *17*, 265. [CrossRef] [PubMed]
12. Hendriks, K.D.W.; Lupi, E.; Hardenberg, M.C.; Hoogstra-Berends, F.; Deelman, L.E.; Henning, R.H. Differences in mitochondrial function and morphology during cooling and rewarming between hibernator and non-hibernator derived kidney epithelial cells. *Sci. Rep.* **2017**, *7*, 15482. [CrossRef] [PubMed]
13. Tolouee, M.; Hendriks, K.D.W.; Lie, F.F.; Gartzke, L.P.; Goris, M.; Hoogstra-Berends, F.; Bergink, S.; Henning, R.H. Cooling of Cells and Organs Confers Extensive DNA Strand Breaks Through Oxidative Stress and ATP Depletion. *Cell Transplant.* **2022**, *31*. [CrossRef] [PubMed]
14. Hendriks, K.D.; Joschko, C.P.; Hoogstra-Berends, F.; Heegsma, J.; Faber, K.-N.; Henning, R.H. Hibernator-Derived Cells Show Superior Protection and Survival in Hypothermia Compared to Non-Hibernator Cells. *Int. J. Mol. Sci.* **2020**, *21*, 1864. [CrossRef] [PubMed]

15. Bayley, J.S.; Winther, C.B.; Andersen, M.K.; Grønkjær, C.; Nielsen, O.B.; Pedersen, T.H.; Overgaard, J. Cold exposure causes cell death by depolarization-mediated Ca^{2+} overload in a chill-susceptible insect. *Proc. Natl. Acad. Sci. USA* **2018**, *115*, E9737–E9744. [CrossRef]
16. Aerts-Kaya, F.S.; Visser, T.P.; Pervin, B.; Mammadova, A.; Özyüncü, Ö.; Wagemaker, G.; Uçkan-Çetinkaya, F.D. SUL-109 Protects Hematopoietic Stem Cells from Apoptosis Induced by Short-Term Hypothermic Preservation and Maintains Their Engraftment Potential. *Biol. Blood Marrow Transplant.* **2020**, *26*, 634–642. [CrossRef] [PubMed]
17. Brinkkoetter, P.-T.; Song, H.; Lösel, R.; Schnetzke, U.; Gottmann, U.; Feng, Y.; Hanusch, C.; Beck, G.; Schnuelle, P.; Wehling, M.; et al. Hypothermic Injury: The Mitochondrial Calcium, ATP and ROS Love-Hate Triangle out of Balance. *Cell. Physiol. Biochem.* **2008**, *22*, 195–204. [CrossRef]
18. Brookes, P.S.; Yoon, Y.; Robotham, J.L.; Anders, M.W.; Sheu, S.-S. Calcium, ATP, and ROS: A mitochondrial love-hate triangle. *Am. J. Physiol. Cell Physiol.* **2004**, *287*, C817–C833. [CrossRef]
19. Aslanidi, K.B.; Aslanidi, G.V.; Vachadze, D.M.; Zinchenko, V.; Yua, L.; Potapova, T.V. A possible role of cold-induced ionic stress in cold-induced cell death. *Membr. Cell Biol.* **1997**, *11*, 57–76. Available online: https://pubmed.ncbi.nlm.nih.gov/9257282/ (accessed on 2 May 2023).
20. Flora, S.J.S.; Pachauri, V. Chelation in Metal Intoxication. *Int. J. Environ. Res. Public Health* **2010**, *7*, 2745–2788. [CrossRef]
21. Knoop, S.; de Groot, H.; Rauen, U. Little evidence for a major role of Ca^{2+} in cold-induced injury of liver cells. *Cryobiology* **2008**, *56*, 103–113. [CrossRef]
22. Rauen, U.; Kerkweg, U.; de Groot, H. Iron-dependent vs. iron-independent cold-induced injury to cultured rat hepatocytes: A comparative study in physiological media and organ preservation solutions. *Cryobiology* **2007**, *54*, 77–86. [CrossRef]
23. Kerkweg, U.; Li, T.; de Groot, H.; Rauen, U. Cold-induced apoptosis of rat liver cells in University of Wisconsin solution: The central role of chelatable iron. *Hepatology* **2002**, *35*, 560–567. [CrossRef]
24. Rauen, U.; Petrat, F.; Li, T.; De Groot, H. Hypothermia injury/cold-induced apoptosis—Evidence of an increase in chelatable iron causing oxidative injury in spite of low O_2^-/H_2O_2 formation. *FASEB J.* **2000**, *14*, 1953–1964. [CrossRef] [PubMed]
25. Cao, J.Y.; Dixon, S.J. Mechanisms of ferroptosis. *Cell. Mol. Life Sci.* **2016**, *73*, 2195–2209. [CrossRef] [PubMed]
26. Dixon, S.J.; Pratt, D.A. Ferroptosis: A flexible constellation of related biochemical mechanisms. *Mol. Cell* **2023**, *83*, 1030–1042. [CrossRef] [PubMed]
27. Dixon, S.J.; Lemberg, K.M.; Lamprecht, M.R.; Skouta, R.; Zaitsev, E.M.; Gleason, C.E.; Patel, D.N.; Bauer, A.J.; Cantley, A.M.; Yang, W.S.; et al. Ferroptosis: An Iron-Dependent Form of Nonapoptotic Cell Death. *Cell* **2012**, *149*, 1060–1072. [CrossRef] [PubMed]
28. Wang, X.; Zhou, Y.; Min, J.; Wang, F. Zooming in and out of ferroptosis in human disease. *Front. Med.* **2023**, *17*, 173–206. [CrossRef]
29. Xie, Y.; Hou, W.; Song, X.; Yu, Y.; Huang, J.; Sun, X.; Kang, R.; Tang, D. Ferroptosis: Process and function. *Cell Death Differ.* **2016**, *23*, 369–379. [CrossRef]
30. Kuang, F.; Liu, J.; Tang, D.; Kang, R. Oxidative Damage and Antioxidant Defense in Ferroptosis. *Front. Cell Dev. Biol.* **2020**, *8*, 586578. [CrossRef]
31. Yang, W.S.; Kim, K.J.; Gaschler, M.M.; Patel, M.; Shchepinov, M.S.; Stockwell, B.R. Peroxidation of polyunsaturated fatty acids by lipoxygenases drives ferroptosis. *Proc. Natl. Acad. Sci. USA* **2016**, *113*, E4966–E4975. [CrossRef]
32. Shintoku, R.; Takigawa, Y.; Yamada, K.; Kubota, C.; Yoshimoto, Y.; Takeuchi, T.; Koshiishi, I.; Torii, S. Lipoxygenase-mediated generation of lipid peroxides enhances ferroptosis induced by erastin and RSL3. *Cancer Sci.* **2017**, *108*, 2187–2194. [CrossRef]
33. Hattori, K.; Ishikawa, H.; Sakauchi, C.; Takayanagi, S.; Naguro, I.; Ichijo, H. Cold stress-induced ferroptosis involves the ASK 1-p38 pathway. *EMBO Rep.* **2017**, *18*, 2067–2078. [CrossRef]
34. Tang, D.; Chen, X.; Kang, R.; Kroemer, G. Ferroptosis: Molecular mechanisms and health implications. *Cell Res.* **2020**, *31*, 107–125. [CrossRef]
35. Galluzzi, L.; Vitale, I.; Aaronson, S.A.; Abrams, J.M.; Adam, D.; Agostinis, P.; Alnemri, E.S.; Altucci, L.; Amelio, I.; Andrews, D.W.; et al. Molecular mechanisms of cell death: Recommendations of the Nomenclature Committee on Cell Death 2018. *Cell Death Differ.* **2018**, *25*, 486–541. [CrossRef]
36. Feng, H.; Stockwell, B.R. Unsolved mysteries: How does lipid peroxidation cause ferroptosis? *PLoS Biol.* **2018**, *16*, e2006203. [CrossRef] [PubMed]
37. Cozza, G.; Rossetto, M.; Bosello-Travain, V.; Maiorino, M.; Roveri, A.; Toppo, S.; Zaccarin, M.; Zennaro, L.; Ursini, F. Glutathione peroxidase 4-catalyzed reduction of lipid hydroperoxides in membranes: The polar head of membrane phospholipids binds the enzyme and addresses the fatty acid hydroperoxide group toward the redox center. *Free. Radic. Biol. Med.* **2017**, *112*, 1–11. [CrossRef] [PubMed]
38. Friedmann Angeli, J.P.; Schneider, M.; Proneth, B.; Tyurina, Y.Y.; Tyurin, V.A.; Hammond, V.J.; Herbach, N.; Aichler, M.; Walch, A.; Eggenhofer, E.; et al. Inactivation of the ferroptosis regulator Gpx4 triggers acute renal failure in mice. *Nat. Cell Biol.* **2014**, *16*, 1180–1191. [CrossRef]
39. Zhang, Y.; Swanda, R.V.; Nie, L.; Liu, X.; Wang, C.; Lee, H.; Lei, G.; Mao, C.; Koppula, P.; Cheng, W.; et al. mTORC1 couples cyst(e)ine availability with GPX4 protein synthesis and ferroptosis regulation. *Nat. Commun.* **2021**, *12*, 1–14. [CrossRef]
40. Seiler, A.; Schneider, M.; Förster, H.; Roth, S.; Wirth, E.K.; Culmsee, C.; Plesnila, N.; Kremmer, E.; Rådmark, O.; Wurst, W.; et al. Glutathione Peroxidase 4 Senses and Translates Oxidative Stress into 12/15-Lipoxygenase Dependent- and AIF-Mediated Cell Death. *Cell Metab.* **2008**, *8*, 237–248. [CrossRef] [PubMed]

41. Miotto, G.; Rossetto, M.; Di Paolo, M.L.; Orian, L.; Venerando, R.; Roveri, A.; Vučković, A.-M.; Travain, V.B.; Zaccarin, M.; Zennaro, L.; et al. Insight into the mechanism of ferroptosis inhibition by ferrostatin-1. *Redox Biol.* **2019**, *28*, 101328. [CrossRef] [PubMed]
42. Xie, B.; Wang, Y.; Lin, Y.; Mao, Q.; Feng, J.; Gao, G.; Jiang, J. Inhibition of ferroptosis attenuates tissue damage and improves long-term outcomes after traumatic brain injury in mice. *CNS Neurosci. Ther.* **2018**, *25*, 465–475. [CrossRef]
43. Vogelaar, P.C.; Roorda, M.; de Vrij, E.L.; Houwertjes, M.C.; Goris, M.; Bouma, H.; van der Graaf, A.C.; Krenning, G.; Henning, R.H. The 6-hydroxychromanol derivative SUL-109 ameliorates renal injury after deep hypothermia and rewarming in rats. *Nephrol. Dial. Transplant.* **2018**, *33*, 2128–2138. [CrossRef] [PubMed]
44. Hajmousa, G.; Vogelaar, P.; Brouwer, L.A.; van der Graaf, A.C.; Henning, R.H.; Krenning, G. The 6-chromanol derivate SUL-109 enables prolonged hypothermic storage of adipose tissue-derived stem cells. *Biomaterials* **2017**, *119*, 43–52. [CrossRef]
45. Lambooy, S.P.H.; Bidadkosh, A.; Nakladal, D.; van Buiten, A.; Girgis, R.A.T.; van der Graaf, A.C.; Wiedenmann, T.J.; Koster, R.A.; Vogelaar, P.; Buikema, H.; et al. The Novel Compound Sul-121 Preserves Endothelial Function and Inhibits Progression of Kidney Damage in Type 2 Diabetes Mellitus in Mice. *Sci. Rep.* **2017**, *7*, 11165. [CrossRef]
46. Petrenko, A.; Carnevale, M.; Somov, A.; Osorio, J.; Rodríguez, J.; Guibert, E.; Fuller, B.; Froghi, F. Organ Preservation into the 2020s: The Era of Dynamic Intervention. *Transfus. Med. Hemotherapy* **2019**, *46*, 151–172. [CrossRef] [PubMed]
47. Parsons, R.F.; Guarrera, J.V. Preservation solutions for static cold storage of abdominal allografts. *Curr. Opin. Organ Transplant.* **2014**, *19*, 100–107. [CrossRef]
48. Compare Belzer UW®Cold Storage Solution to ViaSpan | Bridge to Life. Available online: https://bridgetolife.com/compare-belzer-uw-cold-storage-solution-to-viaspan/ (accessed on 2 May 2023).
49. Belzer, F.O.; Southard, J.H. Principles of Solid-Organ Preservation by Cold Storage. *Transplantation* **1988**, *45*, 673–676. [CrossRef]
50. Ben Abdennebi, H.; Steghens, J.-P.; Margonari, J.; Ramella-Virieux, S.; Barbieux, A.; Boillot, O. High-Na + low-K + UW cold storage solution reduces reperfusion injuries of the rat liver graft. *Transpl. Int.* **1998**, *11*, 223–230. [CrossRef]
51. Southard, J.H.; van Gulik, T.M.; Ametani, M.S.; Vreugdenhil, P.K.; Lindell, S.L.; Pienaar, B.L.; Belzer, F.O. Important components of the uw solution. *Transplantation* **1990**, *49*, 251–257. [CrossRef]
52. Hernández-Gallardo, A.K.; Missirlis, F. Loss of ferritin in developing wing cells: Apoptosis and ferroptosis coincide. *PLoS Genet.* **2020**, *16*, e1008503. [CrossRef] [PubMed]
53. Venema, L.H.; Brat, A.; Moers, C.; Hart, N.A.; Ploeg, R.J.; Hannaert, P.; Minor, T.; Leuvenink, H. Effects of Oxygen During Long-term Hypothermic Machine Perfusion in a Porcine Model of Kidney Donation After Circulatory Death. *Transplantation* **2019**, *103*, 2057–2064. [CrossRef]
54. Jochmans, I.; Brat, A.; Davies, L.; Hofker, H.S.; van de Leemkolk, F.E.M.; Leuvenink, H.G.D.; Knight, S.R.; Pirenne, J.; Ploeg, R.J.; Abramowicz, D.; et al. Oxygenated versus standard cold perfusion preservation in kidney transplantation (COMPARE): A randomised, double-blind, paired, phase 3 trial. *Lancet* **2020**, *396*, 1653–1662. [CrossRef] [PubMed]
55. Bellini, M.I.; Yiu, J.; Nozdrin, M.; Papalois, V. The Effect of Preservation Temperature on Liver, Kidney, and Pancreas Tissue ATP in Animal and Preclinical Human Models. *J. Clin. Med.* **2019**, *8*, 1421. [CrossRef] [PubMed]
56. Garcia, A.; Eljack, N.D.; Sani, M.-A.; Separovic, F.; Rasmussen, H.H.; Kopec, W.; Khandelia, H.; Cornelius, F.; Clarke, R.J. Membrane accessibility of glutathione. *Biochim. Biophys. Acta (BBA) Biomembr.* **2015**, *1848*, 2430–2436. [CrossRef]
57. Cacciatore, I.; Cornacchia, C.; Pinnen, F.; Mollica, A.; Di Stefano, A. Prodrug Approach for Increasing Cellular Glutathione Levels. *Molecules* **2010**, *15*, 1242–1264. [CrossRef]
58. Kelly-Aubert, M.; Trudel, S.; Fritsch, J.; Nguyen-Khoa, T.; Baudouin-Legros, M.; Moriceau, S.; Jeanson, L.; Djouadi, F.; Matar, C.; Conti, M.; et al. GSH monoethyl ester rescues mitochondrial defects in cystic fibrosis models. *Hum. Mol. Genet.* **2011**, *20*, 2745–2759. [CrossRef]
59. Bilzer, M.; Paumgartner, G.; Gerbes, A.L. Glutathione protects the rat liver against reperfusion injury after hypothermic preservation. *Gastroenterology* **1999**, *117*, 200–210. [CrossRef]
60. Czubak, K.; Antosik, A.; Cichon, N.; Zbikowska, H.M. Vitamin C and Trolox decrease oxidative stress and hemolysis in cold-stored human red blood cells. *Redox Rep.* **2017**, *22*, 445–450. [CrossRef]
61. Turell, L.; Zeida, A.; Trujillo, M. Mechanisms and consequences of protein cysteine oxidation: The role of the initial short-lived intermediates. *Essays Biochem.* **2020**, *64*, 55–66. [CrossRef]
62. Schulte, L.; Mao, J.; Reitz, J.; Sreeramulu, S.; Kudlinzki, D.; Hodirnau, V.-V.; Meier-Credo, J.; Saxena, K.; Buhr, F.; Langer, J.D.; et al. Cysteine oxidation and disulfide formation in the ribosomal exit tunnel. *Nat. Commun.* **2020**, *11*, 1–11. [CrossRef] [PubMed]
63. Yangab, N.C.; Hob, W.M.; Chenc, Y.H.; Hu, M.-L. A Convenient One-Step Extraction of Cellular ATP Using Boiling Water for the Luciferin–Luciferase Assay of ATP. *Anal. Biochem.* **2002**, *306*, 323–327. [CrossRef] [PubMed]

Disclaimer/Publisher's Note: The statements, opinions and data contained in all publications are solely those of the individual author(s) and contributor(s) and not of MDPI and/or the editor(s). MDPI and/or the editor(s) disclaim responsibility for any injury to people or property resulting from any ideas, methods, instructions or products referred to in the content.

International Journal of
Molecular Sciences

MDPI

Review

Nitroxidative Stress, Cell—Signaling Pathways, and Manganese Porphyrins: Therapeutic Potential in Neuropathic Pain

Álvaro José Chávez Silva and Mário Sérgio Lima de Lavor *

Department of Agricultural and Environmental Sciences, State University of Santa Cruz (UESC), Ilhéus 45662-900, BA, Brazil; alchasi91@gmail.com
* Correspondence: msllavor@uesc.br

Abstract: Neuropathic pain, a debilitating condition arising from somatosensory system damage, significantly impacts quality of life, leading to anxiety, self-mutilation, and depression. Oxidative and nitrosative stress, an imbalance between reactive oxygen and nitrogen species (ROS/RNS) and antioxidant defenses, plays a crucial role in its pathophysiology. While reactive species are essential for physiological functions, excessive levels can cause cellular component damage, leading to neuronal dysfunction and pain. This review highlights the complex interactions between reactive species, antioxidant systems, cell signaling, and neuropathic pain. We discuss the physiological roles of ROS/RNS and the detrimental effects of oxidative and nitrosative stress. Furthermore, we explore the potential of manganese porphyrins, compounds with antioxidant properties, as promising therapeutic agents to mitigate oxidative stress and alleviate neuropathic pain by targeting key cellular pathways involved in pain. Further research is needed to fully understand their therapeutic potential in managing neuropathic pain in human and non-human animals.

Keywords: chronic pain; manganese porphyrin; neuropathic pain; oxidative stress; UPR

1. Introduction

Chronic pain is a major source of human suffering and disability [1]. Although pain itself and many diseases associated with chronic pain are not immediately fatal, they significantly reduce the quality of life of adults and animals worldwide. This impact is especially severe in patients with neuropathy, in whom sleep disorders, anxiety, and depression are frequent and intense, constituting a risk factor in suicide attempts [2,3]. Neuropathic pain is a specific type of chronic pain, described by persistent or intermittent spontaneous pain, and increased response to mechanical and thermal stimuli, either innocuous or noxious (allodynia or hyperalgesia, respectively), that arises from lesions to the somatosensory nervous system of peripheral or central origin [4], ectopic activity in sensorial fibers, and an imbalance of inhibitory and excitatory neurotransmitters relation, with the involvement of neuroinflammation and nitroxidative stress playing a pivotal role in its instauration and maintenance [5,6]. It is associated with the need for more prescription medications to obtain desired effects, which can lead to adverse effects due to increased doses and frequent visits to health professionals [7,8].

Although neuropathic pain is a multifaceted condition driven by several cellular and molecular mechanisms, those considered key to its development and perpetuation are neuroinflammation, glial cell activation, the release of proinflammatory cytokines, and oxidative stress [5], contributing to altered signaling pathways, which facilitates the expression of further inflammatory mediators and enhanced sensitivity to pain, ultimately, maintaining the cycle of pain where neuroinflammation and oxidative stress influence each other [6,9].

Oxidative stress is known as the imbalance between the cellular oxidative and antioxidant systems, and is widely recognized in chronic pain conditions, since excessive ROS production, exceeding the capacity of cellular antioxidant systems, can result in a dysfunction of the main cellular mitochondrial and endoplasmic reticulum (ER) systems, increasing ROS/PRN concentration during stress conditions [10–13].

Free radicals are molecules highly reactive with unpaired electrons in the outer valence, which makes them inherently chemically unstable and prone to react with surrounding molecules to achieve stabilization [14]. Free radicals can be produced exogenously from ionizing radiation, ozone, cigarette smoking, or by-products of environmental metabolisms. Endogenous sources, such as those derived from oxygen (reactive oxygen species, ROS) or nitrogen (reactive nitrogen species, RNS), are produced through enzymatical or non-enzymatical processes, for instance, in the electron transfer chain (ETC), phagocytosis, cytochrome P-450 system, and prostaglandin (PG) synthesis during normal neuronal functioning or during neuroinflammation processes associated with neuropathic pain, either accepting an electron from or donating an electron to other molecules, consequently, acting also as oxidants [15,16]. Such chemical behavior, when excessive, generates an environment suitable for damage on molecules such as DNA, carbohydrates, lipids, and proteins, leading to homeostasis disruption and cellular damage, and ultimately leading to apoptosis by lowering enzymatic and non-enzymatic antioxidants and compromising cell viability [17,18].

A great amount of information has been developed around free radical chemistry, leading to a better understanding of their biological significance, beyond the concept of harmful uncontrolled molecules. Importantly, it is known that free radicals originate in normal physiological conditions, and have several physiological functions in cellular growth, differentiation, migration, apoptosis, and necrosis, as well as defense against pathogens and species evolution [16,19] Additionally, increasing information sustains the role of reactive species in normal synaptic plasticity and cognitive functions such as memory and learning [20,21], and transducing agonistic modulation during neuropathic pain [22].

Several metalloporphyrins have been identified as potent catalysts for numerous redox reactions; particularly manganese porphyrins (MnPs), which have been found to be efficient antioxidant catalysts [23,24]. These substances dismute superoxide ions at some of the highest rates known for synthetic catalysts and have demonstrated therapeutic effects in conditions where increased ROS/PRN production is a key pathological factor [25], overcoming major conventional monotherapy approaches, usually unsuccessful and accompanied with cumulative adverse effects.

Research into the therapeutic potential of MnPs with antioxidant properties studied in disease models involving oxidative stress [26,27] and in models of chronic pain [28] considering their structural, chemical, pharmacokinetic, and pharmacodynamic properties, is necessary to elucidate the clinical uses of MnPs, based on understanding the pathways and mechanisms that underlie chronic pain of neuropathic origin.

This review examines the nature of ROS/RNS, endogenous antioxidant systems, nitroxidative signaling in neuropathic pain, and the potential of synthetic MnPs in the restitution of neuronal redox homeostasis, which is crucial for understanding scavengers' therapeutic strategies, considering that neuropathic pain conditions still face significant limitation towards effective treatments on human and non-human animals.

2. Reactive Oxygen and Nitrogen Species

Molecular oxygen and nitrogen are undeniably significant in biology, being essential for proper metabolism, cellular function, and the physiological state of eustress in living organisms. While these molecules play critical roles in structural and functional processes, such as signal transduction, gene transcription and expression, and other cellular activities, they can also exert harmful effects on biomolecules in the form of reactive oxygen and nitrogen species, thereby illustrating their dual physiological and pathological nature [29].

Since most of reactive species' effects rely on their compartmental concentration and their chemical nature, it is relevant to understand their main subcellular sources, allowing the compartmentalization of the reactive species gradients, redox-mediated regulation, and role in cell-signaling from several membrane receptors [19,30,31]. In this section, we discuss the chemistry and sources of ROS and RNS in the context of neuropathic pain.

2.1. Reactive Oxygen Species: Chemistry and Sources

Oxygen stands by far as the most vital element, for it bears the basis of biological oxidations, where most of the available energy for aerobic organisms is produced [32]. Yet in excessive amounts, it rends toxic effects, at both cellular and systemic levels [33]. The well-known toxic effect of oxygen is attributed to its chemical structure. Molecular oxygen in its simplest form has a unique electronic configuration, which makes it a radical itself. When an electron is added to the primary structure, the superoxide anion ($O_2\bullet-$), also known as the primary ROS, is formed. $O_2\bullet-$ will interact with other available molecules to generate secondary ROS [34]. Given that ROS are electrically charged, they seek chemical neutralization by reacting with other molecules, causing oxidation, either reversibly or irreversibly, intrinsically increasing the phosphorylation (activity) of various proteins kinases and reducing the phosphatase activity of other proteins; the overall effect appears to enhance neuronal responsivity to sensorial inputs, although, high sustained levels of ROS might excessively oxidize proteins and membrane lipids, causing an abnormal responsiveness of neurons to afferent inputs, inducing central sensitization and pain [35,36]. These deleterious effects occur due to the structural alteration of organic biomolecules, such as protein oxidation, DNA strand breakage through guanine residue oxidation, RNA oxidation, the oxidation of polyunsaturated fatty acids in bilipid membranes, mitochondrial depolarization causing, consequently, apoptosis, and oxidative stress [6,19,37]. Among the most relevant reactive species containing oxygen, exist the hydroxyl radical ($HO\bullet$), superoxide radical anion ($O_2\bullet-$), hydroperoxyl radical ($HOO\bullet$), and peroxyl radicals ($ROO\bullet$), and the non-radicals hydrogen peroxide (H_2O_2) and singlet oxygen ($^1O_2$), which in a series of transformations, involving the gain or loss of electrons, lead to changes in the charge and reactivity of the molecules (Figure 1) [38,39].

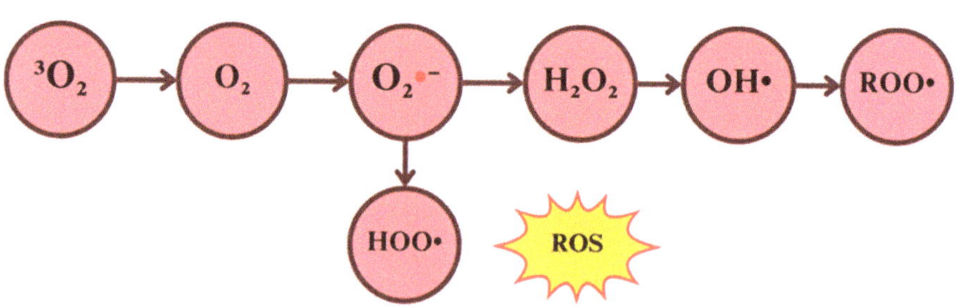

Figure 1. Reactive oxygen species transformation from triplet oxygen ($^3O_2$), singlet oxygen (atmospheric) ($^1O_2$), superoxide anion ($O_2\bullet-$), hydroperoxyl radical ($HOO\bullet$), hydrogen peroxide (H_2O_2), hydroxyl radical ($HO\bullet$), and peroxyl radicals ($ROO\bullet$). ROS: reactive oxygen species.

2.1.1. Singlet Oxygen

O_2 is considered a radical molecule due to the possession of two unpaired electrons with identical spin quantum numbers. The most stable form is triplet oxygen ($^3\Sigma g^-O_2$), also written $^3O_2$, as commonly present in the atmosphere (Figure 2). It acts as a potent oxidant, and its spin arrangement enables it to readily interact with other molecules by accepting electrons individually, rather than releasing them. This specific characteristic clarifies why O_2 tends to react more outstandingly with radicals than non-radicals [40].

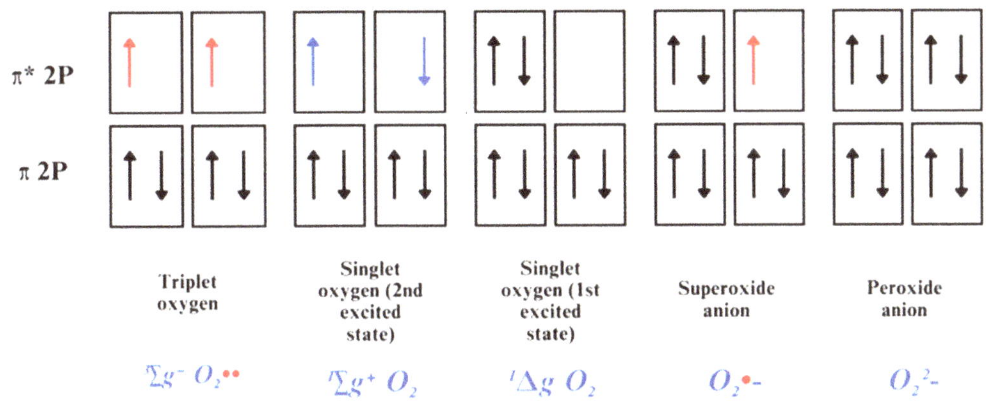

Figure 2. Outer valence and spin distribution of molecular oxygen in different electronic states. π^*: antibonding orbital, π: bonding orbital, 2P: second level main energy of oxygen atomic orbital. Red arrows correspond to unpaired electrons with identical spin, blue arrows correspond to unpaired electrons with opposite spin, black arrows correspond to pair electrons.

The induction of an energy input can lead to the restructuration of electrons' arrangement; once one of the unpaired electrons gains energy, the electron is excited and changes its spin, resulting in the creation of a more reactive dioxygen molecule, known as singlet oxygen [41,42]. This process eliminates unpaired electron spin restrictions, thereby enhancing its oxidative capacity, enabling interactions with proteins, DNA, and lipids [38].

2.1.2. Superoxide Radical Anion

When an electron is added to the outer O_2 orbital, containing an unpaired electron, a less radical molecule is formed, the superoxide radical anion ($O_2\bullet-$). This molecule possesses a single unpaired electron; therefore, it is less reactive than its preceding singlet oxygen [38]. $O_2\bullet-$ is considered a "primary" ROS, produced during the ETC in mitochondria, via electron leakage towards molecular oxygen, in the process of oxidative phosphorylation which is known to be implicated in the pathophysiology of a myriad of diseases [43]. It can readily continue interacting with other molecules to produce "secondary" ROS, by means of direct interaction or via enzyme or metal-catalyzed processes [34,44].

In physiological conditions, most of the superoxide radical is present in its anionic form ($O_2\bullet-$), and only 0.6% is present in its protonated form, the hydroperoxyl radical (HOO•); nonetheless, the catalyzed reaction with HOO•− is a lot faster ($k = 7 \times 10^9$ $M^{-1}{\cdot}S^{-1}$) than the uncatalyzed reaction involving two $O_2\bullet-$ molecules ($k < 0.3\ M^{-1}{\cdot}S^{-1}$) to produce the dismutation of both radical forms, to H_2O_2 and O_2, serving as a great source of H_2O_2 [45], which is of importance in the context of DNA damage, once $O_2\bullet-$ does not react directly with DNA [44,46]. Much of the nuclear DNA damage is restored by the cell, yet, mitochondrial DNA (mtDNA) cannot be fixed in the same manner, which leads to accumulated damage over time, mitochondrial damage, and cell death [44].

Increased levels of $O_2•-$ have been largely observed in neuropathic pain conditions in dorsal horn neurons of rats' spinal cords after nerve ligation [35,47], capsaicin-induced hyperalgesia [48,49], phenyl N-tert-butylnitrone (PBN) and 4-hydroxy-2,2,6,6-tetramethylpiperidine 1-oxyl (TEMPOL) [49–51], and intraplantar injection of $O_2•-$ [51], and have been found to mediate the development and maintenance of central and peripheral sensitization, either increasing nociceptive neurons' excitability, activating calcium/calmodulin-dependent kinase II in glutamatergic spinal neurons, which leads to the presynaptic inhibition of GABAergic interneurons, causing disinhibition and enhancing neuroexcitability in pain signaling pathways and possible excitotoxicity [52,53].

2.1.3. Hydrogen Peroxide

H_2O_2 is a very unstable and slowly decomposing molecule. It comprises an uncharged oxygen–oxygen (O-O) bond peroxide, with a pKa $\pm$ 10.8 at neutral pH, making it able to cross through cellular membranes. It has a high oxidizing power, yet it holds a rather slow rate of reaction with surrounding molecules; such characteristics enable it to exert its radical properties distant from the place it was formed, and confer an elevated capability to accumulate in cells, in relatively excessive concentrations [54].

The main source of H_2O_2 is formed by the dismutation of $O_2•-$ through the SOD-catalyzed reaction. Also, the NADPH-oxidase system is responsible for H_2O_2 production, since it performs the catalysis of NAPDH oxidation by O_2, which yields NAD^+ and H_2O_2 [29], as well as other enzymatic sources, such as glucose oxidase, amino acid oxidase, urate oxidase, and others [46].

Given that H_2O_2 is a by-product of oxidative stress in cell metabolism, for a long time it was considered a strictly harmful and undesired molecule, and it was not until a few decades ago that its role in several biological processes, such as cell differentiation, proliferation, inflammation, wound healing, and more, proved its value as a key redox signaling molecule [55,56]. Additionally, H_2O_2 plays a pivotal role in growth factor-induced signal transduction, thiol redox homeostasis, and mitochondrial function [46].

The physiological or pathological role of the molecule will drastically depend upon the cellular type, subcellular compartment, specific metabolic context, rate of production, and clearance. At micromolar concentrations it appears reactive, whilst at high concentrations it can trigger serious damage, by inactivating the glycolytic enzyme glyceraldehyde-3-phosphate dehydrogenase [54,57]. H_2O_2 is capable of directly oxidizing a significant number of molecules and inactivating enzymes containing thiol groups (RSH) or methionine (Met) residues in their active sites [58]. Moreover, H_2O_2 reacts efficiently with many other radicals, markedly NO, inducing the formation of peroxynitrite ($ONOO^-$) [19].

H_2O_2 can increase the action potential frequency and amplitude in neurons of the dorsal root ganglion (DRG) in neuropathic models [59] and in neurons from within the subnucleus caudalis of the spinal trigeminal nucleus, causing facial pain induced by formalin injection in the lips of rats [60]. Additionally, H_2O_2 can trigger cGMP-dependent kinase Iα (cGKI), causing an increased neurotransmitter release from sensory neurons in the dorsal horn of the spinal cord of neuropathic rats [61], thus, contributing to glutamate homeostasis disturbance, which can potentially cause exacerbated synaptic conductance and amplified Ca^+ influx in neuronal terminals, through the phosphorylation of subunit receptors of N-methyl-D-aspartate (NMDA), inhibition of glutamate transporter (GLT-1), and glutamine synthetase, instigating an advance of excitotoxicity, key in the development of neuropathic pain [62].

After the SOD-catalyzed dismutation reaction of H_2O_2, and under unhinged redox conditions, unbound metal ions, i.e., Fe^{3+}, can catalyze the heterolytic cleavage of H_2O_2, resulting in the formation of hydroxyl radicals (OH•) and anion hydroxide (OH^-), through a well-known reaction referred to as the Fenton reaction ($k = 7.6 \times 10^1\ M^{-1} \cdot S^{-1}$) [34,63].

2.1.4. Hydroxyl Radical

The hydroxyl radical (OH•) is one of the most reactive radicals in biological systems, with a lifespan of 10^{-8}–10^{-9} s. It can be produced by means of redox metal-catalyzed decomposition of H_2O_2, in the presence of transition metal ions, especially when these metal ions are iron or copper [64,65].

OH• may interact with other surrounding molecules by the abstraction of H^+, transfer of electrons, or addition reactions, generating other radicals, although less reactive. Due to this high reactivity, OH• affects molecules' non-selectively, by extracting carbon-bound H^+ on relatively simple molecules, whether abundant or not, by attacking lipidic membranes which triggers oxidizing radical reactions, and by addition to nitrogenous bases, such as the deoxyribosyl backbone of DNA, which leads to oxidation, damaged bases, or strand break [44,66]. A direct link was established between the hydroxyl radical and DNA strand damage to neurons in the context of neuropathic pain induced by chronic constriction injury (CCI) in rats, where an increase in cellular dysfunction and apoptosis in neuropathic animals was observed and promptly reduced in animals treated with an aqueous extract of *Luehea divaricate* containing ROS scavengers. DNA strand break causes an activation of pain pathways that are also directly activated by hydroxyl radicals, and that can further release proinflammatory cytokines, increasing pain sensitization [67]. The hydroxyl radical is relevant for the attenuation of long-term potentiation and facilitation of long-term depression in GABAergic inhibitory interneurons, primarily contributing to pain through GABAergic disinhibition in the dorsal horn leading to enhanced excitatory transmission, hence resulting in neuropathic pain [68].

2.2. Reactive Nitrogen Species: Shemistry and Sources

Nitric Oxide and Peroxynitrite

Nitric oxide (NO•) is endogenously synthetized through a five-electron oxidative reaction involving L-arginine metabolization to L-citrulline, by specific nitric oxide synthases (NOSs). NO• has a role in a variety of physiological processes related to neurotransmission, synaptic plasticity in the CNS, smooth muscle relaxation, vascular tone, platelet aggregation inhibition, leukocyte adhesion inhibition, and antimicrobial immune response [69,70]. Due to its hydrophobicity, NO• can freely migrate through the cytoplasm and cellular membranes, and react distantly from its site of formation, and has a long lifespan (1–10 s) compared to other molecules [71].

Four main isoforms of NOSs have been recognized and located: a mitochondrial NOS (mtNOS) [69]; a neuronal NOS (nNOS), expressed in the postsynaptic terminal of neurons and Schwann cells; an endothelial NOS (eNOS), highly expressed in DRG consistently with allodynia, constitutively expressed and regulated by Ca^{2+}/Calmodulin interaction; and an inducible NOS (iNOS), induced as a response to inflammation, trauma, or infection, which is not regulated by Ca^{2+}, and highly expressed in the cytosol of glial cells [72–74]. The NO• produced by each isoform will perform different and specific functions, depending on the place where they are produced, e.g., the regulation of mitochondrial oxygen consumption and transmembrane potential via a reversible reaction with cytochrome c oxidase (mtNOS), communication among neurons (nNOS), the relaxation of blood vessels and maintenance of blood pressure in endothelium (eNOS), and contribution to defensive mechanisms of macrophages (iNOS) [73].

In the context of oxidative stress, the cytotoxic potential of NO• is due to the secondary production of oxidants derived from NO•−, which are by far more reactive, rather than the oxidative potential itself [75]. NO_2• can interact with NO• to yield dinitrogen trioxide (N_2O_3), which is even more reactive than the priors, with a high ability to cause protein nitrosation and deamination reactions [76]. Therefore, NO• oxidizing potential is mainly related to its ability to interact with O_2•− once they react to producing significant amounts of a much more oxidative molecule, the peroxynitrite anion ($ONOO^-$) [77].

Regarding ONOO , it is a short-lived (10 ms) one- or two-electron oxidant radical, produced by the interaction of oxygen and NO•, serving as the main NO•-derived oxidant. This process occurs in a competitive environment of O_2•− against the enzymatic SOD-catalyzed dismutation to H_2O_2, NO• migration across the cell, NO• preferential reaction with guanylate cyclase, and finally, the reaction of O_2•− with NO•, to ultimately yield $ONOO^-$ in a potentially faster reaction when NO• concentrations increase (Figure 3) [76].

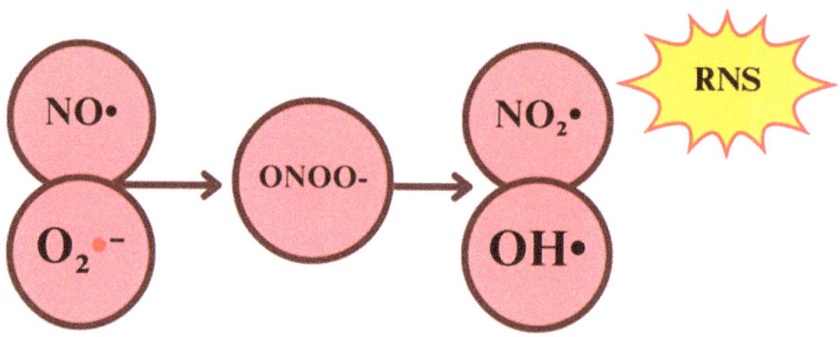

Figure 3. Peroxynitrite ($ONOO^-$) formation from nitric oxide (NO•) and anion superoxide O_2•−), and subsequent formation of hydroxyl radical (OH•) and nitrogen dioxide (NO_2•) from peroxynitrite. RNS: reactive nitrogen species.

The ratio of the competitive environment determines the switch from physiological functions to pathological pathways signaling, once $ONOO^-$ can act either as a cytotoxic effector molecule against pathogens or as an endogenous cytotoxic molecule towards host cells, based on the kinetics setting [78].

$ONOO^-$ is able to engage with biomolecules like DNA, proteins, and lipids that lead to structural modifications in biomolecules, resulting in dysfunction such as interference with signaling pathways through protein amino acid nitration. The two major direct ways of $ONOO^-$ acting on biomolecules are through one-electron (with transition metals) or two-electron (with thiols and CO_2) oxidation reactions in an aqueous milieu [79].

On the other hand, $ONOO^-$ can indirectly induce oxidation or nitration in amino acids including cysteine (Cys), tyrosine (Tyr), and methionine (Met) [76], as well as freely diffuse though lipidic membranes (ONOOH), and react via its secondary radicals with proteins and lipids in the hydrophobic milieu, thus, negatively impacting protein structural and functional roles, or inducing lipid peroxidation [80].

Both $ONOO^-$ and NO• play significant roles in the settlement and perpetuation of neuropathic pain, through several molecular mechanisms, as described above; nevertheless, they can specifically contribute to the process through neuroexcitability enhancement of nociceptive neurons, activating Calcium/calmodulin-dependent protein kinase II (CaMKII) in excitatory glutamatergic neurons, potentiating synaptic transmission, increasing calcium influx, or causing GABAergic interneurons inhibition, leading to neuronal disinhibition [62,81]. Moreover, both species can cause transient receptor potencial (TRP) channel activation [82], critical for pain processing, preceding membranal protein and lipid nitrosative modifications, and Toll-like receptors (TLR) activation (iNOS) [37], consequently inducing nociceptive hypersensitivity. Mitochondrial stress is a major contributor to neuropathic pain conditions, and it has been seen that elevated levels of NOS- and NOX-derived species can cause mitochondrial homeostasis disturbance, and neuronal degeneration (BENNETT, 2014). As an overall effect, both species contribute to central and peripheral sensitization following nerve injury.

2.3. Endogenous Sources of Reactive Oxygen and Nitrogen Species (ROS/RNS)

Several sources of ROS and RNS have been identified in neurons during normal and neuropathic pain processing, giving its fundamental role in signaling pathway modulation, discussed later in this paper. Although, as previously stated, ROS and RNS are exogenously and endogenously formed, the focus will be on the endogenous sources. Thus, at the cellular level, the majority of them come from enzymatic activity (Figure 4) [76].

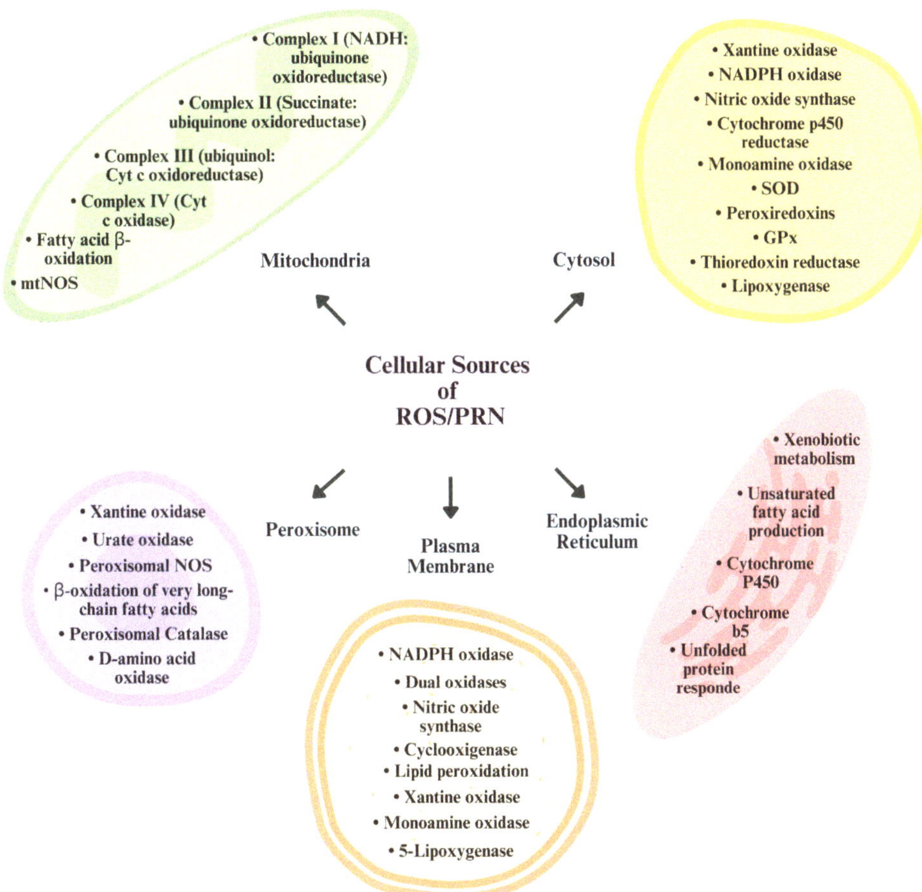

Figure 4. Endogenous sources of reactive oxygen and nitrogen species. Figure illustrates the major cellular organelles and enzymatic sources involved in the endogenous production of reactive oxygen and nitrogen species (ROS/RNS) during physiological cell functioning.

2.3.1. Endoplasmic Reticulum

The endoplasmic reticulum (ER) is formed as an intricately linked net with numerous tubules and sacs tightly interconnected and is thought to occupy around 10% of the cellular space and has a pivotal role in lipid and protein biosynthesis [83]. More specifically, it is involved in biosynthesis, folding, transport, and posttranslational modifications of proteins, whether of the plasma membrane, Golgi apparatus, or lysosomes, as well as calcium storage, and drug detoxification, in certain cell types [72,84,85].

In the smooth ER, reactive species are produced by any of two systems in charge of xenobiotic metabolism and unsaturated fatty acid production. Regarding xenobiotic metabolism, through a two-phase process, an initial monooxygenase reaction (phase I) using an NADPH-cytochrome P450 reductase and a complex of cytochrome P450 adds a polar group (OH^-) in a lipophilic substrate, using O_2 as a co-substrate, typically replacing a hydrogen atom, which introduces polarity into the molecule, and increases its water solubility. Following oxidation, the xenobiotic with added polarity undergoes a conjugation reaction (phase II) with endogenous molecules (glucuronic acid, sulfate, or glutathione), increasing further the hydrophilicity of the conjugate, facilitating its elimination [72].

In relation to the unsaturated fatty acid production, reactive species are produced by two mechanisms of action of cytochrome b5, which acts in electron transfer. First, cytochrome b5 can function as part of a multienzyme complex consisting of cytochrome b5, NADH cytochrome b5 reductase, and a specific desaturase. In this process, cytochrome b5 transfers electrons from NADH cytochrome b5 reductase to the desaturase. With the aid of molecular oxygen (O_2), the desaturase inserts carbon-carbon double bonds into fatty acids, producing two molecules of water (H_2O) [86]. Secondly, it can act through a direct electron transfer from the NADH cytochrome b5 reductase to the catalytic site of a fused cytochrome b5/desaturase, adding the same carbon-carbon bond, previously mentioned. In both cases, cytochrome b5 acts positively over the cytochrome P450 monooxygenase reaction, contributing to the availability of the second of two needed electrons for O_2 to activate cytochrome P450, accelerating the catalytic process and increasing the formation of H_2O_2 and O•−, via the leakage of electrons to O_2 by NADH cytochrome b5 reductase [72,87].

Altogether, and both present in the ER, cytochrome P450 and cytochrome b5 can contribute to reactive species production through diverse mechanisms. Given that cytochrome P450 enzymes participate in the metabolism of xenobiotics and endogenous molecules, some of the reactions involved occur with electron transfer to O_2, which subsequently contributes to ROS production. Moreover, cytochrome P450 enzymes suffer redox-cycling processes; in such processes, electron transfer between heme-iron centers and additional molecules occurs, which can result in ROS generation, more precisely $O{\bullet}^-$ and H_2O_2, as derivatives [88–90].

Into the bargain, it is believed that redox homeostasis in the ER can induce ER-associated stress, which triggers redox signaling mediators involved in ROS formation [46]. As an example, unfolded or misfolded proteins that accumulate in the ER lumen can elicit such a response, associated with chronic pain of neuropathic nature [91] As the first of two possible mechanisms, ROS can be formed during electron transfer from protein thiols to O_2 by the Endoplasmic Reticulum Oxidoreductin-1/Protein Disulfide Isomerase (ERO-1/PDI). Next, during the protein misfolding process due to the depletion of glutathione reductase (GSH), ROS are formed; when GSH is used, thiols are restored and can readily interact with ERO-1/PDI, so as to be re-oxidized. In this cyclic and repetitive process, disulfide bond formation and scission generate ROS as derivatives [91–93].

Since ER stress in increasingly recognized as a relevant component in the development of neuropathic pain, ROS produced in the process play a significant role in it. Under high protein traffic in ER stress conditions, great quantities of misfolded proteins and ROS can be generated, triggering the unfolded protein response (UPR), which might lead to deviant cellular signaling, neurotoxicity, neuroinflammation, the disruption of cellular functioning, and apoptosis, particularly relevant in neuropathic conditions, where ROS levels are boosted [94,95].

ER stress and pain behavior were normalized in neuropathic animals undergoing nerve ligation, and BiP and *XBP1* expression in the spinal cord were reduced after treatment with tauroursodeoxycholic acid, an ER chaperone that aids the ER in protein folding, indicating the participation of ER stress in the development of neuropathic pain [96].

2.3.2. Mitochondria

Among the most important functions of mitochondria are the production of energy in the form of ATP, either by the tricarboxylic acid cycle or oxidative phosphorylation, intracellular calcium concentration balance, and β-oxidation [72]. Almost all of these cellular functions are associated with reactive species production, and the mitochondria stands as one of the most important sources of reactive species with oxidant potentials. This amount of reactive species contributes to cellular damage and, at the same time, accounts for relevant redox signaling pathways [31].

It is widely accepted that oxidative phosphorylation and the ETC promote mitochondrial ROS formation in higher concentrations than any other mitochondrial activity; even so, the relevance of the other sites should not be disregarded, solely based on the percentage of production, given that there exist at least eleven different sites associated with substrate oxidation and oxidative phosphorylation electron leakage to O_2 that also produce O• and H_2O_2 [97]. In general terms, if electrons leak individually or in pairs, they will generate O•− or H_2O_2, respectively. If they are transferred four-at-a-time, as expected, they will form H_2O by inflowing to oxidative phosphorylation [97].

The mitochondrial ETC is responsible for ATP formation, through the process of oxidative phosphorylation. The oxidation of metabolites contained in energy sources undergoes a transfer of electrons to electron carriers, such as nicotinamide adenine dinucleotide (NAD^+) and flavin adenine dinucleotide (FAD); the reduced form of these carriers transfers the electrons to the respiratory chain, and finally to oxygen [46,98]. During this process, as electrons pass through the mitochondrial complexes, some may leak out and react with O_2, which then suffers either a reduction mediated by NADPH and xanthine oxidase, yielding O•− or by ubiquinol, a reduced form of ubiquinone [99].

Yet, it is notable that the production of O•− and H_2O_2 in the mitochondria will drastically depend on the protonmotive force, the $NADH/NAD^+$ and the $CoQH_2/CoQ$ ratios, and the local concentration of molecular oxygen, which are unsteady and difficult to measure in in vivo biological systems. Thus, O•− and H_2O_2 levels will raise or lower depending on the redox state and the specific location of the electron donor: if they act upstream, causing an oxidation of the site, they will lead to reduced electron leak; if they act downstream, causing a reduction of the site, this will lead to increased electron leak. Altogether, this makes it troublesome to assume that any mitochondrial dysfunction will necessarily lead to increased O•− and H_2O_2 levels, or to oxidative stress; instead, a wider frame must be taken into consideration [97,100].

ROS levels are found to be increased in neuropathic pain, more specifically in microglia, astrocytes, and neurons of the dorsal horn in the spinal cord [17,48]. In the event of nerve injury, mitochondria dysfunction emerges as a consequence of excessive reactive species production, Exceeded by endogenous antioxidant systems, triggering mitochondrial fission and an unpaired expression of mitochondrial transcription factor A (TFAM) and mitofusin 2 (MFN2) proteins, responsible for mitochondrial biogenesis and fusion, respectively, and hence, contributing to oxidative stress and increased mitochondrial damage [101]. These events seem to be significantly regulated by the cellular energy sensor (adenosine 5′-monophosphate (AMP)-activated protein kinase) AMPK, which acts upon oxidative stress conditions, triggering peroxisome proliferator-activated receptor-gamma coactivator-1alpha (PGC-1α), which promotes mitochondrial gene expression related to biogenesis and mDNA replication and transcription, ultimately restoring mitochondrial function [102]. Furthermore, excessive ROS produced by mitochondria can trigger mitochondrial damage and exacerbate neuropathic pain through TRP channel activation, NMDA receptor upregulation, and GABAergic gating mechanisms inhibition, inducing NLRP3 inflammasome proinflammatory responses, and causing oxidative damage to neurons [50,103,104].

2.3.3. Plasma Membrane

Given its direct contact with the extracellular space, a high exposure to oxidizing agents and reactions are plausible, leading to reactive species production [72]. Several metabolic processes can be disturbed due to oxidatively damaged lipids of the plasma membranes, which will precede transmembrane ion gradient alterations and the interruption of secretory and signaling functions [105].

The main source of reactive species generation is owed to the effect of NOX, which will indorse O•− production. NOX are a group of enzymes bounded to the plasma membrane present in a myriad of cell types [44]. NOX are constituted by six subunits: one Rho-GTPase, either Rac1 or Rac2, and five phagocytic oxidases (phox): $gp91_{phox}$, $p22_{phox}$, $p40_{phox}$, $p47_{phox}$, and $p67_{phox}$. The first two are restricted to the membrane, and the last three are free in the cytoplasm [20]. Upon stimulation, when proper conditions allow the cytoplasmic subunit to migrate to the membrane subunits, all the components bound together, leading to allosteric modification and subsequent NOX activation. Once activated, NOX will catalyze the oxidation of cytoplasmic NADPH into NAD^+, releasing an electron to O_2, to produce O_2•− onto the plasma membrane or in its outer side [106,107].

Out of the NOX family, NOX1, NOX2, and NOX4 have been linked to the neuroinflammation process and neuropathic pain development [53,108,109]. NOX1 is expressed in DRG, spinal cord neurons, and glial cells [110], and its inhibition by ML171 significantly reduced pain behavior through TRPA1-mediated ROS production in a model of neuropathic pain induced by peripheral nerve injury in rats [111]. NOX2 is expressed in microglia from the spinal cord, and in macrophages and neurons in DRG in neuropathic pain models induced by nerve injury; thus, its suppression was associated with decreased ROS production, microglia activation, and TNF-α and IL-1β, therefore, improving thermal and mechanical hypersensitivity [112,113]. Finally, NOX4 was detected in dorsal horn neurons and interneurons of the spinal cord, as well as in the injured nerve and DRG in peripheral nerve injury and diabetic neuropathy [114,115], contributing to ROS production, peripheral myelin integrity impairment through myelin protein zero (MP) and peripheral myelin protein 22 (PMP22) degradation, associated with apoptosis, and consequently, the exacerbation of the manifestations of neuropathic pain, all of which were reduced when NOX4 activity was inhibited in knock-out mice [116].

2.4. *ROS/RNS-Mediated Cell Signaling*

RONS play pivotal roles in cell functioning, by eliciting or regulating changes in signaling cellular messages, enzymatic activity, response to growth factor, the induction of inflammatory response [117], immune response, cellular adhesion, differentiation, proliferation, autophagy and apoptosis [118], gene transcription [119], and protein and membrane integrity [19,120], to name a few.

Given that the cellular activity is modified as a response to stimuli, mainly through the activation of gene transcription, the two recognized ways by which ROS participate in such activation, are by (1) the direct modification of transcription factors, which will be addressed to particular DNA promoters of target genes, or (2) via the activation of mitogen-activated protein kinase (MAPK) pathways, like extracellular signal-regulated kinases (ERK), c-Jun N-terminal kinases (JNK), and p38, which will lead to the phosphorylation and activation of other transcription factors, such as activator protein 1 (AP-1) and cAMP response element-binding protein (CREB) [121,122]. The two mechanisms by which cell signaling happen, by means of ROS, can be explained as (1) modifications of target protein molecules, once reactive species oxidize main Cys residues, or by (2) changes in intracellular redox state by the redox-buffering capacity of the cells [72]. The molecular alteration suffered by the former mechanism acts over enzymes, transporters, receptors, transcription factor regulatory sites, and allosteric and macromolecular sites [123], and can alter their localization, activity, and DNA-binding ability of the transcription factors, thus repressing or enhancing the transcription of target genes [120,124].

Although some reactive species can cause irreversible structural modifications over molecules, with low specificity (ex.: OH•), permanently damaging them, other moderate oxidants, such as H_2O_2, will primarily target Cys residues of proteins, more precisely thiol groups (-SH) [72]. Thiol groups from Cys residues can undergo several oxidative modifications, and the process typically occurs as follows: initially, ROS oxidize the -SH group of Cys to form sulfenic acid (-SOH); sulfenic acid (-SOH) can further react with a neighbor -SH group, either from the same protein or from another protein, yielding the formation of a disulfide bond (-S-S-). Under certain conditions, such as excessive reactive species, sulfenic acid (-SOH) can be further oxidized to sulfinic acid ($-SO_2H$) and then to sulfonic acid ($-SO_3H$). These two last higher oxidation states are usually irreversible. On the other hand, even though sulfenic acid formation has been seen as damaging, it has its own relevance as a redox-sensing signal for cellular functions, due to the reversible oxidation by ROS [125,126].

Other targets of redox signaling exist, such as the iron–sulfur (Fe–S) cluster or Tyr residues, but the list could be vastly extensive for the purpose of this paper. Therefore, this section takes into consideration those redox-signaling targets deemed more sensitive to nitroxidative effects and/or implicated in the regulation of protein function, associated with transcription factor control and well-characterized mechanisms, and that contribute to the regulation of oxidative stress, offering insights to how cells sense and respond to oxidative stress and neuropathic pain.

2.4.1. NRF2–ARE Pathway

Nuclear factor erythroid 2-related factor 2 (NRF2) is a redox-sensitive leucine zipper (bZIP) transcription factor [127] that has a significant role in neuropathic pain modulation through the inflammatory response and oxidative stress management by hyperalgesia and allodynia regulation, regulating the expression of HO-1 and IL-10 [127,128]. It is expressed in microglia in the spinal cord, DRG, and sciatic nerve [129]. NRF2 binds to specific DNA sequences to control transcription from DNA to RNA [130]. It is activated in response to oxidative and electrophilic stress [131]. Its major function is to initiate the expression of a wide range of cytoprotective genes once it has a role in maintaining redox homeostasis [56]. In quiescent conditions, in the cytoplasm, NRF2 is bound to the Kelch-like ECH-associated protein 1 (KEAP1) dimer, which acts as an adaptor protein that links NRF2 to the Cullin 3 (Cul3)-based E3 ubiquitin ligase complex, which causes its ubiquitination, by attaching to lysine residues on NRF2, through a process involving ubiquitin-activating enzymes 1 (E1), ubiquitin-conjugating enzymes (E2), and E3 [132]; this tags NRF2 for subsequent degradation by the 26S proteosome into small peptides and this interaction maintains low NRF2 cellular levels, as expected in normal conditions (Figure 5A) [133].

When an increase in oxidative stress occurs, NRF2 is released from KEAP1, accumulates in the cytoplasm, translocate to the nucleus, and heterodimerizes with specific regions of DNA encoding antioxidant response elements (ARE), detoxification, and cytoprotective proteins, such as heme-oxygenase 1 (HO-1) [134], CAT, SOD, and others (Figure 5B) [130,135].

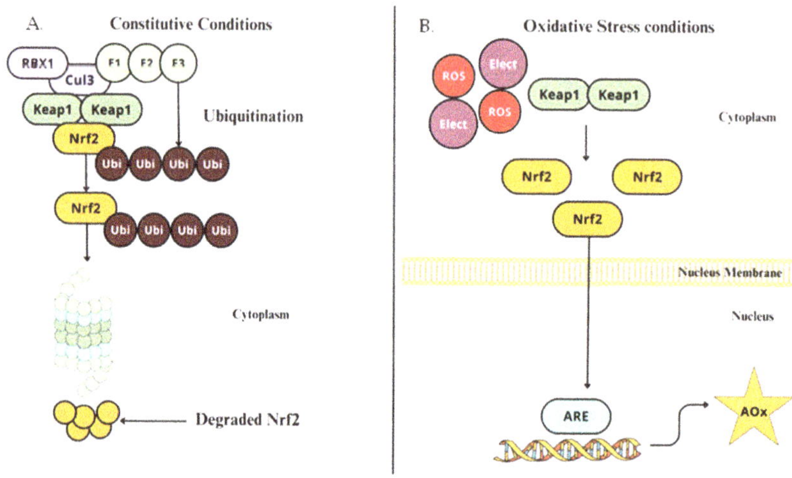

Figure 5. NRF2–ARE pathway. (**A**). Constitutive conditions of normal Nrf2 degradation by ubiquitination process in the cytoplasm. (**B**). Oxidative stress conditions of Nrf2 release by KEAP1 and translocation to the nucleus towards DNA regions responsible for the ARE encoding. AOx: antioxidant; Elect: electrolytic stressors; Ubi: ubiquitin.

As previously stated, KEAP1 is found to have a double function, as an oxidative and electrophilic stress sensor, and negative regulator of NRF2, by degradation via ubiquitination [132]. Out of 25 cysteine residues of KEAP1, three are among the most critical for redox sensing, responsible for NRF2 dissociation (Cys^{151}) and degradation (Cys^{273} and Cys^{288}) [136]. KEAP1 is highly reactive and able to undergo modifications in response to alterations in the inner cellular redox state; these Cys can be oxidized or covalently modified, inducing the formation of disulfide bonds [137].

Thus, conformational variations reduce the affinity for NRF2, ultimately preventing KEAP1 from enabling the ubiquitination and degradation of NRF2. This allows NRF2 to stabilize, and move forward with its transcriptional inducing capacity, which then triggers cellular protective antioxidant [138], anti-neurotoxic [139], neuroprotection [140], and anti-inflammatory responses [141].

Decreased NRF2 and HO-1 levels were observed in neuropathic rats after vincristine administration, where oxidative stress, DNA damage, neuronal cell damage, and inflammation, evidenced by increased 8OH-dG, diminished glial fibrillary acidic protein (GFAP), and nuclear factor kappa-B (NF-κB) were present, and promptly reversed after quercetin administration, showing the role played by the NRF2 pathway, and the protective effect of quercetin against neuropathic pain caused by vincristine [128]. To evaluate the role of NRF2 in neuropathic pain, NRF2 siRNA was administered to CCI-induced rats, which prominently decreased the expression of *NRF2* RNA and NRF2 and HO-1 protein and worsened the hyperalgesia behavior (withdrawal threshold and latency) [127]. Also, a study investigating the involvement of NRF2 expression in spared nerve injury in rats with our without an anhedonia phenotype, classified after a hierarchical cluster analysis of a sucrose preference test, showed that NRF2 expression was significantly decreased in the medial prefrontal cortex, hippocampus, and spinal cord of anhedonian rats, compared to non-anhedonian ones. After the NRF2 activator sulforanphane was administered, a reduction in mechanical withdrawal threshold was observed, while sucrose preference remain unchanged, due to NRF2 level normalization in the animals with the anhedonia phenotype [142].

2.4.2. NF-κB Signaling Pathway

The nuclear factor κ-light-chain-enhancer of activated B cells (NF-κB) is a pleiotropic inducible transcription factor implicated in pro-survival regulatory inflammation and immune functions, cellular adhesion, differentiation, and cell proliferation. It is activated in a wide variety of neuroinflammation-associated stimuli via pro-apoptotic pathways [143,144]. It is composed by homo- and heterodimers of five structural protein members, of which Rel-A(p65), Rel-B, and c-Rel contain C-terminal transactivation domains (TADs), and p50/p105 and p52/p100, which also serve inhibitory functions [120].

Conversely, in the nervous system, it is composed mainly by p50/Rel-A heterodimers [145], but the most widely studied dimers are those of Rel-A and p50, also referred to as NF-κB$_{can}$, once they participate in the canonical activation pathway. On the other hand, Rel-B and p52 form a dimer referred to as NF-κB$_{non}$ for they participate in the non-canonical activation pathway, discussed later [146]. Additionally, NF-κB contains Rel-1-homology domains (RHD) that bear a nuclear localization signal, accountable for dimerization, detection, and binding to DNA, as well as communication with the NF-κB inhibitor (IκB) [147].

As previously stated, the NF-κB pathway is mainly activated by either of two kinase-dependent pathways, the canonical and the non-canonical. However, in the nervous system, the most studied and well known is the canonical pathway, which includes a series of cytokines and Toll-like receptors (TLR) [144]. In normal conditions, NF-κB dimers form complexes with IκB, to retain them in the cytosol, by masking their nuclear localization signal (NSL), and also preventing NF-κB from binding to DNA by masking its DNA-binding sites [147]. There, H_2O_2, HOCl, $_1O^2$, and peroxynitrite act via the oxidation or nitration of specific critical Cys residues on the IκB–kinase (IKK) complex, which terminates the joining of IκB to NF-κB. Once IκB is structurally modified, activated NF-κB is released and translocated to the nucleus (Figure 6) [19,148].

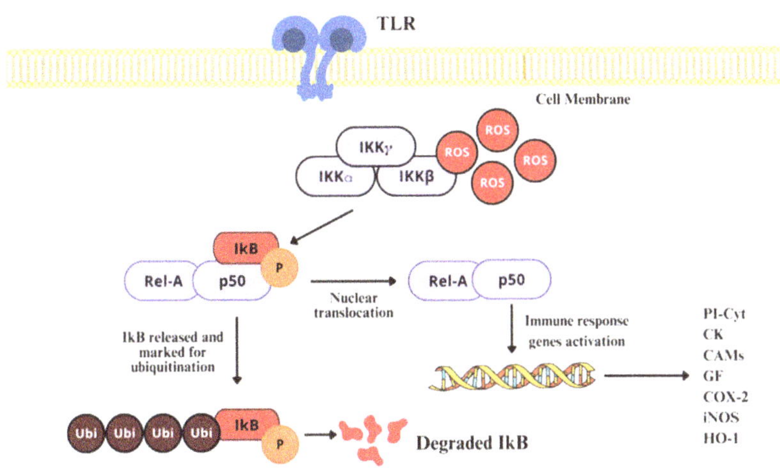

Figure 6. Canonical pathway of NF-κB activation via TLR (Toll-like receptors) and IKK complex oxidation/nitration leads to IκB–NF-κB complex separation and leads to NF-κB nuclear translocation. PI-Cyt: pro-inflammatory cytokines; CK: chemokines; CAMs: cell adhesion molecules; GF: growth factors; COX-2: Cyclooxygenase-2; iNOS: Inducible Nitric oOxide Synthase; HO-1: Heme-Oxygenase-1.

When phosphorylated at a specific regulatory aminoacidic residue by the IKK complex, IκB is ubiquitinated and marked for 26S proteosome degradation, still in the cytosol [149]; NF-κB is freed for nuclear translocation, to activate the transcription of target genes [46], where it binds to enhancer regions of specific genes, such as proinflammatory cytokines, chemokines, adhesion molecules, growing factors (GF), and promoters of COX-2, iNOS, and heme-oxygenase [149,150]. On the other side, H_2O_2 can regulate NF-κB by direct interaction with the Cys of the DNA-binding region (Figure 7) [151]. In that sense, higher than normal levels of RONS in the nucleus can prevent NF-κB from interacting with DNA, reducing its transcriptional activity, although increased H_2O_2 can induce peroxiredoxin 1 activity, which is a H_2O_2 scavenger.

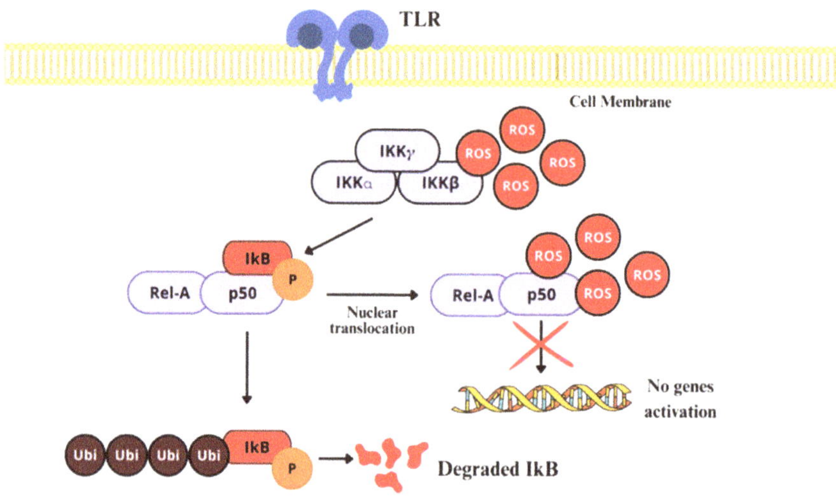

Figure 7. NF-κB pathway inactivation via reactive oxygen and nitrogen species direct interaction with the DNA-binding region's Cys leads to inhibited gene activation in the nucleus under increased levels of ROS/RNS.

A diverse array of genes with sometimes differing functions respond to NF-κB_{can}, i.e., pro-survival genes such as inhibitors of apoptosis (IAP), *Bcl-2*, *Bcl-x*, *Bcl-w*, *Bfl-1*, *BIRCS*, and *MnSod*; yet NF-κB_{can} can also regulate pro-apoptotic genes like *p53*, *TNF-α*, *Bax*, *c-myc*, *APO-1*, and *FasL* [146]. Given the extremely specialized types and subtypes of neuronal cells, it is a major challenge to study NF-κB, due to traditional techniques limitations in their ability to mirror realistic and accurate locations and activities of proteins that are regulated by specific dimer combinations and several post-translational modifications.

NF-κB is widely implicated in spinal microglial activation to M1 phenotype in the establishment of neuropathic pain, as seen in the examination of growth and differentiation factor 11 (GDF11), an inhibitor of macrophage activation, and its impact on microglia polarization, thermal and mechanical hyperalgesia in sciatic nerve injury in mice, showing a marked reduction of pain behaviors, enhancing the switch from M1 to M2 phenotype through the modulation of the transforming growth factor beta receptor type 1 (TGF-βR1)/SMAD2/NF-κB pathway [152,153]. Another study aimed to investigate the participation of Pellino 1 (Peli1), a component of the E3 ubiquitin ligase, in the activation of microglia in neuropathic pain induced by sciatic nerve injury. It was found that higher levels of Peli1 and NF-κB expression, as well as pain behavior after spared nerve injury (SNI), induced neuropathic pain. Also, the administration of intrathecal Peli1 shRNA prior to SNI and siRNA after the establishment of SNI, reduced the expression of Peli1 and

NF-κB in the spinal dorsal horn of mice, alleviating the thermal and mechanical threshold through NF-κB pathway modulation, as Peli1 is needed for the ubiquitination of proteins regulating NF-κB activation [154].

2.4.3. MAPK/AP-1 Signaling Pathway

The activator protein 1 (AP-1) is a heterogenous family constituted by the basic-region leucine zipper (bZIP) transcription factor, accountable for the regulation of cellular responses to numerous extracellular stimuli, including ROS [155]. This vast group is represented by the members of the Jun (c-Jun, JunB, and JunD), Fos (c-Fos, FosB, Fra-1, and Fra-2), Maf (c-Maf, MafA, MafB, MafG/F/K, and Nrl) and ATF (ATF2, ATF3, B-ATF, JDP1, and JDP2) protein subfamilies [156,157]. They harbor a C-terminal leucine zipper domain that allows them to form hetero- and homodimers, which subsequently bind to specific DNA sequences known as 12-O-tetradecanoylphorbol-13-acetate (TPA)-responsive elements (TRE) or cAMP response elements (CRE) in the promoter regions of target genes, where they regulate gene expression, mediated by basic N-acid-rich domains, parallel to the leucine zipper [157]. The specific arrangement of the dimers composing the AP-1 complex establishes the interaction patterns, molecular affinities, and functional specificity that represent their tasks in the cell [158,159].

AP-1 transcription factor can be redox activated by ROS, ROS-dependent factors, cytokines, GF, neurotransmitters, UV light, and ionizing radiation [160]. The increased exposure to H_2O_2, OH•, and O•− can directly determine the redox state of several conserved cysteine residues included in the AP-1 pathway [161,162]. Also, AP-1 can be indirectly regulated by ROS signaling pathways, like NF-κB and interferon-γ. The AP-1 transcription factor family is a target of the ROS-sensitive MAPK cascade for phosphorylation, this leads to enhanced transcriptional activation [163,164]. ROS activate ERK1/2, JNK, and p38 MAPKs (Figure 8), which are crucial mediators for signal transduction pathways leading to AP-1 activation, through the phosphorylation of c-Jun on serine and threonine residues, enhancing its stabilization and c-Fos transcriptional activity over genes involved in neuron, microglia, and astrocyte activation, in a sequential manner [165], particularly relevant in dorsal horn neurons of the spinal cord and DRG, and playing distinct roles in different cells across the development of neuropathic pain [166]. The accurate redox control of AP-1 expression and activation is essential for homeostasis, and its alteration has been implicated in altered neurons states, for it is a key molecular switch that controls the expression of downstream proinflammatory factors [167]. Nevertheless, the regulation of AP-1 activity is complex as it depends on severa factors such as, changes in *Jun* and *Fos* gene transcription and mRNA expressionthe effects on Jun and Fos protein turnover, post-translational modifications of Jun and Fos proteins that modulate its transactivation capacity, and finally, on the interaction with other transcription factors that can positively or negatively affect AP-1 activity [168].

An oxidative shift of the intracellular thiol/disulfide redox state, or ROS, are accountable for significant activation of the MAPK cascade elements, as seen in neurons [169,170]. They consist in a series of protein kinases that activate each other in specific sequences. MAPKs are regulated by the phosphorylation and dephosphorylation of threonine and serine residues, and are triggered by tyrosine kinase receptor activation, protein tyrosine kinases, cytokine receptors, heterotrimeric G protein-coupled receptors, and GF [39]. The major MAPK cascades include ERK1/2 (extracellular signal-regulated kinase 1 and 2), JNK (c Jun N-terminal kinase), the p38 pathway, and ERK5, each of which has specific upstream activators, downstream targets, and biological actions [171].

Each cascade is composed by three central kinases (MAP3K, MAP2K, and MAPK); an additional upstream MAP4K and downstream MAPKAPK components are included.

Respectively, each cascade has its own signal transmitted by the successive phosphorylation and activation of sequential kinases, that at a given point lead to the phosphorylation of the target regulatory proteins by the MAPK and MAPKAPK constituents [172]. With the aim of executing their functions, MAPKS and MAPKAPKs phosphorylate and regulate their substrates, so as to induce and regulate the de novo gene expression of other transcription factors and suppressors, and chromatin remodeling proteins, at the nucleus level, after being physically transported across the nuclear membrane [173].

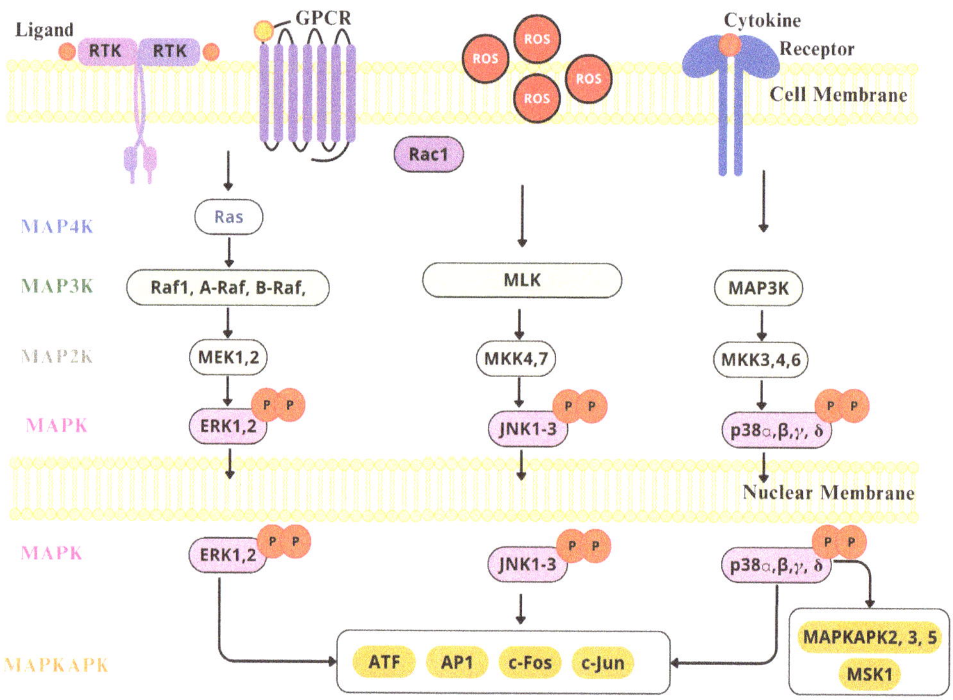

Figure 8. MAPK/AP-1 signaling pathway. MAPKs are regulated by a cascade of phosphorylation and dephosphorylation of threonine and serine residues, triggered by tyrosine kinase receptors (RTK) activation, protein tyrosine kinase (TK), cytokine receptors (CR), heterotrimeric G protein-coupled receptors (GPCR), growth factors (GF), or ROS/RNS. ATF: activating transcription factor; AP-1: Activator Protein—1; MLK: Mixed-Lineage Kinase; MEK: MAPK/ERK Kinase; MKK: Mitogen-Activated Protein Kinase; ERK: Extracellular Signal-Regulated Kinase; JNK: c Jun N-terminal Kinase.

The ERK1/2 cascade is activated through various membrane receptors, but mostly via GTPase Ras, which receives extracellular signals at the plasma membrane, that in turn conscript the Raf-1 and B-Raf (MAP3K) tier of the cascade to the plasma membrane, probably through homo- or hetero-dimerization and phosphorylation [174]. Therefore, the kinases activating the Raf components are considered MAP4Ks. Afterward, the signal is transmitted to the MEK1/2 (MAP2K), through the phosphorylation of serine residues in their activation loop. The activated MEK1/2 transmit the signal to ERK1/2 (MAPK) also by the phosphorylation of the threonine and tyrosine regulatory residues in the Thr-Glu-Tyr domain, in the activation loop [175]. Ultimately, the signals pass on to the MAPKAPK elements, RSKs, MNKs, and MSKs, and other substrates located in the cytoplasm or the nucleus. Both ERK1/2 and MAPKAPK can phosphorylate substrates involved in cellular proliferation, differentiation, neuroplasticity, stress response, pro-survival, and pro-apoptotic processes [176]. As a matter of fact, ERK is phosphorylated in dorsal horn neurons of the spinal cord, in DRG neurons, and in peripheral nerve terminals after nerve injury, contributing to peripheral and central sensitization [165,177]. ERK activation occurs in a sequential manner, starting in the dorsal horn neurons after spinal nerve ligation (SNL) within 10 min through the transcriptional modulation of NK-1 and prodynorphin [178,179], followed by microglia for some days, having its most robust expression in 3 days, and finally, in astrocytes for weeks, supporting the role of ERK in astrocytes in the maintenance of neuropathic pain, whose inhibition reduced behavior associated with neuropathic pain [165]. Isoliquiritigenin reduced mechanical and thermal hyperalgesia by a CCI in mice, inhibiting the activation of spinal microglia and neuroinflammation through the ERK pathway inhibition [180].

The JNK cascade is usually activated by stress signals such as UV radiation and cytokines [181]. It includes MLK (MAP3K), MKK4/7 (MAP2K), and JNK (MAPK). Once activated by stimuli dependent and independent of stress and mitogens, it transmits their signals to GTPases, such as Rac1, that activate the MAP3K-level kinases directly or via MAP4Ks [182]. The MAP3K tier transmits the signals through the phosphorylation of the threonine and serine residues in the activation loop, which consequently activates the MKK4/7; finally, these kinases activate any of the components of the MAPK level by the direct phosphorylation of tyrosine and threonine residues in the activation loop. As expected, the JNKs and their respective MAPKAPKs phosphorylate substrates in the cytoplasm and nucleus that modulate the transcription of genes mediating apoptosis, immunological responses, and neuronal activity [183]. JNK-1 is activated in early stages (three days) and late stages (21 days) in GFAP- expressing astrocytes in neuropathic pain caused by SNL, and transiently (12 h) in DRG neurons after SNL, specifically in small C-fibers; therefore, this supports the idea that spinal astrocytes contribute to the continuance of neuropathic pain [166]. The neuroprotective effect of Orexin B was evaluated in a model of neuropathic pain induced by CCI in rats. Orexin was found to be effective in improving mechanical and thermal hyperalgesia, preventing microglia activation reducing Iba1 levels and preventing JNK/NF-κB signaling pathway activation [184].

As for the p38 cascade, it is activated by stress signals and environmental stressors, but it also responds to normal cellular processes, like inflammatory cytokines [185]. It involves MKK3/4/6 (MAPKK) and p38 (MAPK) [172]. After activation by stress conditions or receptors, the signals are transmitted via GTPases, MAP4K, and MAP3K, in a similar manner to the JNK cascade. The MAP3K level of the cascade phosphorylates and activates MKK3/6, and under certain conditions, MKK4 (MAP2K) components of the p38 cascade [186]. At this point, the p38 isoforms and functional unconventional spliced forms at the MAPK level are activated via the phosphorylation of the threonine and tyrosine residues in the regulatory domain in their activation loop. Lastly, the signals are transmitted by the p38, or MAPKAPKs to target motifs accountable for regulatory processes [172]. p38 MAPKs are particularly expressed in the superficial laminae of dorsal horn neurons in the spinal cord, whose increase correlates with lower mechanical withdrawal and thermal latency thresholds in SNL-induced neuropathic pain [166]. Increasing evidence points to the role of p38 MAPK in the activation of spinal microglia after nerve injury and its contribution to neuropathic pain, via downstream mechanisms of proinflammatory mediators after nuclear translocation, where it influences the gene expression of TNF-α, IL-1β, and IL-6, critical mediators of neuroinflammation and pain that also exacerbate other pain signaling pathways [187]. p38 MAPK involvement in the pathophysiology of neuropathic pain mediating microglial activation was evaluated in SNL rat models treated with electroacupuncture to assess the analgesic effect of such treatment; it was found that a significant decrease in p38 MAPK activation and enhanced mechanical withdrawal threshold, both in the ipsilateral and contralateral dorsal horn of the spinal cord, not only demonstrated the beneficial effect of electroacupuncture, but the potential of targeting the p38 MAPK pathway to treat neuropathic pain [188].

2.4.4. Phosphoinositide 3-Kinase (PI3K)/Akt Pathway

The PI3K pathway constitutes a conserved family of kinases present in the inner side of the plasma membrane, capable of acting specifically on the D-3 position of the inositol ring of inositol phospholipids [189]. Such phosphorylation produces phosphatidylinositol-3,4,5-triphosphate (PIP3), phosphatidylinositol-3,4-biphosphate (PIP2), and phosphatidylinositol-3-phosphate (PIP) [190]. The PI3K enzymes, when activated, transmit intracellular signaling cascades concerning signal transduction, protein synthesis, vesicular traffic, cytoskeletal reorganization, cell growth, proliferation, survival, metabolism, apoptosis, autophagy [189,191], drug resistance in response to GF, such as epidermal growth factor (EGF), platelet-derived growth factor (PDGF), nerve growth factor, (NGF), vascular endothelial growth factor (VEGF), and brain-derived neurotrophic factor (BDNF) [122,192]. PI3K has also been found to be essential in the initiation and maintenance of neuropathic pain [193].

Although in mammalians cells, the PI3Ks exists in three classes (I, II, and III), and each of these have different structural characteristics, functional homologies, and substrate specificity, the most prevalent and significant in the nervous systems is the class I PI3Ks, where they play a crucial role in neuronal development, synaptic plasticity, and neuroprotection [194]. Class I PI3K are further divided into class IA and class IB subgroups [191,195,196]. The heterodimeric class IA PI3Ks are composed of a regulatory subunit p85, including five polypeptides p85α, p85β, p55α, p50α, and p55γ, and a catalytic subunit p110 (p110α, p110β, or p110δ), which are activated by receptor tyrosine kinase (RTK) and G-protein-coupled receptors (GPCRs), small G protein Ras, and cytokines [197]. The p110α and p110β are ubiquitously expressed in all cell types, while p110δ expression is more restricted to leukocytes [198]. The catalytic p110 subunit contains a C-2 region for membrane anchoring; a Ras-binding region (RBD); an adaptor-binding region (ABD), which holds an N-terminal region for interaction with the regulatory p85 subunit; a helical region; and a catalytic region. The p85 regulatory subunits have in common a p110 binding region (inter-Src-homology-2 region, iSH2) and two SH2 regions for p110 ABD binding, constitutively [189,199]. On the other hand, class IB PI3Ks enzymes are formed by the p110γ polypeptide, associated with an adaptor subunit p101 or p84 and p87, to form heterodimers, p101/p110γ, or p84/p110γ, which are primarily activated by GPCRs, due to the interaction of its regulatory subunit with the subunit of trimeric G protein [200]; this isoform is present in the nervous system.

Either of these PI3Ks catalyzes the phosphorylation of PIP2 to produce PIP3 [201–203]. Typically, its activation starts with the union of a ligand to RTKs or GPCRs, through its regulatory p85 subunit. In non-activated states, the p85 subunit remains bonded to the catalytic subunit p110 ABD region, through its iSH2 region, which stabilizes the p110 catalytic subunit. Once properly activated through cytokines, GF, insulin, and LPS, the SH2 regions bind to their respective receptors or adapter proteins, leading to the allosteric activation of the p110 catalytic subunit of the PI3K [194,204]. When activated, PI3K transforms the PIP2 to PIP3, which serves as a membrane-embedded second messenger, inducing the repositioning and activity of pleckstrin-homology (PH) region-containing proteins, in the inner side of the plasma membrane [192]. Among these proteins, can be included Akt (protein serine/threonine kinase), also knowns as PKB, and phosphoinositide dependent kinase 1 (PDK1) [202]. Akt activation permits the transportation on protein kinase to the cytoplasm and nucleus, where it downstream modulates pro-apoptotic proteins, such as Bcl-2, caspase 9, fork head transcription factors (FOXOs), NF-κB, mTOR, and glycogen synthase kinase 3 (GSK3) (Figure 9) [192,197,205].

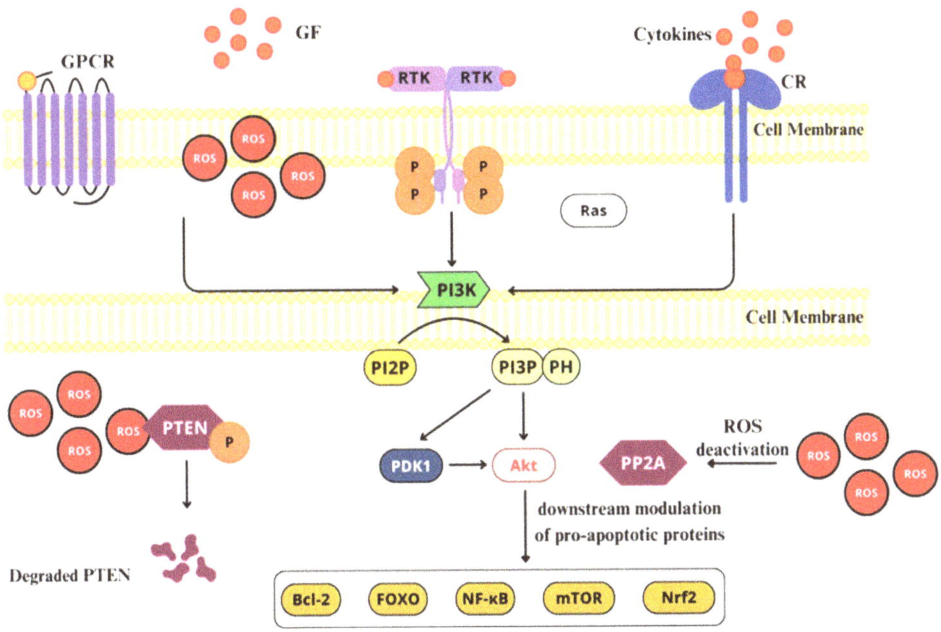

Figure 9. Phosphoinositide 3-kinase (PI3K)/Akt pathway. PI3K is activated by RTKs or GPCRs, GF, CR, and ROS. Once activated, PI3K transforms the PIP2 to PIP3. This serves as a second messenger for Akt (protein serine/threonine kinase) and phosphoinositide dependent kinase 1 (PDK1) activation. Akt activation allows downstream modulation of pro-apoptotic proteins such as Bcl-2, caspase 9, fork head transcription factors (FOXOs), NF-κB, and mTOR. ROS can inhibit phosphatase and tensin homolog (PTEN), which negatively regulates PIP3 synthesis and inhibits Akt activation, leading to further activation of the PI3K/Akt signaling pathway. Phosphatase 2A (PP2A), a Akt/PKB inhibitor, can be deactivated by ROS.

ROS and RNS interact with specific amino acids residues on tyrosine phosphatases, protein tyrosine kinase, and protein kinase C, that can trigger kinase cascades, MAPKs, and PI3Ks [206]. ROS can directly activate PI3K, increasing its downstream signaling pathway [207]. At the same time, H_2O_2 can oxidize and inhibit phosphatase and tensin homolog (PTEN), which negatively regulates PIP3 synthesis and inhibits Akt activation, hence, leading to a further activation of PI3K/Akt signaling pathway [208]. Additionally, ROS enhance the phosphorylation of PTEN through casein kinase II, allowing PTEN to undergo proteolytic degradation [209]. Moreover, phosphatase 2A (PP2A), an Akt/PKB inhibitor, can be deactivated by ROS [209,210]. Nevertheless, low ROS levels cause the oxidation of Akt/PKB -S-S- bridges, promoting PP2A association with the Akt/PKB complex, promoting the short-term activation of the Akt/PKB pathway [211,212].

The dysregulation of this pathway has major implications in neuropathic pain [193]. PI3K/Akt inhibition or alteration leads to elevated states of ROS production, membrane depolarization, mitochondrial membrane stability impairment, the diminishment of oxidative phosphorylation, and ATP production [213,214]. The class I PI3K catalytic subunit p110α has a role in controlling ROS levels at a physiologic rate through the NRF2/ARE-dependent pathway [215], which, if not regulated, could lead to activation of the autophagy machinery [206]. The dysregulation of the PI3K/Akt pathway has a relevant role in the establishment of chronic pain conditions derived from traumatic lesions [192,193,216,217]. PI3K has also been linked to TRK systems NGF/TrkA and BDNF/TrkB activation, displaying central sensitization and hyperalgesia mediation induced by G-CSF [218], and mediating pain induced by plantar incision in mice [216]. Additionally, PI3K mediates peripheral and central sensitization and hyperalgesia caused by the intraplantar injection of EphrinB1-Fc, triggering the EphrinBs (ligand) and EphBs (receptor) pathway activation [219], affecting neuronal excitability, leading to an increased firing of pain signals, contributing to pain hypersensitivity [220]. PI3K was proved to be crucial for the DRG and dorsal horn neurons' upregulation of the kinesis family member 1a (Kif1a), following a positive correlation between increased levels of PI3K/Akt/CREB, and Kif1a protein and mRNA expression in a chronic pain model induced by chronic constriction injury. The process starts with CREB phosphorylation by PI3K, which recruits the DNA demethylation TET1 towards the *Kif1a* promoter region, inducing an upregulation of its expression, which then leads to pain sensation. Such insight offers a novel target for managing neuropathic pain, since the inhibition of PI3K led to decreased Kif1a expression, and alleviating pain-associated behaviors [221].

2.4.5. Calcium Signaling

Divalent calcium ions (Ca^{2+}) are widely implicated in a plethora of pro-survival and pro-apoptotic cell processes, considering that it is one of the most relevant second messengers for intracellular signaling in most cells, but particularly in neurons and astrocytes, where it serves intricate and integrated functions like dendritic response to neurotransmitters, gene expression through nucleus signaling, and neurotransmitter release from presynaptic terminals [222]. Ca^{2+} aids in communicating depolarization status and synaptic activity to neurons [223]. The biochemically elegant mechanism through which it exerts signaling properties relies on the calcium concentration gradient relation of the extracellular space and the cytosolic space, as well as the relation of the cytosolic space and intracellular compartments, such as the endoplasmic reticulum and mitochondria [223]. In this context, given that Ca^{2+} is a main element for maintaining and controlling neuronal excitability, it is not absurd to consider the high energetic demand that this implies for neurons, as all of the Ca^{2+} that enters the cytosolic space must be removed from within, through plasma membrane calcium ATPase (PMCA), sodium-calcium exchanger (NCX), mitochondrial calcium uniporter (MCU), mitochondrial efflux systems, endoplasmic reticulum calcium ATPase (SERCA), calcium release channels, and cation-dependent calcium pumps (Figure 10) [224].

In the mitochondria, Ca^{2+} is transported through the MCU, against the concentration gradient, by means of electrochemical gradient–mitochondrial membrane potential; once in the mitochondria, Ca^{2+} forms inactive precipitates bound to inorganic phosphates [225]. When Ca^{2+} is not bound, it is transported back to the cytosol by the NCX. Nevertheless, when mitochondrial levels of Ca^{2+} are excessively high, this will force its way out through the mitochondrial permeability transition pores (mPTP); this marks the initial steps towards the release of pro-apoptotic proteins and the beginning of cell death through necrosis or apoptosis [226,227]. Cytosolic Ca^{2+} level increases mediated by ROS induce the AP-1 proteins c-Fos and c-Jun [39,228], as well as PKC-α activation through oxidative stress [229].

As formerly stated, the mitochondria is a paramount producer of reactive species, through the ETC as by-products, mainly $O\bullet-$ and H_2O_2, as well as through the TCA cycle, α-ketoglutarate dehydrogenase, or pyruvate dehydrogenase complexes, which mainly produce $O\bullet-$ and H_2O_2, or through enzymes not related to the ATP production, yet residents of the mitochondrial membrane, such as monoamine oxidase (MAO), cytochrome b5 reductase (Cb5R), and glycerol–3–phosphate dehydrogenase [230]. Altogether, the mitochondria can be considered a producer and a target itself of ROS actions, due to its intricate interaction, mutual regulation, and complementary functions, either in physiological or pathological conditions [231].

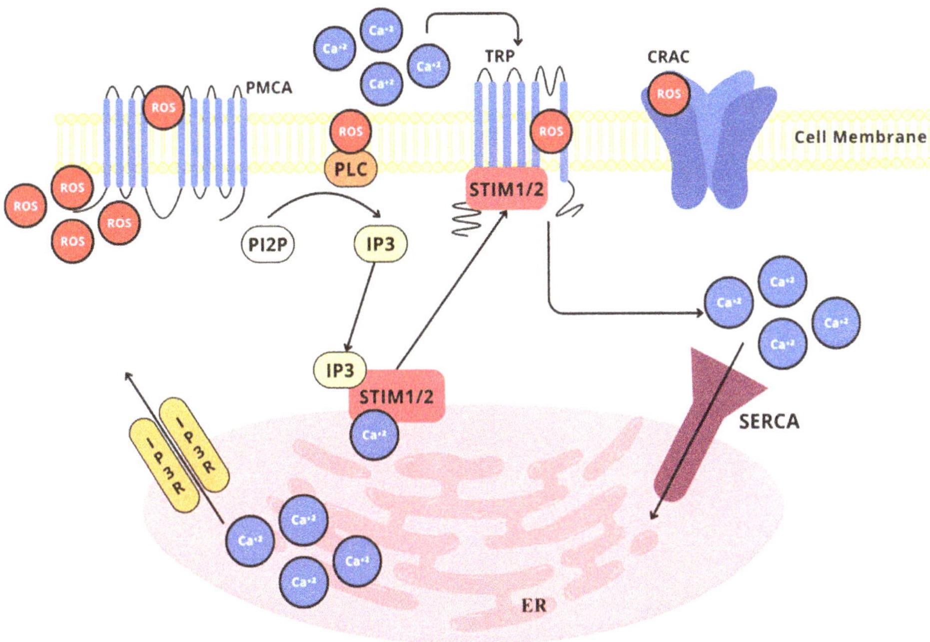

Figure 10. Calcium signaling induced by reactive species. ROS can stimulate calcium signaling in cells of the nervous system, initiating lipid peroxidation and activation of phospholipase C (PLC) and inositol 1,4,5—triphosphate (IP3) receptor (IP3R)-mediated calcium signaling. Also, ROS/RNS can directly oxidize or nitrosylate cysteine residues in Ca^{2+} channels and can influence Ca^{2+} channel gene expression, affecting their quantity and proportion in the cellular membrane, open-time probability, and trafficking. PMCA: plasma membrane calcium ATPase; STIM1/2: Stromal Interaction Molecule 1/2; CRAC: calcium release-activated calcium channel; SERCA: endoplasmic reticulum calcium ATPase.

ROS can stimulate calcium signaling in cells of the nervous system, as seen in cases where astrocytic MAO produced H_2O_2, initiating lipid peroxidation and the activation of phospholipase C and inositol 1,4,5—triphosphate (IP3) receptor (IP3R)-mediated calcium signaling [232]. Astrocytes in hypoxia conditions have been shown to trigger Ca^{2+} signaling as a response to ROS; also, RONS can directly oxidize or nitrosylate cysteine residues in Ca^{2+} channels, specifically in the pore-forming α-1-subunit [233], altering its conformational structure and function, modifying neurotransmitter release and synaptic plasticity, essential for learning, memory, and chronic pain [234]. Additionally, RONS can influence Ca^{2+} channels' gene expression, affecting their quantity and proportion in the cellular membrane, open-time probability, and trafficking [235]. Two major types of channel proteins seem to be involved in receptor-induced Ca^{2+} signals, the store-operated Ca^{2+} channels (SOC) and transient receptor potential channels (TRP). Both are of major importance for cell functions and thus have been accounted for in the entrance and release of Ca^{2+} through plasma membrane channels, and from its sites of storage, respectively. TRP, which are known to transport mainly Ca^{2+} ions, have six non-voltage sensitive transmembrane channels, divided based on their activation mechanism and N-termini or C-termini regulatory domain, the canonical (TRC1–TRC7), melastatin-related TRP (TRPM1–TRPM8), and vanilloid receptor-related TRP (TRPV1–TRPV4) [233].

All of the TRP family members are redox sensitive and participate in redox regulation and oxidative stress, either through activation by H_2O_2 mediated by ADP-ribose or cyclic

ADP-ribose reacting in the binding cleft of the C-terminal [236], by other oxidizing agents that sensitize TRP to pH alterations [237], or by the modulation of SOC, which is regulated by the translocation of the ER Ca^{2+} sensors stromal interaction molecule 1/2 (STIM1/2) to the plasma membrane, where they activate Orai channels to ignite calcium entry and storage [238]. The TRP channels are of paramount relevance in the processing and discrimination of heat, cold, pain, and stress components and in the establishment of neuropathic pain derived from chemotherapy treatment. Paclitaxel can cause severe peripheral neuropathic pain, demonstrated by a mechanical withdrawal threshold decrease in rats, which was inhibited after intrathecal and intraperitoneal administration of the TRPV1 antagonist, capsazepine; TRPM8 protein expression was also reduced in DRG neurons. The topical application of menthol increased TRPM8 activity and reduced TRPV1 activity. In that sense, TRPV1 upregulation and TRPM8 inhibited activity are related to neuropathic pain induced by paclitaxel [239]. Following nerve injury or neuron dysfunction, TRP channel expression is altered affecting neuronal excitability, neuroplasticity, and neuroinflammation, leading to neuropathic pain, as evidenced by TRPC1, TRPC4, TRPC5 and TRPC6 increased expression associated to mechanical hypersensitivity in SNI [240–242].

In this sense, both mitochondrial ROS production and Ca^{2+} have physiologic roles in normal cell functioning. Nevertheless, dysfunction in mitochondrial ROS production and Ca^{2+} equilibriums also play a major role in the pathogenesis of chronic pain states and cell death, once neurons are vulnerable to redox and energetic disturbances [243].

2.4.6. Unfolded Protein Response (UPR) Pathway

As formerly mentioned, when the ER undergoes stressing conditions, such as alterations in redox homeostasis, hypoxia, altered Ca^{2+} regulations, sustained hyperglycemia, and the accumulation of unfolded/misfolded proteins in the ER lumen [244,245], a series of adaptative/protective or maladaptive responses initiates, named the unfolded-protein response (UPR), which directly or indirectly affects the ER itself, or other organelles [92]. The primary objective of this elicited response is to reestablish normal ER function and homeostasis; nevertheless, when stress conditions are indeed critical and chronic, the cell is driven towards dysfunctionality, oxidative stress, ER stress, and ultimately, cell death [246]. ER stress is widely implicated in neuronal injury [247,248]. SNL induces an increased expression of ER stress markers such as *sXBP1* and BiP, as well as inducing the activation of the ATF6 and PERK/eiF2 pathway in dorsal horn neurons of the spinal cord [249]. The use of chaperones such as tauroursodeoxycholic acid has been effective in reversing ER stress, elicited by the UPR in neuropathic conditions like CCI, diabetic peripheral neuropathy, and that induced by formalin [248–250]. The influence of the UPR towards neuropathic pain involves central and peripheral sensitization, either by increased ER stress markers, or impairing calcium signaling, for the trigger of nociceptive pathways, leading to hyperalgesia.

The initiation and development of the UPR involves three proximal ER stress-sensor transmembrane kinases/transcription factor, namely inositol-requiring kinase 1 (IRE-1α), activating transcription factor 6 (ATF-6), and double stranded RNA-activated protein kinase (PKR)-like endoplasmic reticulum kinase (PERK), each of which has its own ER branch, but also works complementarily to restore normal ER functions, which are, under non-stressed conditions, attached to the immunoglobulin heavy chain binding protein (GRP78/BIP) through their amino terminals at their ER luminal ends (Figure 11) [92,251]. The ER embodies a highly oxidizing-folding environment to enable disulfide bond formation as one of the fundamental steps in normal protein structural formation; thus, 25% of the ROS produced by cells comes from ER activity [245].

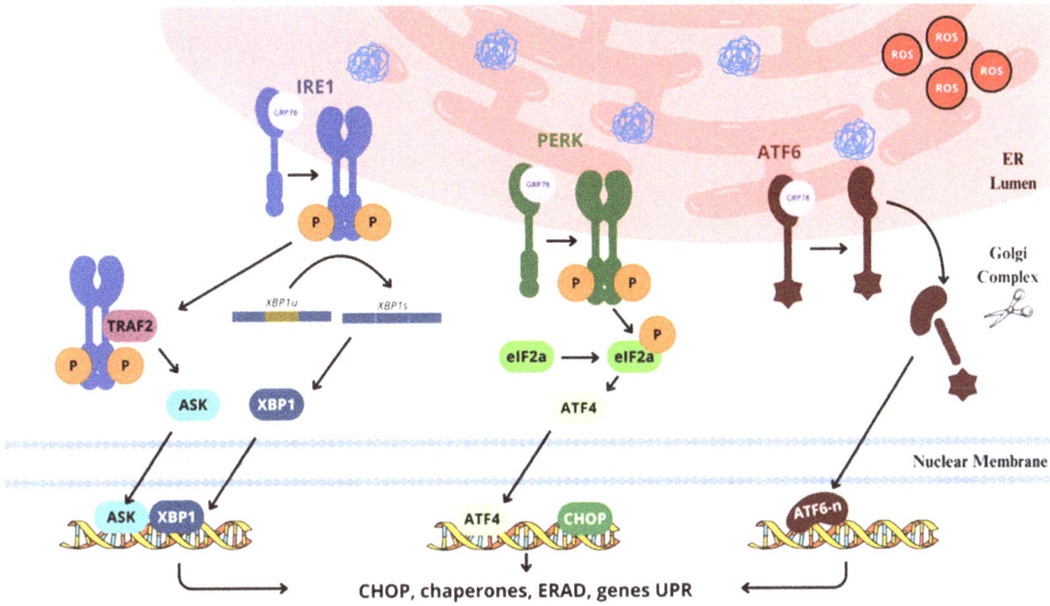

Figure 11. Unfolded protein response (UPR) pathway. Under non-stressed conditions, GRP78 attaches to unfolded protein response (UPS) sensors, IRE-1α, ATF-6, and PERK. In the presence of unfolded/misfolded proteins, GRP78 is detached from the UPR sensors; once activated, GRP78 induces complex pro-survival autophagy pathways, the antioxidant ERAD pathway, and ER regeneration, or pro-apoptotic and ferroptosis mechanisms. Also, ROS produced serves as a signal for activation or modification of the cellular stress response, once ROS-oxidated cysteine residues in the UPR sensors can mediate UPR signals. IRE-1α: Inositol-Requiring Kinase 1; ATF-6: Activating Transcription Factor 6; PERK: RNA-activated protein kinase (PKR)-like Endoplasmic Reticulum Kinase; eiF2a: Eukaryotic Initiation Factor 2 Alpha; XBP1: X-box Binding Protein 1; CHOP: C/EBP Homologous Protein; ASK1: Apoptosis Signal-regulating Kinase 1; TRAF2: Tumor Necrosis Factor Receptor-Associated Factor 2.

This oxidative process of intramolecular disulfide bond formation, known as oxidative protein folding (OPF) is one of the most usual posttranslational protein modifications [252]. The catalyzation of the disulfide bond formation is primarily carried out by the thioredoxin protein disulfide isomerase (PDI), through thiol-disulfide oxidation. PDI are formed by four thioredoxin (Trx) domains, namely a, a′, b, and b′, and a KDEL ER retention group of c-domain [253]. The redox state of the catalytic a-domains in the PDI enzyme establishes the oxidase or isomerase activity, depending on the CGHC (cysteine-glycine-histidine-cysteine) active site motifs; thus, PDI in its reduced state, due to the active site cysteines within the CGHC motif present as thiols (-SH), allows PDI to disrupt incorrect disulfide bonds, acting as disulfide reductase, as well as rearrange incorrect disulfide bonds into correct ones, acting as isomerase, while in the oxidized form, the active site cysteines within the CGHC are present as disulfide (-S-S-) bonds, allowing PDI to oxidize substrates, and to form new disulfide bonds between cystine residues, working as an electron acceptor [254].

On the other hand, b′ domains are responsible for the determination of misfolded or unfolded protein, through the identification of hydrophobic exposed terminals of the proteins [255]. H_2O_2, in addition to being the main source for oxidation, produced by O_2, is also used by peroxidases to shape disulfide bonds to re-oxidize PDI. The cyclic reoxidation of PDI in the active sites help maintain the redox balance within the ER, which is more oxidative than the cytosolic environment. Yet, under highly oxidizing conditions, the CGHC motif can be hyperoxidized to its cysteine sulfinic acid or sulfonic acid form, turning it inactive, and unable to catalyze further reactions, leading to an accumulation of unfolded or misfolded proteins [254].

Successively, the ER membrane-associated oxidoreductin (ERO-1), that possesses two cysteine pairs on a flexible loop and on the CXXC motif, transfers sulfhydryl electrons from the reduced PDI to O_2, yielding H_2O_2, which adds to maintaining the oxidizing environment [246,256]. In this sense, the ER redox state is tightly related to ER homeostasis, which depends on ROS production during disulfide bond formation in protein folding, to participate in the pro-oxidizing environment in the ER [257]. Additionally, quiescin sulfhydryl oxidase (QSOX), an FAD domain- and Trx domain-containing enzyme, catalyzes the formation of disulfide bonds, via free thiol group oxidation on cysteine residues of substrate proteins, transferring electrons from the cysteines of substrates to the cysteines in the Trx domain of QSOX, which are then shifted to its FAD domain. The reduced $FADH_2$ domain acts as an electron acceptor, ready to be re-oxidized, conveying the electrons to O_2, yielding H_2O_2, allowing QSOX to reenter a new cycle of disulfide bonds formation [258].

Further increase in ROS formation is triggered by Nox4 activity after UPR [259]. Additionally, another source of ER-stress is attributed to ROS-dependent Ca^{2+} depletion from the ER, which restrains the ability of Ca^{2+} to activate ER chaperones, such as calreticulin and calnexin. This can be explained due to the fact that IP3R, the ryanodine receptor (RyR), and Ca^{2+} pumps are redox regulated. For example, SERCA activity is inhibited when the Cys674 residue undergoes sulfoxidation and is solely activated after the same cysteine residue undergoes glutathionylation by NO; IP3R's sensitivity to IP3 and RyR activity is enhanced by ROS, leading to augmented Ca^{2+} release from the ER [260].

However, excessive RONS can lead to the dysregulation of these receptors, contributing to ER and cellular stress and death through disrupted calcium homeostasis. As another source of ER oxidizing environment, when misfolded proteins appear, disulfide bonds are corrected by reduction through GSH, resulting in higher oxidized glutathione (GSSH) rates; this also participates in redox ER homeostasis [92].

In this process, paradoxically, the UPR induces increased ROS levels, while at the same time, ROS-mediated ER stress is expected to be restored by UPR. In this context, ROS ER levels increase due to higher ERO-1-PDI activity, but the ROS produced also serve as a signal for the activation or modification of the cellular stress response, once ROS-oxidated cysteine residues in the UPR sensors, IRE-1α, ATF-6, and PERK can mediate UPR signals, being activated to induce complex pro-survival autophagy, the antioxidant ERAD pathway, and ER regeneration, or pro-apoptotic and ferroptosis mechanisms [94]. Similar to ROS, RNS induce nitrosative stress, being linked to UPR-mediated cytotoxicity via cysteine residue inactivation in the active sites of PDI [261].

3. Antioxidant Enzymes

The nervous system exhibits a high oxidative metabolic activity rate with a consequent preeminent amount of reactive species production via enzymatic and non-enzymatic reactions as by-products of normal neuronal functioning [20], being particularly elevated during neuron and glia activation by glutamate and ATP [262]. Moreover, the relatively low levels of antioxidants [263], the high membrane surface to cytoplasm rate, and the high lipid content make neurons particularly vulnerable targets to oxidative damage over macromolecules and lipid peroxidation [264]. Although, as mentioned above, redox signaling works as an intrinsic sensor for oxidative stress, serving several physiological functions, and when the cellular antioxidant systems fail to prevent the overproduction of such reactive species, detrimental pathways can be activated, inducing cellular and tissue damage.

The cellular antioxidant systems, that is, those compounds present at a lower concentration in relation to oxidizable substrates that are able to hinder or preclude the oxidation of the substrate [265], are composed by enzymatic and non-enzymatic molecules and are chemically structured for and capable of reducing other compounds with an oxidizing potential of different chemical natures [65]. The first line of endogenous defense is constituted by an enzymatic antioxidant system that includes (superoxide dismutase) SOD, (catalase) CAT, and (glutathione peroxidase) GPx, whereas the second line of defense is formed by non-enzymatic antioxidants, such as antioxidant enzyme cofactors, reactive species scavengers, oxidative enzyme inhibitors, and Cu, Fe, Zn, and Mn transition metals, by participating in reactions that neutralize ROS directly or catalyzing the formation of less reactive species (Figure 12) [263,266]. Further processes of the restoration of damage by a repair system can be activated, eliminating oxidized nucleic acids by proteolytic enzymes, and lipids, either by phospholipases or peroxidases [16,267].

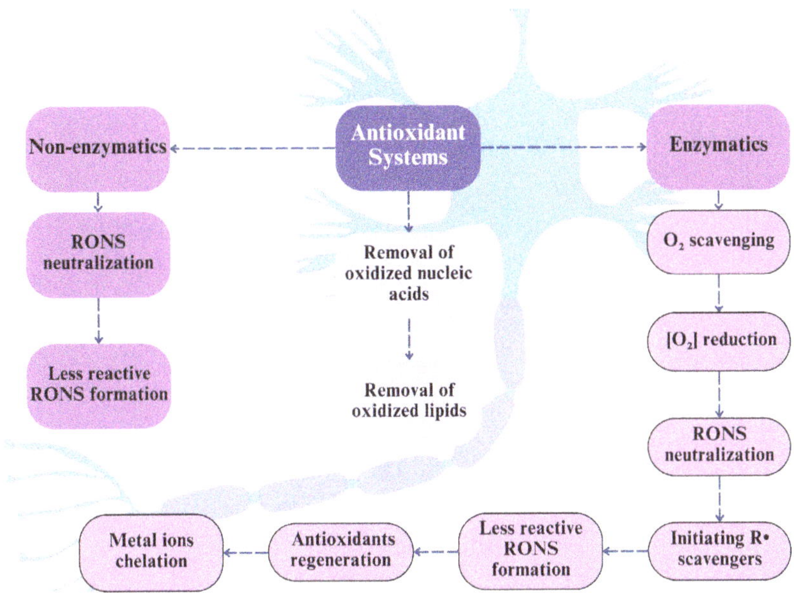

Figure 12. Enzymatic and non-enzymatic antioxidant endogenous systems' mechanisms of action.

For the purpose of this review, the focus will be on preventative enzymatic antioxidants, as these are considered the most important among the endogenous antioxidant systems for their compelling contribution to the normal redox state, and for their role in promoting the synthesis or regeneration of further antioxidant or pro-survival mechanisms [46,268]. In addition to SOD, CAT, and GPx, which are considered to be the first front line of the endogenous antioxidant defense system, the Trx enzyme and thioredoxin reductase (TrxR) will also be included for their enzymatic nature, as opposed to those considered by other authors [269].

3.1. Superoxide Dismutase

The superoxide dismutase family (SODs) includes transition metal-containing enzymes present in all living organisms existing in the presence of oxygen. SODs are considered one of the most powerful antioxidant enzymes against biological oxidants and possess the ability to transform two O•− molecules into H_2O_2 and O_2 in a pH-independent medium [268]. In mammals, three forms of SOD are acknowledged. The Cu, Zn-containing

SOD-1 is a homodimer constituted by eight beta strands, with one Cu atom serving for catalytic function and one Zn atom relevant for structural stability, in each of the two subunits, respectively; it is ubiquitously expressed, and found predominantly in the cytoplasm, but also expressed in the mitochondrial intermembrane space and the nucleus [270,271]. SOD-2 is a homotetrameric enzyme that contains Mn in its active site. It is found in the mitochondrial matrix where the pH is considerably higher (pH approx. 7.8), compared to that of the intermembrane space (approx. 7.0–7.4); it is expressed at different amounts depending on the type of cell, being significantly higher in those cells that encompass elevated proportions of mitochondria [268].

They perform their dismutation power over the O•− molecules produced in the ETC, providing electrons from the metal core directly to the negatively charged O•− molecules and a protonation to yield H_2O_2 [271]. Finally, the Cu- and Zn-containing enzyme SOD-3, produced intracellularly and found predominantly in the extracellular space [272], is homotetrameric in humans and mice, but dimeric in rats (Figure 13) [273]. As opposed to SOD-1 and SOD-2, which are ubiquitously expressed, SOD-3 is diversely expressed across different types of cells, and found to be highly expressed in heart, lungs, and pancreas, but low in the brain [273]. Additionally, SOD-3 is generally less responsive to direct induction by O•− molecules or other oxidants, due to its presence and role in the extracellular matrix, where exposure to regulatory signals other than those encountered in the intracellular space are found. SOD3 is regulated in conjunction with signaling molecules and cytokines, such as interleukin-1 (IL-1β) and TNF-α, as part of a coordinated response to inflammation, contributing to modulate oxidative stress in the extracellular environment [274].

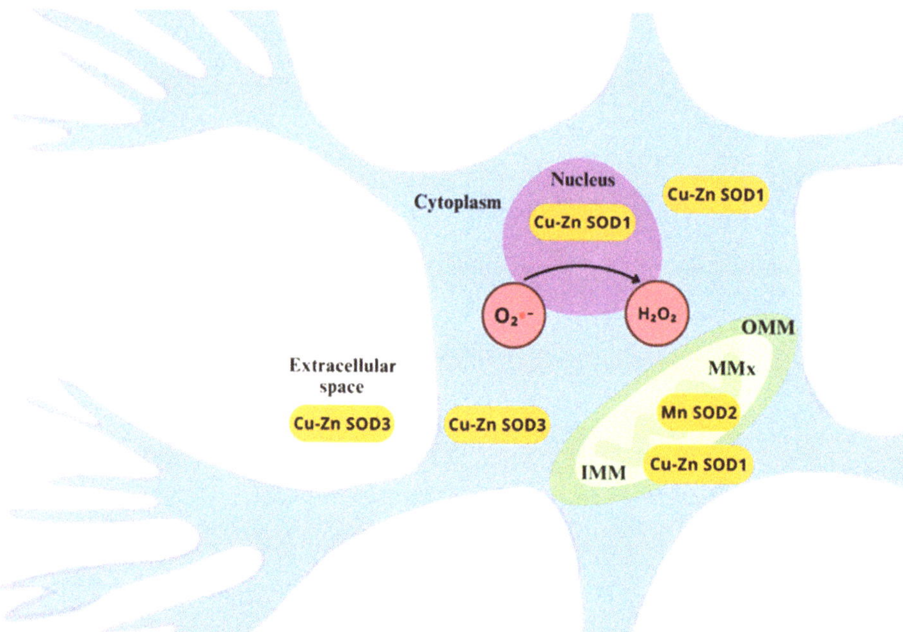

Figure 13. Superoxide dismutase (SOD) sites of expression in neuronal cell. IMM: internal mitochondrial membrane; OMM: outer mitochondrial membrane; MMx: mitochondrial matrix.

The dismutation reaction catalyzed by SOD is considered to be highly efficient given that it happens at a diffusion-limited rate of ~2 × 10^9 $M^{-1} \cdot s^{-1}$, that is ~10^4 the rate constant during spontaneous dismutation [275]. In the process, an electron reduction and protonation of the O•− radical results in the formation of H_2O_2, by means of an initial step of O•− radical binding to the core Cu(II) ion, resulting in the oxidative transfer of an electron to O•− and the production of reduced Cu(I); then, a second O•− radical is bound to a partly protonated Arg143 in the anion-binding site of the SOD that oxidizes the O•− radical. Moreover, His63 donates a proton and an electron from reduced Cu(I) to O•− radical, producing H_2O_2, and the oxidation of Cu(I) to Cu(II), to pursue further dismutation [276].

Changes in SOD expression or function are linked to chronic pain. Maropitant treatment increases SOD expression, reducing pain threshold in peripheral nerve injury, often marked by neuronal loss [248,271]. The enhanced expression of SOD-1 was observed to reduce lipopolysaccharide-mediated superoxide production, suppress cyclooxygenase (COX)-2 and iNOS expression, and confer protection against microglia-mediated inflammatory responses [277]. Compromised SOD-2 function is a common characteristic of pathogenesis associated with oxidative stress in brain injury [278]. Moreover, an increase in SOD-2 expression avoided neuronal death from oxidative stress [279].

3.2. Catalase

Catalase (CAT) is a tetrameric heme-containing enzyme, formed by four tetrahedrally arranged subunits of 60 kDa each, containing an active heme group and NADPH bound to each subunit [280]. It is part of the front-line defense antioxidant system, commonly found in mammalian cells as part of the peroxisome [263]. CAT is produced due to oxygen exposure and is characterized by the ability to catabolize the rupture of two H_2O_2 molecules to form two molecules of H_2O and one of O_2, through a two-step reaction, using Fe as a cofactor (Figure 14) [281]. It is expressed ubiquitously in central nervous system cells, including neurons and glia, where it protects cells by the detoxification of H_2O_2, allowing the body to tolerate and adapt to oxidative stress, as an adaptative response [282]. In addition to its protective effects against ROS, CAT offers influence over several pathways and metabolic processes such as growth, proliferation, and apoptosis [283] and influence over redox-sensitive signaling molecules, like protein kinases and transcription factors, through the modulation of H_2O_2 levels, impacting cellular responses to environmental or endogenous stressors [284].

The reaction is described as a two-step reaction, as follows: initially, a compound I, the covalent oxyferryl species Fe(IV) possessing a porphyrin π cation radical, is formed from Fe(III), via the reduction of one H_2O_2 molecule; following this, the Fe(IV) compound is decomposed through redox reactions via two-electron transfer from the second H_2O_2 molecule, which works as an electron donor to the reaction, therefore, producing O_2 and H_2O, and restoring the compound I to its resting state Fe(III) [285]. After this the final products, O_2 and H_2O, are released from the catalase, and the latter returns back to its original state, to pursue further reactions, binding and decomposing another H_2O_2 molecule. CAT is maintained chemically and structurally available after repeated cycles of chemical reductions of H_2O_2, contributing to steady O_2 concentrations states, and redox balance [285].

Also, it is considered one of the most powerful antioxidant molecules in biological systems, as it has the highest turnover of all enzymes, degrading millions of H_2O_2 per second [268]. The inhibition of CAT activity induced increased cytotoxicity and higher ROS rates, denoting a pivotal role in oxidative balance [263].

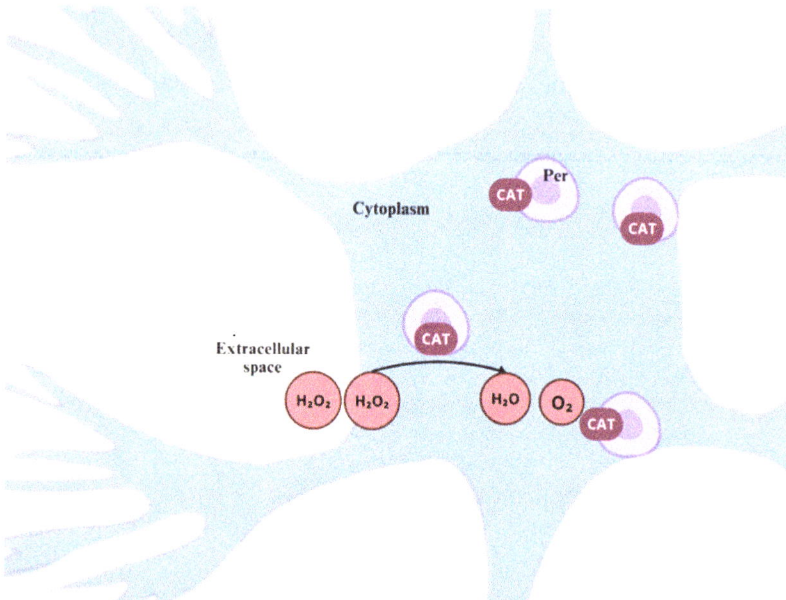

Figure 14. Catalase (CAT) sites of expression in neuronal cell. CAT is characterized by the ability to catabolize the rupture of two H_2O_2 molecules to form two molecules of H_2O and one of O_2. Per: peroxisome.

Therefore, CAT dysfunction or deficiency is thought to cause impaired glucose metabolism and lipid accumulation [286], as well to contribute to oxidative stress and neuronal damage in paclitaxel-induced neuropathic pain, where it is decreased together with NRF2, SOD2, and HO-1, and can be reversed by cannabidiol (CBD) and tetrahydrocannabivarin (THCV) by the upregulation of catalase, but also other protective proteins such as p-AMPK [287], contributing to pain relief and neuronal protection.

CAT is increasingly recognized as a therapeutic target for diseases, founded on CAT-based therapeutic strategies providing neuroprotective effects through integrity and function neuronal preservation, for treating or averting neuroinflammatory events, using viral-cultured CAT or gene therapy, delivering *CAT* genes, and enhancing endogenous *CAT* gene expression. Ongoing innovative studies are utilizing protein engineering to enhance CAT-based therapies, aiming for improved stability, activity, and specificity, and demonstrating promising potential to expand therapeutic applications [288].

3.3. Glutathione Peroxidase

The family of antioxidant enzymes glutathione peroxidase (GPx) consist of a group of eight isoforms (GPx1–GPx8) identified to date and present in all living organisms. Among them, GPx1 is the most abundantly found in almost all tissues and is highly expressed in the cytoplasm and mitochondria; GPx2 is found mostly in the gastrointestinal tract and human liver; GPx3 is found in the plasma and in the extracellular space; GPx4 is expressed ubiquitously, but especially in the brain, testis, and sperm cells where cell membranes need to be protected from lipid peroxidation; GPx5 is specific to the epididymis; GPx6 is expressed in the olfactory epithelium and embryonic tissues; and GPx7-8 are expressed predominantly in the ER, where they play a role in protection from oxidative stress and maintaining the oxidative balance for oxidative protein folding and the secretory pathway (Figure 15) [289]. Chemically, GPx are oligomeric enzymes, formed by subunits arranged with either a selenocysteine or cysteine amino acid redox-sensitive active site, both playing main functions in the catalytic reduction of hydroperoxides and H_2O_2, which are transformed to alcohols (ROHs) or H_2O, respectively [290].

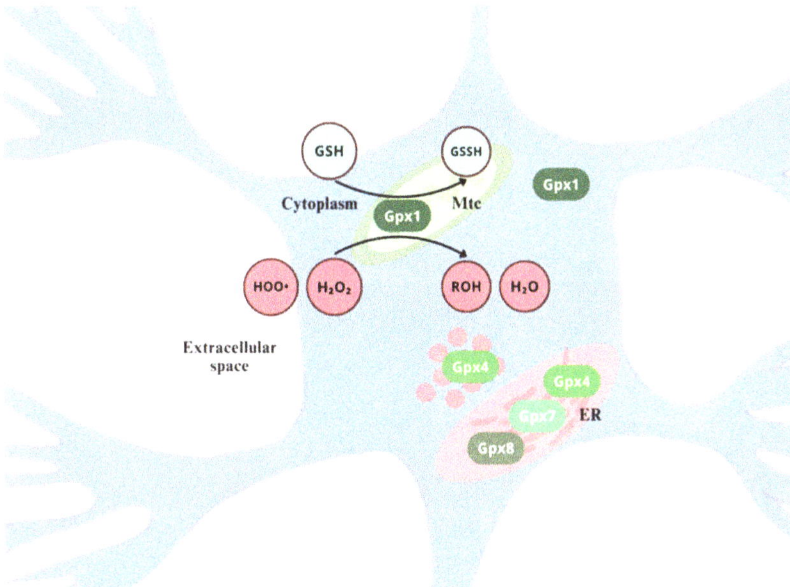

Figure 15. Glutathione peroxidase (Gpx) sites of expression in neuronal cell. Gpx drives the enzymatic reduction of H_2O_2 and hydroperoxides via oxidation of reduced glutathione (GSH) to its oxidized form (GSSH) in a ping-pong mechanism involving oxidation of the GPx followed by a two-step reduction.

The GPx family has a vast number of physiological functions, although the most important role played is related to the enzymatic reduction of H_2O_2 and hydroperoxides via the oxidation of reduced glutathione (GSH) to its oxidized form (GSSH) in a ping-pong mechanism involving the oxidation of the GPx followed by a two-step reduction [291]. As a first phase, the catalytic site of GPx consisting of a selenocysteine or a cysteine amino acid is oxidized to selenenic acid or sulfenic acid, respectively, by hydroperoxides or H_2O_2 which successively are transformed to their corresponding reduced forms, alcohols (ROHs) or H_2O [292]; thereafter, the second phase involves the return of GPx to its initial reduced state, through a series of reduction reactions by the GSH and NADPH as a cofactor. Here, the two-step reduction process starts with selenenic acid and sulfenic acid return to their oxidized states, selenocysteine and cysteine, with the help of two GSH molecules, which become oxidized (GSSG) in the process [289]. The return of the oxidized GSSG form to the reduced GSH form involves the enzyme glutathione reductase (GR); for such, NADPH binds to the active site of the GR that has an FAD cofactor. Here, the NADPH donates electrons to FAD, reducing it to $FADH_2$, which in turn transfers these electrons to the disulfide bond in GSSG, disrupting the bond and yielding two molecules of GSH. As FAD is reduced, NADPH is oxidized to NADP+, which is released from the enzyme, finishing the cycle [289].

The intracellular depletion of GSH and decreased GPx4 activity caused by ferroptosis, an iron-dependent cell death pathway seen in pathological processes, results in the accumulation of lipid peroxides, the Fe(II)-catalyzed oxidation of lipids, and an increased production of reactive species [293]. In a spinal cord injury model in rats, GPx was upregulated following treatment with L-carnitine and atorvastatin, as opposed to non-treated animals, improving neuronal function [294]. Gpx is also upregulated in SNL- induced neuropathic pain [295]. Gpx4 was downregulated in a model of spared nerve injury in rats; after methyl ferulic acid administration, Gpx4 levels were increased, accompanied by a significant NOX4, ferroptosis- related protein ACSL4 decrease, and consequent mechanical withdrawal threshold and thermal withdrawal latency improvement [296]. Another study assessing whether sirtuin 2 (SIRT2) attains a neuroprotective effect in spared nerve injury-induced neuropathic pain, found enhanced levels of Gpx4 and ferroportin 1 (FPN1) after SIRT2 recombinant adenovirus intrathecal administration, reducing intracellular iron accumulation and oxidative stress, hence, attenuating mechanical hypersensitivity [297]. It has also been demonstrated that Gpx4 upregulation can reverse and inhibit ferroptosis-induced stress, blocking neurons and glia activation from the dorsal horn of spinal cord in neuropathic pain induced by chronic constriction injury in rats [298,299]. Altogether, GPx is a relevant target for therapeutic approach towards neuropathic pain [300].

3.4. *Thioredoxin*

The thioredoxin (Trx) system is formed by two oxidoreductase enzymes, Trx and TrxR, with NADPH as an electron donor. Trx are highly conserved across living cells, including mammals. The Trx system has greatly important functions protecting cells against oxidative stress and damage, being particularly important in the CNS cells, where it assists correct protein synthesis and the modulation of apoptosis [301]. Trx serves as a redox homeostasis regulator through the modulation of signaling pathways, induction of transcription factors, and pro-survival mechanisms [302]. Trx is depicted with a Cys-Gly-Pro-Cys active site vital for protein disulfide reductase of enzymes like peroxidredoxins (Prxs), ribonucleotide reductase (RNRs), and methionine sulfoxide reductase (MSRs), as it serves as an oxidoreductase and an electron donor [303]. Trx has the ability to catalyze the reductive conversion of disulfide (S-S) bonds into dithiol (-SH) bonds on substrates, due to its reductase power, acting through disulfide–dithiol bonds exchange [263]. The whole process happens as follows: Trx binds to the protein substrate that contains a disulfide bond (S-S); then, one thiol (-SH) group of the N-terminal cysteine in the active site (Cys-Gly-Pro-Cys) of the Trx performs a nucleophilic attack on the disulfide (S-S) bond of the substrate by electron donation, yielding an intermolecular mixed S-S between the Trx and the substrate. Afterwards, the second thiol (-SH) group in the C-terminal cysteine in the active site (Cys-Gly-Pro-Cys) of the Trx attacks the intermolecular mixed S-S bond, releasing the reduced substrate and leading to the creation of a disulfide (S-S) bond in the Trx itself. Finally, the disulfide (S-S) bond in the oxidized Trx is reduced back to its original free thiol (-SH) groups state, through the enzyme TrxR, using NADPH as an electron donor, ready to engage in the next catalytic cycle (Figure 16) [304].

In mammals, there are two primaries strict Trx proteins: Trx1 and Trx2. Trx1 is the most widely studied and present mainly in the cytosol but can be found in the nucleus upon oxidative/nitrosative stress [305] or released through the cellular membrane [306]; Trx1 is coupled to Prx1 and Prx2, and MSR, being critical for cellular growth and apoptosis, and is expressed in neuronal cells (Figure 17). Thus, neurons with dysfunctional mitochondria induced by complex IV inhibition show increased H_2O_2 vulnerability due to low Trx expression [307]. The Trx2 isoform is localized in the mitochondria and is exceptionally relevant for maintaining mitochondrial redox balance, as well as for the protection of mitochondrial DNA and proteins from oxidative damage [308]. Similarly to Trx1, Trx2 is expressed in neurons, and shows lower expression levels after complex IV inhibition, making the cells more susceptible to H_2O_2 attack [307]. Both Trx1 and Trx2 are primarily expressed in sensitive neurons in the DRG and in the spinal cord and sciatic nerve and are particularly altered in nociceptive signaling sites after sciatic nerve injury, participating in the regulation of oxidative stress and redox homeostasis, relevant in the pathophysiology of neuropathic pain [309].

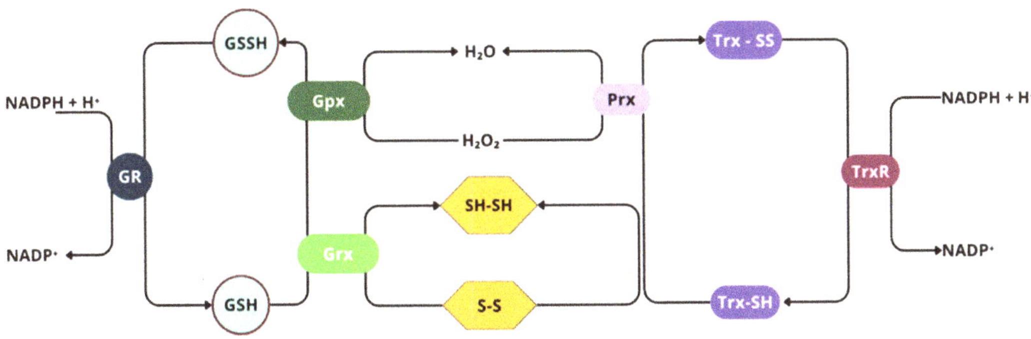

Figure 16. The thioredoxin system in neuronal antioxidant defense. Trx has the ability to catalyze the reductive conversion of disulfide (S-S) bonds into dithiol (-SH) bonds on substrates, due to its reductase power, acting through disulfide–dithiol bond exchange. The disulfide (S-S) bond in the oxidized Trx is reduced back to its original free thiol (-SH) group state, through the enzyme TrxR, using NADPH as an electron donor, ready to engage in the next catalytic cycle. GPx catalyzes the reduction of hydrogen peroxide (H_2O_2) to water. To do this, GPx utilizes glutathione (GSH) as an electron donor. Here, two molecules of GSH are oxidized to form a disulfide bond (S-S), resulting in GSSG (oxidized glutathione). Now, the oxidized glutathione (GSSG) is recycled back to its reduced form (GSH) to maintain its antioxidant capacity. GSSH: oxidized glutathione; GSH: glutathione; Gpx: glutathione peroxidase; Grx: glutathione reductase; GR: glutathione reductase; Trx: thioredoxine; TrxR: thioredoxine reductase; SH: dithiol bonds; SS: disulfide bonds.

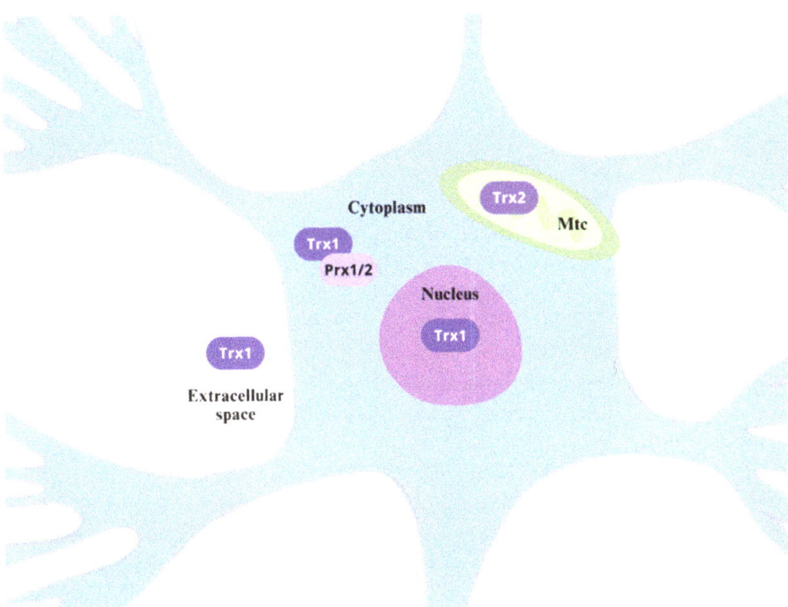

Figure 17. Glutathione peroxidase (Gpx) sites of expression in neuronal cell. Gpx drives the enzymatic reduction of H_2O_2 and hydroperoxides.

The above-mentioned antioxidant enzymatic proteins play critical roles as regulators of various redox signaling pathways, which profoundly depend on reversible tissue- and target-specific modifications of main thiols in proteins. The CNS, a highly oxidizing environment, is acknowledged as very vulnerable to reactive species effects carried out by nitrosative or oxidative damages. While each enzyme has a specific role, they work together through synergistic interactions, redundancy, and compartmentalization to ensure a robust, correct defense against reactive species and to guarantee cellular redox equilibrium.

4. Nitroxidative Stress in Neuronal Cells

Pain can be considered, in a wide sense, as a protective mechanism to alert to actual and imminent danger to tissue integrity, or as a warning to potential lesions, eliciting a myriad of responses across emotional, sensory, social, cellular, and molecular domains. When the sensory process that provides triggering signals for pain occurs, nociception initiates, comprising the detection of harmful or potentially harmful stimuli by nociceptors, as a normal response to such stimuli. That is, after nerve injury, the activation of macrophages, circulating neutrophils, and T cell recruitment to the site of the lesion, along with cytokines, chemokines, neurotransmitters, and ROS production, prompts peripheral and central sensitization [310]. In some cases of injury, it often leads to abnormal sensory phenotype expression in the nervous system, causing hypersensitivity due to peripheral nerves, dorsal root ganglia (DRG), or central nervous system damage, ultimately leading to neuropathic pain [311], which is defined as a pain caused by a lesion or a disease of the somatosensory system [312] estimated to affect approximately 7–10% of the global population [313], characterized by paresthesia, dysesthesia (tingling, numbness, pins, and needles sensation), associated with non- painful sensory neurological deficits in the disturbed area, together with motor and cognitive deficits, depending on the affected region [314].

Glial cells are important for homeostasis maintenance in the CNS, providing structural and nutritional assistance for neurons through hydro-electrolyte balance and neuronal synapsis (astrocytes), or surveying and protecting the neuronal milieu against potential disruptions of homeostasis and astrocyte activation (microglia) [315]. Immune cells of the CNS play a critical role in the development and continuance of neuropathic pain, as nerve insult leads to the neuroinflammation response, characterized by increased local vascular permeability, the activation of glial cells and inflammatory mediator release, enhancing pain sensitization [316]. Once microglia are activated, phenotypical and functional modifications occur, leading to proinflammatory cytokine release, such as TNF-α, IL-1β, and IL-6, cluster differentiation molecule 11b (CD11b), CD68, and ionizing calcium binder adapter molecule (Iba1) [9,317], all of which contribute in the early phase of neuropathic pain stablishment, and also causing astrocyte activation, supported by an increased expression of glial fibrillary acidic protein (GFAP) and CD11b, indicating a shift in astrocyte function, which in turn activates microglia, thus, creating cross-talk for neuropathic pain pathophysiology [194, 318]. While microglia contribute initially to the establishment of neuropathic pain via the inflammatory response initiation, astrocytes are accountable for the long-term maintenance of such state [319,320].

Activated glia release chemokines, RONS, and neurotransmitters that modulate inhibitory and excitatory synaptic transmission, relevant for pain amplification [5]. Glutamate, the primary excitatory neurotransmitter is upregulated in astrocytes in response to inflammatory signals, contributing to central sensitization and pain amplification; aspartate, similarly to glutamate can be influenced by glial activation, contributing to excitatory signaling in neuropathic pain; and (Gamma-amynobutyric acid) GABA, the primary inhibitory neurotransmitter, responsible for reducing neuronal excitability and neuronal firing, can be downregulated in neuropathic pain, causing diminished inhibitory signaling, and consequently, increasing pain sensitization [316]. Additionally, purinergic receptors (P2X and P2Y), TLRs, Src kinase, and MAPK (JNK, ERK, and P38) pathways can be upregulated in conditions of nerve injury. This leads to increased expression of transcription factors in activated glia through various mechanisms, including further GF, proinflammatory cytokine and chemokine, and RONS production, contributing to pain sensitization, as described above [321]. In this sense, RONS act as redox sensors for signal transduction pathways modulating the neuroinflammatory process in neuropathic pain [37]. In case the prompting offense to glial cells is not resolved, chronic activation arises, reflecting in a continuum of glial activation states [322].

The molecular mechanisms underlying the neuropathic pain shift the microglial phenotype from a resting, small, highly branched, compact cell body and surveillance state to an activated bigger cell body, with a less branched, and ameboid state, causing an altered expression of cell surface proteins, growth factors, intracellular signaling molecules, and redox state consistent with inflammatory mediators release, such as cytokines and proteases contributing to the development of the neuroinflammation and neuropathic pain sensitization [311].

Morphologically, three states are described in the microglial cells, similar to that used in peripheral macrophages: M0 phenotype for the resting state, M1 phenotype for the classical activation state, and M2 phenotype for the alternative activation state [323]. The first state is characterized by the expression of genes related to neuronal function and development, whereas M1 and M2 phenotypes are induced by CNS insults and chronic pain, portrayed as mixed temporally changing and lacking equilibrium in the activation continuum. Thus, the M1 state is thought to lead to excessive RONS production and neuroinflammation, and M2 hampers an adequate immune response [323]. Also, astrocytes express antioxidant proteins to maintain overall redox homeostasis in the CNS under normal or pathological conditions [324,325]. In this sense, microglia and astrocytes can influence direct structural and molecular modifications in neuronal synapses, neuron differentiation, and the axon growth of mature neuronal cells, both physiologically, as part of normal CNS functioning, and pathologically to cause oxidative and nitrosative damage and an imbalance of redox signaling [308].

Microglial cells are of prominent relevance regarding the inflammatory process typical of neuropathic pain [326], as they release a large number of neuroactive substances, which in turn contribute to the RONS production. Among these RONS, it can be pointed out that Nox- and iNOS-derived radicals in neurons, such as $O\bullet-$ and NO, contribute to the cell death of dopaminergic neurons in neuropathic pain [17,114]. Microglia under oxidative stress produce increased inflammatory mediators that can move through the cell membrane, working as signaling molecules to other cells or inducing peroxynitrite formation causing DNA fragmentation, lipid oxidation, and neuronal death in a faster manner than that perceived in astrocytes [327]. In microglia cells producing $O\bullet-$ catalyzed by nitrate and nitrites, O_2 and H_2O_2 levels are rapidly imbalanced, disturbing microglial functions, which in turn facilitates adjacent neuronal cell damage induced by ROS [328]. Elevated levels of RONS have been proved to increase proinflammatory cytokines in microglial cells, whereas lower RONS levels are associated with anti-inflammatory cytokine upregulation [114]. ROS generated by Nox2 contributes to neuropathic pain inducing the activated microglia phenotype, further enhancing the development of the chronic glial activation state in the spinal cord, and assisting the interaction of neurons and macrophages in the DRG [112,329,330], pointing the central and peripheral role in neuropathic pain, whereas Nox4's role is primarily related to peripheral demyelination, causing peripheral nerve sensitization, and therefore adding up to the overall effect of neuropathic pain [114,116].

In addition to microglial cells, other CNS cells can produce O•− and NO wherever NOS and Nox are expressed, contributing to radical-mediated damage, as seen in β-amyloid-induced neurotoxicity in astrocytes [331]. Astrocytes serve key functions in neuronal recovery but can, under certain circumstances, be neurotoxic. When activated, they also release NO, TNF, and ROS to the extracellular space, which can negatively affect axon regeneration and nerve development, and lead to neuronal injury and death through mitochondrial stress, DNA strand breaks, and lipid peroxidation [332]. Moreover, when astrocytes suffers sustained oxidative stress, glutamate transporters, astrocyte–neuron communicating enzymes [333], and connexins undergo substantial modifications, which are implicated in the establishment of neuropathic pain [334]. Mitochondria dysfunction, NADPH-derived ROS, and RNS in astrocytes are the main factors accountable for oxidative and nitrosative stress, which is distributed across the long and thin processes and cell bodies, thus leading to astrocyte degeneration, SOD1 aggregation, and ischemic/reperfusion injuries [335,336].

In a 1-methyl-4-phenyl-1,2,3,6-tetrahydropyridine (MPTP) model of cell death of dopaminergic neurons, researchers aimed to document whether chronic exposure to small doses of MPTP were capable of affecting neurons from the dorsal horn of the spinal cord, responsible for the sensory component of pain. Considering that MPTP inhibits mitochondrial complex I, this leads to an augmented production of ROS, oxidative stress, lipoperoxidation, and mitochondrial membrane rupture, leading to apoptosis pathways and neuronal damage, evidenced by decreased calbindin D28K, calretinin, and parvabulmin-positive neurons in the laminae II; a reduction of SP interneurons and Met—enkephalin-positive fibers in laminae I and II; and finally, increased alpha-synuclein in spared neurons and fibers, establishing a clear connection between mitochondrial dysfunction, increased ROS production, and neuronal damage [337]. Also, Nox-deficient animals in an inflammatory model of dopaminergic cells death were found to be less affected than the wild type of animals, suggesting a direct role of Nox-derived O•−in the mechanisms of microglial cell activation and neurotoxicity [338].

Furthermore, TRPA1, TRPM2, and TRPV1 are activated by nitroxidative species, mediating the integration of endogenous and exogenous sensory stimuli, at the peripheral and central terminals and somas of afferent neurons [37]. TRPA1 and TRPV1 are expressed by peptidergic C fibers, activated directly by peroxidized and nitrated phospholipids and carbonylated proteins, as well as aldehyde formation, due to oxidative stress [82,339]; TRPA1 is primarily related to nociceptive hypersensitivity development, while TRPV1 is crucial for nociceptive hypersensitivity and thermal hyperalgesia [82]. TRPM2 is expressed in neurons and microglia, activated directly via H_2O_2 and cytosolic ADP—ribose [340,341], which are by-products of mitochondrial damage and are associated with spinal cord microglia activation, the local migration of macrophages after nerve injury, proinflammatory response, and promoting nociceptive hypersensitivity and are also crucial in the activation process of the MAPK pathway, causing NF-κB nuclear translocation, consequently increasing the proinflammatory cytokine and chemokine production [340,342]. In this context, TRP plays a critical role in the development and maintenance of neuropathic pain through polymodal activation, including chemical, thermal, and mechanical signals, inducing ionic changes across neuronal membranes, and initiating downstream signaling pathways comprising: AMPK, MAPK, NF-κB, and TGF- β. TRP also enhance hyperexcitability via the increased expression and activity of TRP, leading to spontaneous pain and disturbed sensitivity [343]. Additionally, such a state of increased expression and activity of TRP leads to the hyperexcitability, apoptosis, hyperactivation, and infiltration of immune cells, amplified levels of proinflammatory cytokines, mitochondria dysfunction, and autophagia as seen in models of diabetic neuropathy in peripheral nerves [344], cancer-induced pain [345], chemotherapy-induced peripheral pain [346], and trigeminal neuropathic pain [22]

It is widely known that proteins can undergo reversible modifications through oxidative actions in amino acids. Amongst the most redox-sensitive amino acids it can be cited that the -SH group cysteine, and the aromatic ring of tyrosine [21], displaying different predispositions to endure RONS alterations depending on the specific oxidants in the redox reaction, the proportion and distribution of thiol groups, transition metal ion presence as prosthetic groups, and motifs and amino acid residue exposure on the molecular surface of the proteins, as well as different levels of chemical modifications ranging from reversible to irreversible, and specific or non-specific union, which depends on the selective attachment to residues and their location, or multiple sidechain and backbone sites affected, respectively [347].

From a redox perspective, the reversible and specific oxidation of a few or at least one single redox-sensitive residue in a protein by an electrophile molecule can be enough to induce chemical modification in a switch manner for modulating the protein activity [348]. Some of the plausible amino acid residue modifications bearing redox switch capacities can be elicited through cysteine residue S-sulfenylation, S-glutathionylation, S-nitrosylation, and S-persulfidation, given that cysteine residues are key redox-sensitive molecules due to the presence of -SH groups, bestowing structural and functional flexibility to proteins (Figure 18) [349,350].

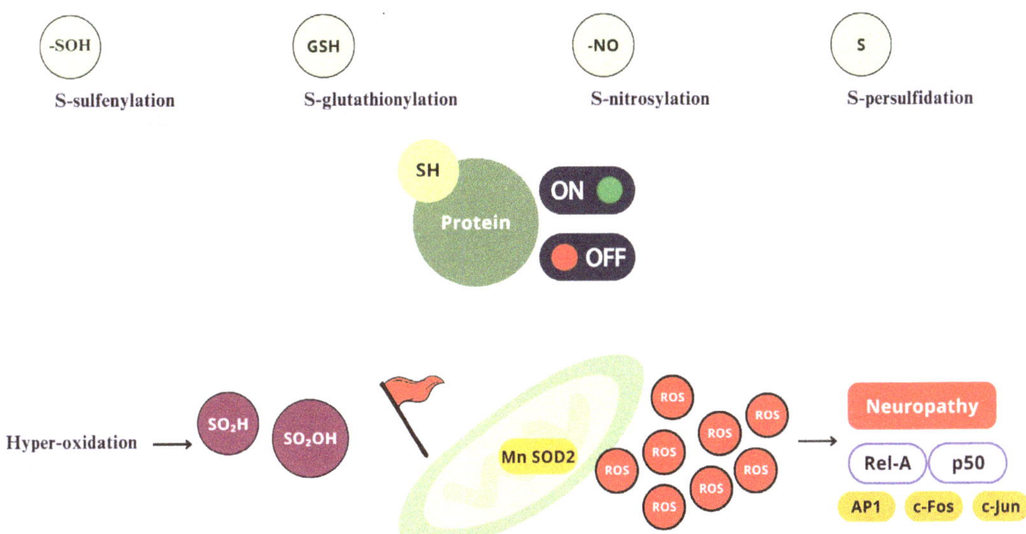

Figure 18. Reversible and irreversible proteins change under ROS/RNS. Reactive species serve as redox-sensitive switch molecules due to amino acid -SH group cysteine, or to aromatic rings of tyrosine, for modulating protein activity. Nevertheless, when RS• interact with NO to form S-nitrosylated cysteine residues in proteins, or undergo hyper-oxidation, to produce sulfinic and sulfonic cysteine after interaction with H_2O_2, $ONOO^-$, and HOCl, changes can become irreversible, impairing antioxidant defense and leading to neuronal damage.

Cysteine residues are usually found in catalytic and regulatory sites of enzymes, signaling pathway effectors, and transcription factors, and they can initially undergo the ionization to thiolate (R-S^-) which is the deprotonated form of -SH, and further undergo one- or two-electron oxidations, to yield thiyl radicals (RS•), or cysteine sulfenic acid (-SOH), respectively, both reversible. Then, RS• can interact with NO to form S-nitrosylated cysteine residues in proteins, or undergo hyper-oxidation, to produce sulfinic and sulfonic cysteine after interaction with H_2O_2, $ONOO^-$, and HOCl, which are irreversible (Figure 18) [351], as seen in the irreversible oxidation of SOD1 Cys111 to its sulfinic and sulfonic states, present in ALS mice models that presented signs of peripheral pain [352–354]. Accordingly, in spite of ALS being a neurodegenerative disease characterized by upper and lower motor neuron degeneration, it has been recently found to have a sensory neuropathy component observed in up to 20% of ALS human patients, pain among them, affecting unmyelinated small and myelinated large fibers as part of the process of primary neurodegeneration [355,356], potentially leading to neuropathic pain, and relevant to understand the full spectrum of ALS, associated with SOD1 mutation. Furthermore, cysteine residues of proteins and peptides exposed to NO and $ONOO^-$ can suffer S-nitrosylation (-SNO groups formation) and S-glutathionylation (SS and GSH mixed groups formation); such reactions are relevant to cell signaling, protein stability and activity, and redox homeostasis through GST, Grx, and Trx [357].

Tyrosine residues oxidation leading to 3–hydroxytyrosine, 3-nitrotyrosine, halogenated tyrosine, and intramolecular tyrosine cross-links are deeply associated with interference with cellular functions, as seen in endogenous ROS exposure, yielding inter- and intramolecular cross-links in proteins, accompanied by spontaneous fragmentation of the polypeptide chains involved; some of these cross-links, like 3,3′-dityrosine, are biomarkers of oxidative stress, as seen in a study using a neuropathic pain model of chronic constriction injury, where they found dityrosine-containing protein cross-links exposition to activate satellite glial cells (SGCs) in the DRG, and a subsequent increase in proinflammatory mediators and ROS, mediated by the NF-κB pathway and the receptors for advanced glycation end products, exacerbating hyperalgesia [358]. As for tyrosine and tryptophane nitration, once they react with OH•, and subsequently form tyrosyl and tryptophanyl radicals, both can further react with NO_2• after covalent attachment to produce 3-nitrotyrosine and 6-nitrotryptophan, respectively [359].

Proteotoxic stress has been linked to neuropathic pain through nitration reactions, which profoundly trigger chemical and structural changes in proteins, leading to function alterations and cell signaling modulation [36,99]; as such, peroxynitrite, a cytotoxic proinflammatory and pronociceptive agent is portrayed as a main factor in the development of peripheral and central sensitization associated with neuropathic pain [360,361]. The mitochondrial MnSOD can be negatively affected by nitration, rendering it inactive, and thus contributing to the perpetuation of a vicious cycle of free radical formation. This, in turn, leads to further nitrosative damage to relevant enzymes unable to fight back against the development and maintenance of hyperalgesia and central sensitization [36]. The accumulation of ROS and RNS, increased during oxidative and nitrosative stress conditions, can induce the activation of transcription factors such as AP-1, NF-κB, and MAPKs, which can activate and increase the expression and production of COX enzymes and PGs [362,363].

Whenever antioxidant systems are exceeded by the production of ROS or RNS, the outcome is likely to cause an oxidative or nitrosative stress state, in which subsequent damage to biomolecules like proteins, membrane lipids, and nucleic acids leads to cellular and tissue dysfunction and cell death via necrosis, apoptosis, and autophagy, implicated in chronic pain of neuropathic origin [364]. Nevertheless, the final outcome or the biological impact of the cross-links between antioxidants and pro-oxidants will depend on the specific type interacting, physicochemical properties, cellular and subcellular location and gradient, the rate of formation and degradation, and quantity. Thus, in recent years, not only antioxidants, but also RONS have gained more attention from the scientific community, as the regulatory roles of RONS have emerged as adaptive mechanisms involved in cellular homeostasis, in low steady concentrations, for reversible redox-sensitive modifications in functional motifs of several proteins related to cell signaling.

5. MnPs: Mechanisms of Action and Potential Therapy for Neuropathic Pain

The role of oxidative and nitrosative stress has been stablished for chronic pain of neuropathic origin, through the effects of hyper-physiological levels of reactive species on neuronal and glial cells [365]. Antioxidant therapies from diverse origins, aiming for the restoration of normal cell functioning, have been applied for various chronic pain disorders, with different, sometimes conflicting, and specific pain type-dependent outcomes, either positive, negative, or not clear [366,367]. Among the most studied synthetic therapeutic antioxidant compounds, there is a group of proteins named SOD mimics. By designation, SOD mimics are compounds with the ability to catalyze the oxidation and reduction of O•−. Yet, it has been proved that in addition to reaction with O•−, SOD mimics can also catalyze redox reactions with H_2O_2, $ONOO^-$, thiols, and others under specific conditions [368], working as oxidants and reductants.

More specifically, there is a class of metalloproteins of low molecular weights with a central metal ion of manganese contained in the porphyrin ring of synthetic origin, exhibiting different oxidation states of manganese Mn(II), Mn(III), and Mn(IV), which contributes to their redox applications, and these are named manganese porphyrins (MnPs) [369]. MnPs were initially, in a very narrow way, seen as simply scavengers of O•−, but the new knowledge about redox chemistry applied to cellular physiology proved its worth based on evidence of their potential role as therapeutics, due to their physicochemical properties, longer half-life, lower cost of production compared to SOD enzymes, and lack of immunogenic reactions in biological systems, hence, affecting signaling pathways and cellular functions such as proliferation, differentiation, and cell death [26], through the modulation of the redoxome [370].

In addition to the specific reactive species scavenged by MnPs, even more importantly, there are other aspects intrinsic to their structure that must be taken into consideration when applying MnSOD mimic therapeutic strategies for neuropathic pain. First, the half-wave redox potential ($E_{1/2}$(O•−)), which determines whether a compound can be readily oxidized and/or reduced by the substrate within the redox active molecular targets in the cell. $E_{1/2}$(O•−) is a reliable measure of physiological relevance, as it compares the electrochemical potency at half of the wave of the compound to that of the SOD enzyme (+300 mV vs. normal hydrogen electrode—NHE) in relation to O•−; the closer it is, the more biologically compatible the compound is for electron interchange among reactive species and signaling proteins in the cellular context. Potentials that are too high or too low can be toxic or inefficient, respectively [368]. A MnP within a compatible range will also allow a more specific action of the compound over the target molecules, without interfering with other biological processes, and more stability, avoiding fast degradation or inactivation [371].

Secondly, the intrinsic catalytic activity of the SOD mimic protein (Kcat) for the catalysis of O•− is another reliable measure of the therapeutic potential of the compound, as it quantifies the efficiency to catalyze the redox reaction, expressing the number of substrate molecules converted into the reaction product for each molecule of the compound by the unit of time, when the compound is fully saturated with the substrate (i.e., SOD Kcat(O•−) ~2 × 109 $M^{-1} \cdot s^{-1}$) [372]. Such factors establish not only the catalytic efficiency, but the rate of the reaction, which is relevant for reaching the therapeutic outcomes in a reasonable time, and the compatibility with the cellular redox conditions, that is the pH, temperature of the reaction, and substrate concentrations [373]. Notwithstanding these factors, the in vivo efficacy of metalloporphyrin-based SOD mimics is not only associated with their SOD-like activity, but with other factors like the shape, size, and bulkiness of the molecule [374], bioavailability, biodistribution in different cell kinds and subcellular spaces, and reactivity toward other cellular biomolecules [375], which will dictate the overall therapeutic effect of the treatment. Assuming the rich environment of redox-based signaling processes occurring in the cellular metabolism through cellular respiration, glycolysis, cyt P450 detoxifying system, ETC, NO synthesis, and others, an approach targeting these redox-signaling pathways for therapeutic purpose would make more sense when assessing the effects of MnP on redoxome alterations, than on separate scavenging activity toward specific reactive species [23].

The main consideration is that, under impaired conditions of neurons' ability to bear peroxides elimination, the excessive accumulation of H_2O_2 as a byproduct of the O•− dismutation through MnP might reach toxic levels. That is, the overall antioxidant power of the synthetic enzyme will only be efficient if H_2O_2 is also removed [368]. Cationic MnPs are not potent scavengers of H_2O_2, but once reduced in vivo, may bind O_2 to O_2•− and H_2O_2 [376]. Yet, cationic MnPs have higher Kcat(O•−) in comparison to neutral or anionic MnP, since O•− is negatively charged; therefore, they are prone to be attracted to the positively charged cationic MnPs, facilitating a more efficient interaction and better catalytic activity. This translates into a more effective MnP in the detoxification and utilization of the O•− radical for antioxidant mechanisms [377,378]. In addition, when MnP holds a highly positive reduction potential, it is very likely that they will tend to accept electrons by cellular reductants, rather than give electrons away; in this process, Mn will be reduced to Mn(II), which will reduce O•− to H_2O_2, working as superoxide reductase, rather than superoxide dismutase [379].

The electrophilic nature of MnP, especially Mn(III) explains its tendency to react with or bind to anionic substrates, normally electron-rich, such as HOO^-, RS^-, ClO^-, $ONOO^-$, etc. When MnP enters the cell, it can straightaway suffer reduction while inducing the oxidation of thiols, thus, working as an oxidant and antioxidant [368]. As proved by previous research, MnPs can work as suppressors of glycolysis and mitochondrial respiration [379], exhibit significant antihyperglycemic activity, and enhance neuronal death [380]; additionally, MnP can induce NF-κB inactivation [381] and protective responses against oxidative damage to neuronal and glial cells [248,365], as well as reduced NF-κB expression, thus, resulting in a significative decrease in proinflammatory cytokines such as IL-1β and TNF-α [382].

MnPs with potent log Kcat(O•−) are also potent $ONOO^-$ scavengers. As previously stated, electron-deficient MnPs react or bind to electron-rich anionic substrates, such as $ONOO^-$. When Mn(III) is oxidized to Mn(IV), either through an O•− or $ONOO^-$ reaction, reducing itself back to Mn(III) with cellular reductants present in the electron pool, yielding the highly oxidizing radical nitrogen dioxide (NO•$_2$), or through a two-electron approach to Mn(II) yielding nitrite (NO_2^-), this latter reaction being the most likely to occur in cells, as MnPs are maintained in the Mn(II) state by cellular reductants [383]. In addition, MnPs can interact with deprotonated reductants like GSH, cysteine, and protein thiols [26]; in this sense, the MnP works as an oxidant, accepting electrons from the substrates, and being reduced to either the Mn(III) or Mn(II) lower oxidation states as a result, while oxidizing the reductants.

A significant degree of interaction between MnP and thiols has been reported, specifically towards cysteine residues present in the p50 subunit of the NF-κB protein, resulting in disulfide formation [384]. The S-glutathionylation of the p65 NF-κB subunit, which is responsible for the κB attachment to certain DNA regions, thereby initiating the transcription of target genes, has been demonstrated in hyaluronan degradation-induced inflammation in mice [385]. In such cases, the MnP functioned as an oxidant, yielding a

GS radical, which combined with additional GS to produce GSSG radicals; GSSG, when oxidized by O_2, produces O•−. However, glutathionylation will only be significant if GSH and H_2O_2 are present, and GSSG can be safely removed if glutathione reductase (GR) is present in sufficient levels [381]. The antioxidative effect of MnP, as previously mentioned is only effective when physiological levels of peroxide enzymes are present to remove H_2O_2. Hence, the cycling of MnP with GSH or cysteine would lead to an accumulation of H_2O_2, contributing to the further oxidation of biomolecules of significant importance in the homeostasis complex process [368]. In a chemotherapy-induced neuropathic pain model in rats, mitotoxicity influenced by reduced ATP production in mitochondria due to respiratory complex I and II inhibition in primary nerve sensory axons was thwarted by the administration of Mn(III) 5,10,15,20-tetrakis(N-n-hexylpyridinium-2-yl)porphyrin (MnTE-2-PyP(5+)), without compromising anti-tumor effect, and mechanical hypersensitivity was reversed [386].

Although evidence for manganese-based porphyrins' therapeutic effects over neuropathic pain is scarce, acknowledged evidence of iron-based porphyrins exists [387,388]. Manganese and iron porphyrins can both dismutate superoxide anion, although their efficiency and pathway might differ in favor of manganese porphyrins, which have an overall mechanism more promising regarding interaction with cellular reductants and flavoenzymes, and the production of •OH after interacting with H_2O_2, as MnPs have a very limited capacity to interact with H_2O_2, contrary to iron porphyrins, which can produce •OH in a Fenton-like reaction, contributing to reactive species production in the cell [28,374]. Yet, it is important to acknowledge that iron porphyrin complexes perform remarkable peroxynitrite decomposition catalyst activity, highlighting the potential of porphyrin-based research for therapies in the context of neuropathic pain.

Structural and functional neuropathic alterations in diabetic type 2 rats were evaluated and the therapeutic potential of peroxynitrite decomposition catalysts Fe(III) tetrakis-2-(N-triethylene glycol monomethyl ether)-pyridyl porphyrin (FP15) and Fe(III) tetramesitylporphyrin octasulfonate (FeTMPS) were assessed. Mechanical and sensorial deficits and thermal allodynia, as well as nitrotyrosine and poly(ADP-ribose) levels were restored with both treatments, compared to untreated animals [389]. Ischemia/reperfusion-induced neuropathic pain has also been evaluated regarding the peroxynitrite decomposition catalyst effect of intraperitoneal FeTMPyP [5,10,15,20-tetrakis (N-methyl-4′-pyridyl)porphyrinato iron (III)]. A significant enhancement of mechanical withdrawal threshold and phosphorylated NMDA subunit 1 were observed in treated animals, consequently reducing mechanical allodynia and central sensitization [390]. The anti-nociceptive and antioxidant effect of peroxynitrite decomposition catalyst FeTMPyP was tested in chronic constriction of the sciatic nerve-induced neuropathic pain in rats. High iNOS, NF-κB, IL-6, and TNF-α expression levels in sciatic nerves, together with increased poly (ADP) ribose DRG levels were completely reversed after treatment, reducing PARP overactivation, oxidative stress, and normalizing behavioral and functional pain parameters [391].

6. Final Considerations

This review evaluated the role of reactive species in cell signaling, focusing on their redox-sensitive characteristics and their link to neuropathic pain. Reactive oxygen and nitrogen species, significant for cellular signaling pathways, act as both mediators of cellular damage and modulators of physiological processes. Their dysregulation, especially under oxidative stress, has been increasingly linked to the sensitization of pain pathways, highlighting the potential of targeting redox-sensitive mechanisms for novel therapeutic strategies. Enzymatic antioxidant systems play a crucial role in maintaining redox homeostasis. Enhancing these defenses offers a promising strategy to reduce oxidative and nitrosative damage associated with neuropathic pain. Among emerging therapeutic agents, manganese porphyrins (MnPs) stand out due to their mimetic antioxidant properties, offering significant potential for addressing redox imbalances and neuropathic pain. This review underscores the significance of understanding the intricate dynamics of reactive species, subcellular compartment gradients, and MnP-mediated redox chemistry in the context of neuronal dysfunction. Further research is essential to optimize the design of biocompatible, target-specific MnP molecules and to elucidate their safety and efficacy in clinical settings. Advancing knowledge in this field may pave the way for innovative treatments for neuropathic pain, ultimately improving the quality of life for affected individuals.

Author Contributions: Conceptualization, Á.J.C.S. and M.S.L.d.L.; methodology, Á.J.C.S. and M.S.L.d.L.; investigation, Á.J.C.S.; resources, M.S.L.d.L.; data curation Á.J.C.S. and M.S.L.d.L.; writing—original draft preparation, Á.J.C.S.; writing—review and editing, Á.J.C.S. and M.S.L.d.L.; supervision, M.S.L.d.L.; project administration, M.S.L.d.L.; funding acquisition, M.S.L.d.L. All authors have read and agreed to the published version of the manuscript.

Funding: This research was supported by Conselho Nacional de Desenvolvimento Cientifico e Tecnologico (CNPq), Fundação Coordenação de Aperfeiçoamento de Pessoal de Nível Superior (CAPES), and the Universidade Estadual de Santa Cruz (UESC) (number: 073.6762.2024.0001737-36).

Institutional Review Board Statement: Not applicable.

Informed Consent Statement: Not applicable.

Data Availability Statement: Data will be made available on request.

Acknowledgments: The authors extend their gratitude to Marcia Souza from the UESC veterinary hospital for her constant support, and to CNPq and CAPES for funding and kindly providing resources for the author's (Á.J.C.S.) doctorate scholarship.

Conflicts of Interest: The authors declare no conflicts of interest. The funders had no role in the design of the study; in the collection, analyses, or interpretation of data; in the writing of the manuscript; or in the decision to publish the results.

Abbreviations

AP-1	Activator Protein 1
ATF6	Activator Transcription Factor
BDNF	brain-derived neurotrophic factor
CAT	catalase
CHOP	C/EBP Homologous Protein
Cys	cysteine
EGF	epidermal growth factor
eNOS	Endothelial Nitric Oxide Synthase
ER	endoplasmic reticulum
ERK	Extracellular Signal-Regulated Kinase
ERO-1	Endoplasmic Reticulum Oxidoreductin-1
ETC	electron transfer chain
FAD	flavin adenine dinucleotide
GH	glutathione
GLUT1	Glucose Transporter 1
GPx	glutathione peroxidase
GR	glutathione reductase
GRP78	Glucose-Regulated Protein 78
GSH	reduced glutathione
GSSG	oxidized glutathione
H_2O_2	hydrogen peroxide

HIF-1α	Hypoxia-Inducible Factor 1α
HO•	hydroxyl radical
HOCL	hypochlorous acid
HOO•	hydroperoxyl radical
IL	interleukin
iNOS	Inducible Nitric Oxide Synthase
IRE-1α	Inositol-Requiring Kinase 1α
IRI	ischemic reperfusion injury
JNK	c-Jun NH2 quinase terminal
KEAP1	Kelch-like ECH-associated protein 1
LOOH	lipidic peroxides
M	Molar
MAPK	Mitogen-Activated Protein Kinase
MCU	Mitochondrial Calcium Uniporter
MDA	malondialdehyde
Met	methionine
MnP	manganese porphyrin
mtNOS	Mitochondrial Nitric Oxide Synthase
NAD	nicotinamide adenine dinucleotide
NCX	Na^+/Ca^{2+} exchangers
NF-κB	Nuclear Factor kB
NGF	nerve growth Factor
nNOS	Neuronal Nitric Oxide Synthase
NO	nitric oxide
NOX	NADPH oxidase
NRF2	Nuclear factor (erythroid-derived 2)-like 2
O_2•−	superoxide radical anion
$ONOO^-$	peroxynitrite
PDI	Protein Disulfide Isomerase
PERK	Protein Kinase (PKR)-Like Endoplasmic Reticulum Kinase
PG	prostaglandin
PI3K	phosphatidylinositol 3 kinase
PIP	phosphatidylinositol phosphate
PIP2	phosphatidylinositol biphosphate
PIP3	phosphatidylinositol triphosphate
PMCA	the plasma membrane calcium ATPase
PPP	pentose phosphate pathway
RNS	reactive nitrogen species
ROO•	peroxyl radical
ROS	reactive oxygen species
RS•	thiyl radical
SERCA	Endoplasmic Reticulum Calcium ATPase
SOD	superoxide dismutase
Tyr	tyrosine
VGF	vascular growth factor
XO	xanthine oxidase

References

1. Burma, N.E.; Leduc-Pessah, H.; Fan, C.Y.; Trang, T. Animal Models of Chronic Pain: Advances and Challenges for Clinical Translation. *J. Neurosci. Res.* **2017**, *95*, 1242–1256. [CrossRef] [PubMed]
2. Goldberg, D.S.; McGee, S.J. Pain as a Global Public Health Priority. *BMC Public Health* **2011**, *11*, 770. [CrossRef]
3. Racine, M. Chronic Pain and Suicide Risk: A Comprehensive Review. *Prog. Neuropsychopharmacol. Biol. Psychiatry* **2018**, *87*, 269–280. [CrossRef] [PubMed]

4. Finnerup, N.B.; Attal, N.; Haroutounian, S.; McNicol, E.; Baron, R.; Dworkin, R.H.; Gilron, I.; Haanpää, M.; Hansson, P.; Jensen, T.S.; et al. Pharmacotherapy for Neuropathic Pain in Adults: A Systematic Review and Meta-Analysis. *Lancet Neurol.* **2015**, *14*, 162–173. [CrossRef] [PubMed]
5. Teixeira-Santos, L.; Albino-Teixeira, A.; Pinho, D. Neuroinflammation, Oxidative Stress and Their Interplay in Neuropathic Pain: Focus on Specialized pro-Resolving Mediators and NADPH Oxidase Inhibitors as Potential Therapeutic Strategies. *Pharmacol. Res.* **2020**, *162*, 105280. [CrossRef]
6. Mobasheri, A.; Simon, F.; Carrasco, C.; Nazirŏglu, M.; Nazirŏglu, N.; Rodríguez, A.B.; Pariente, J.A. Neuropathic Pain: Delving into the Oxidative Origin and the Possible Implication of Transient Receptor Potential Channels. *Front. Physiol.* **2018**, *9*, 95. [CrossRef]
7. IASP. IASP Terminology. Available online: https://www.iasp-pain.org/terminology?navItemNumber=576#Pain (accessed on 10 April 2021).
8. Colloca, L.; Ludman, T.; Bouhassira, D.; Baron, R.; Dickenson, A.H.; Yarnitsky, D.; Freeman, R.; Truini, A.; Attal, N.; Finnerup, N.B.; et al. Neuropathic Pain. *Nat. Rev. Dis. Primers* **2017**, *3*, 17002. [CrossRef]
9. Ji, R.R.; Nackley, A.; Huh, Y.; Terrando, N.; Maixner, W. Neuroinflammation and Central Sensitization in Chronic and Widespread Pain. *Anesthesiology* **2018**, *129*, 343–366. [CrossRef]
10. Singh, P.; Bansal, S.; Kuhad, A.; Kumar, A.; Chopra, K. Naringenin Ameliorates Diabetic Neuropathic Pain by Modulation of Oxidative-Nitrosative Stress, Cytokines and MMP-9 Levels. *Food Funct.* **2020**, *11*, 4548–4560. [CrossRef]
11. Zamani, K.; Fakhri, S.; Kiani, A.; Abbaszadeh, F.; Farzaei, M.H. Rutin Engages Opioid/Benzodiazepine Receptors towards Anti-Neuropathic Potential in a Rat Model of Chronic Constriction Injury: Relevance to Its Antioxidant and Anti-Inflammatory Effects. *Naunyn-Schmiedeberg's Arch. Pharmacol.* **2025**, 1–15. [CrossRef]
12. Hashemi, B.; Fakhri, S.; Kiani, A.; Abbaszadeh, F.; Miraghaee, S.; Mohammadi, M.; Echeverría, J. Anti-Neuropathic Effects of Astaxanthin in a Rat Model of Chronic Constriction Injury: Passing through Opioid/Benzodiazepine Receptors and Relevance to Its Antioxidant and Anti-Inflammatory Effects. *Front. Pharmacol.* **2024**, *15*, 1467788. [CrossRef] [PubMed]
13. Ilari, S.; Giancotti, L.A.; Lauro, F.; Gliozzi, M.; Malafoglia, V.; Palma, E.; Tafani, M.; Russo, M.A.; Tomino, C.; Fini, M.; et al. Natural Antioxidant Control of Neuropathic Pain-Exploring the Role of Mitochondrial SIRT3 Pathway. *Antioxidants* **2020**, *9*, 1103. [CrossRef] [PubMed]
14. Di Meo, S.; Venditti, P. Evolution of the Knowledge of Free Radicals and Other Oxidants. *Oxid. Med. Cell. Longev.* **2020**, *2020*, 9829176. [CrossRef] [PubMed]
15. Teleanu, D.M.; Niculescu, A.G.; Lungu, I.I.; Radu, C.I.; Vladâcenco, O.; Roza, E.; Costăchescu, B.; Grumezescu, A.M.; Teleanu, R.I. An Overview of Oxidative Stress, Neuroinflammation, and Neurodegenerative Diseases. *Int. J. Mol. Sci.* **2022**, *23*, 5938. [CrossRef]
16. Lobo, V.; Patil, A.; Phatak, A.; Chandra, N. Free Radicals, Antioxidants and Functional Foods: Impact on Human Health. *Pharmacogn. Rev.* **2010**, *4*, 118. [CrossRef]
17. Kim, D.; You, B.; Jo, E.K.; Han, S.K.; Simon, M.I.; Lee, S.J. NADPH Oxidase 2-Derived Reactive Oxygen Species in Spinal Cord Microglia Contribute to Peripheral Nerve Injury-Induced Neuropathic Pain. *Proc. Natl. Acad. Sci. USA* **2010**, *107*, 14851–14856. [CrossRef]
18. Osmanlıoğlu, H.Ö.; Nazıroğlu, M. Resveratrol Modulates Diabetes-Induced Neuropathic Pain, Apoptosis, and Oxidative Neurotoxicity in Mice Through TRPV4 Channel Inhibition. *Mol. Neurobiol.* **2025**, *61*, 7269–7286. [CrossRef]
19. Sies, H.; Jones, D.P. Reactive Oxygen Species (ROS) as Pleiotropic Physiological Signalling Agents. *Nat. Rev. Mol. Cell Biol.* **2020**, *21*, 363–383. [CrossRef]
20. Massaad, C.A.; Klann, E. Reactive Oxygen Species in the Regulation of Synaptic Plasticity and Memory. *Antioxid. Redox Signal.* **2011**, *14*, 2013. [CrossRef]
21. Moldogazieva, N.T.; Mokhosoev, I.M.; Feldman, N.B.; Lutsenko, S.V. ROS and RNS Signalling: Adaptive Redox Switches through Oxidative/Nitrosative Protein Modifications. *Free Radic. Res.* **2018**, *52*, 507–543. [CrossRef]
22. Trevisan, G.; Benemei, S.; Materazzi, S.; De Logu, F.; De Siena, G.; Fusi, C.; Fortes Rossato, M.; Coppi, E.; Marone, I.M.; Ferreira, J.; et al. TRPA1 Mediates Trigeminal Neuropathic Pain in Mice Downstream of Monocytes/Macrophages and Oxidative Stress. *Brain* **2016**, *139*, 1361–1377. [CrossRef] [PubMed]
23. Batinic-Haberle, I.; Spasojevic, I. Complex Chemistry and Biology of Redox-Active Compounds, Commonly Known as SOD Mimics, Affect Their Therapeutic Effects. *Antioxid. Redox Signal.* **2014**, *20*, 2323–2325. [CrossRef] [PubMed]
24. Carballal, S.; Valez, V.; Alvarez-Paggi, D.; Tovmasyan, A.; Batinic-Haberle, I.; Ferrer-Sueta, G.; Murgida, D.H.; Radi, R. Manganese Porphyrin Redox State in Endothelial Cells: Resonance Raman Studies and Implications for Antioxidant Protection towards Peroxynitrite. *Free Radic. Biol. Med.* **2018**, *126*, 379–392. [CrossRef] [PubMed]
25. Ali, D.K.; Oriowo, M.; Tovmasyan, A.; Batinic-Haberle, I.; Benov, L. Late Administration of Mn Porphyrin-Based SOD Mimic Enhances Diabetic Complications. *Redox Biol.* **2013**, *1*, 457–466. [CrossRef]

26. Batinic-Haberle, I.; Spasojevic, I.; Tse, H.M.; Tovmasyan, A.; Rajic, Z.; Clair, D.K.S.; Vujaskovic, Z.; Dewhirst, M.W.; Piganelli, J.D. Design of Mn Porphyrins for Treating Oxidative Stress Injuries and Their Redox-Based Regulation of Cellular Transcriptional Activities. In *Amino Acids*; Springer: Vienna, Austria, 2012; Volume 42, pp. 95–113. [CrossRef]
27. Ratcliffe, P.; Koivunen, P.; Myllyharju, J.; Ragoussis, J.; Bovée, J.V.; Batinic-Haberle, I.; Vinatier, C.; Trichet, V.; Robriquet, F.; Oliver, L.; et al. Hypoxia Dovepress Update on Hypoxia-Inducible Factors and Hydroxylases in Oxygen Regulatory Pathways: From Physiology to Therapeutics. *Hypoxia* **2017**, *5*, 11–20. [CrossRef]
28. Rausaria, S.; Ghaffari, M.M.E.; Kamadulski, A.; Rodgers, K.; Bryant, L.; Chen, Z.; Doyle, T.; Shaw, M.J.; Salvemini, D.; Neumann, W.L. Retooling Manganese(III) Porphyrin-Based Peroxynitrite Decomposition Catalysts for Selectivity and Oral Activity: A Potential New Strategy for Treating Chronic Pain. *J. Med. Chem.* **2011**, *54*, 8658–8669. [CrossRef]
29. Singh, A.; Kukreti, R.; Saso, L.; Kukreti, S. Oxidative Stress: A Key Modulator in Neurodegenerative Diseases. *Molecules* **2019**, *24*, 1583. [CrossRef]
30. Bedard, K.; Krause, K.H. The NOX Family of ROS-Generating NADPH Oxidases: Physiology and Pathophysiology. *Physiol. Rev.* **2007**, *87*, 245–313. [CrossRef]
31. Murphy, M.P. How Mitochondria Produce Reactive Oxygen Species. *Biochem. J.* **2009**, *417*, 1–13. [CrossRef]
32. Miller, D.M.; Buettner, G.R.; Aust, S.D. Transition Metals as Catalysts of "Autoxidation" Reactions. *Free Radic. Biol. Med.* **1990**, *8*, 95–108. [CrossRef]
33. Zheng, M.; Storz, G. Redox Sensing by Prokaryotic Transcription Factors. *Biochem. Pharmacol.* **2000**, *59*, 1–6. [CrossRef] [PubMed]
34. Valko, M.; Morris, H.; Cronin, M. Metals, Toxicity and Oxidative Stress. *Curr. Med. Chem.* **2005**, *12*, 1161–1208. [CrossRef] [PubMed]
35. Park, E.S.; Gao, X.; Chung, J.M.; Chung, K. Levels of Mitochondrial Reactive Oxygen Species Increase in Rat Neuropathic Spinal Dorsal Horn Neurons. *Neurosci. Lett.* **2006**, *391*, 108–111. [CrossRef] [PubMed]
36. Salvemini, D.; Little, J.W.; Doyle, T.; Neumann, W.L. Roles of Reactive Oxygen and Nitrogen Species in Pain. *Free Radic. Biol. Med.* **2011**, *51*, 951–966. [CrossRef]
37. Grace, P.M.; Gaudet, A.D.; Staikopoulos, V.; Maier, S.F.; Hutchinson, M.R.; Salvemini, D.; Watkins, L.R. Nitroxidative Signaling Mechanisms in Pathological Pain. *Trends Neurosci.* **2016**, *39*, 862–879. [CrossRef]
38. Halliwell, B. Reactive Species and Antioxidants. Redox Biology Is a Fundamental Theme of Aerobic Life. *Plant Physiol.* **2006**, *141*, 312–322. [CrossRef]
39. Dröge, W. Free Radicals in the Physiological Control of Cell Function. *Physiol. Rev.* **2002**, *82*, 47–95. [CrossRef]
40. Klotz, L.-O.; Briviba, K.; Sies, H. Signaling by Singlet Oxygen in Biological Systems. In *Antioxidant and Redox Regulation of Genes*; Academic Press: Cambridge, MA, USA, 2000; pp. 3–20. [CrossRef]
41. Al-Shehri, S.S. Reactive Oxygen and Nitrogen Species and Innate Immune Response. *Biochimie* **2021**, *181*, 52–64. [CrossRef]
42. Turrens, J.F. Mitochondrial Formation of Reactive Oxygen Species. *J. Physiol.* **2003**, *552*, 335. [CrossRef]
43. Kovacic, P.; Pozos, R.S.; Somanathan, R.; Shangari, N.; O'Brien, P.J. Mechanism of Mitochondrial Uncouplers, Inhibitors, and Toxins: Focus on Electron Transfer, Free Radicals, and Structure-Activity Relationships. *Curr. Med. Chem.* **2005**, *12*, 2601–2623. [CrossRef]
44. Valko, M.; Izakovic, M.; Mazur, M.; Rhodes, C.J.; Telser, J. Role of Oxygen Radicals in DNA Damage and Cancer Incidence. *Mol. Cell. Biochem.* **2004**, *266*, 37–56. [CrossRef]
45. De Grey, A.D.N.J. $HO_2\bullet$: The Forgotten Radical. *DNA Cell Biol.* **2004**, *21*, 251–257. [CrossRef] [PubMed]
46. Jomova, K.; Raptova, R.; Alomar, S.Y.; Alwasel, S.H.; Nepovimova, E.; Kuca, K.; Valko, M. Reactive Oxygen Species, Toxicity, Oxidative Stress, and Antioxidants: Chronic Diseases and Aging. *Arch. Toxicol.* **2023**, *97*, 2499–2574. [CrossRef]
47. Siniscalco, D.; Fuccio, C.; Giordano, C.; Ferraraccio, F.; Palazzo, E.; Luongo, L.; Rossi, F.; Roth, K.A.; Maione, S.; de Novellis, V. Role of Reactive Oxygen Species and Spinal Cord Apoptotic Genes in the Development of Neuropathic Pain. *Pharmacol. Res.* **2007**, *55*, 158–166. [CrossRef] [PubMed]
48. Schwartz, E.S.; Lee, I.; Chung, K.; Chung, J.M. Oxidative Stress in the Spinal Cord Is an Important Contributor in Capsaicin-Induced Mechanical Secondary Hyperalgesia in Mice. *Pain* **2008**, *138*, 514. [CrossRef] [PubMed]
49. Lee, I.; Kim, H.K.; Kim, J.H.; Chung, K.; Chung, J.M. The Role of Reactive Oxygen Species in Capsaicin-Induced Mechanical Hyperalgesia and in the Activities of Dorsal Horn Neurons. *Pain* **2007**, *133*, 9. [CrossRef] [PubMed]
50. Kim, H.K.; Park, S.K.; Zhou, J.L.; Taglialatela, G.; Chung, K.; Coggeshall, R.E.; Chung, J.M. Reactive Oxygen Species (ROS) Play an Important Role in a Rat Model of Neuropathic Pain. *Pain* **2004**, *111*, 116–124. [CrossRef]
51. Yowtak, J.; Lee, K.Y.; Kim, H.Y.; Wang, J.; Kim, H.K.; Chung, K.; Chung, J.M. Reactive Oxygen Species Contribute to Neuropathic Pain by Reducing Spinal GABA Release. *Pain* **2011**, *152*, 844–852. [CrossRef]
52. Gwak, Y.S.; Hassler, S.E.; Hulsebosch, C.E. Reactive Oxygen Species Contribute to Neuropathic Pain and Locomotor Dysfunction via Activation of CamKII in Remote Segments Following Spinal Cord Contusion Injury in Rats. *Pain* **2013**, *154*, 1699–1708. [CrossRef]
53. Kallenborn-Gerhardt, W.; Schröder, K.; Schmidtko, A. NADPH Oxidases in Pain Processing. *Antioxidants* **2022**, *11*, 1162. [CrossRef]

54. Andrés, C.M.C.; Pérez de la Lastra, J.M.; Juan, C.A.; Plou, F.J.; Pérez-Lebeña, E. Chemistry of Hydrogen Peroxide Formation and Elimination in Mammalian Cells, and Its Role in Various Pathologies. *Stresses* **2022**, *2*, 256–274. [CrossRef]
55. Lennicke, C.; Rahn, J.; Lichtenfels, R.; Wessjohann, L.A.; Seliger, B. Hydrogen Peroxide—Production, Fate and Role in Redox Signaling of Tumor Cells. *Cell Commun. Signal.* **2015**, *13*, 39. [CrossRef] [PubMed]
56. Sies, H. Hydrogen Peroxide as a Central Redox Signaling Molecule in Physiological Oxidative Stress: Oxidative Eustress. *Redox Biol.* **2017**, *11*, 613–619. [CrossRef] [PubMed]
57. Hyslop, P.A.; Chaney, M.O. Mechanism of GAPDH Redox Signaling by H_2O_2 Activation of a Two-Cysteine Switch. *Int. J. Mol. Sci.* **2022**, *23*, 4604. [CrossRef]
58. Day, A.M.; Brown, J.D.; Taylor, S.R.; Rand, J.D.; Morgan, B.A.; Veal, E.A. Inactivation of a Peroxiredoxin by Hydrogen Peroxide Is Critical for Thioredoxin-Mediated Repair of Oxidized Proteins and Cell Survival. *Mol. Cell* **2012**, *45*, 398–408. [CrossRef]
59. Sözbir, E.; Nazıroğlu, M. Diabetes Enhances Oxidative Stress-Induced TRPM2 Channel Activity and Its Control by N-Acetylcysteine in Rat Dorsal Root Ganglion and Brain. *Metab. Brain Dis.* **2016**, *31*, 385–393. [CrossRef]
60. Viggiano, A.; Monda, M.; Viggiano, A.; Viggiano, D.; Viggiano, E.; Chiefari, M.; Aurilio, C.; De Luca, B. Trigeminal Pain Transmission Requires Reactive Oxygen Species Production. *Brain Res.* **2005**, *1050*, 72–78. [CrossRef]
61. Lorenz, J.E.; Kallenborn-Gerhardt, W.; Lu, R.; Syhr, K.M.J.; Eaton, P.; Geisslinger, G.; Schmidtko, A. Oxidant-Induced Activation of CGMP-Dependent Protein Kinase Iα Mediates Neuropathic Pain After Peripheral Nerve Injury. *Antioxid. Redox Signal.* **2014**, *21*, 1504–1515. [CrossRef]
62. Chen, Z.; Muscoli, C.; Doyle, T.; Bryant, L.; Cuzzocrea, S.; Mollace, V.; Mastroianni, R.; Masini, E.; Salvemini, D. NMDA-Receptor Activation and Nitroxidative Regulation of the Glutamatergic Pathway during Nociceptive Processing. *Pain* **2010**, *149*, 100–106. [CrossRef]
63. Schieber, M.; Chandel, N.S. ROS Function in Redox Signaling and Oxidative Stress. *Curr. Biol.* **2014**, *24*, R453–R462. [CrossRef]
64. Halliwell, B.; Clement, M.V.; Ramalingam, J.; Long, L.H. Hydrogen Peroxide. Ubiquitous in Cell Culture and in Vivo? *IUBMB Life* **2000**, *50*, 251–257. [CrossRef] [PubMed]
65. Kadiiska, M.B.; Mason, R.P. In Vivo Copper-Mediated Free Radical Production: An ESR Spin-Trapping Study. *Spectrochim. Acta A Mol. Biomol. Spectrosc.* **2002**, *58*, 1227–1239. [CrossRef] [PubMed]
66. Zastawny, T.H.; Altman, S.A.; Randers-Eichhorn, L.; Madurawe, R.; Lumpkin, J.A.; Dizdaroglu, M.; Rao, G. DNA Base Modifications and Membrane Damage in Cultured Mammalian Cells Treated with Iron Ions. *Free Radic. Biol. Med.* **1995**, *18*, 1013–1022. [CrossRef] [PubMed]
67. Kroth, A.; do Carmo Quevedo Santos, M.; da Silva, T.C.B.; Silveira, E.M.S.; Trapp, M.; Bezzerra, R.M.N.; Simabuco, F.; Niero, R.; Partata, W.A. Aqueous Extract from *Luehea divaricata* Mart. Leaves Reduces Nociception in Rats with Neuropathic Pain. *J. Ethnopharmacol.* **2020**, *256*, 112761. [CrossRef] [PubMed]
68. Bittar, A.; Jun, J.; La, J.H.; Wang, J.; Leem, J.W.; Chung, J.M. Reactive Oxygen Species Affect Spinal Cell Type-Specific Synaptic Plasticity in a Model of Neuropathic Pain. *Pain* **2017**, *158*, 2137. [CrossRef]
69. Ghafourifar, P.; Cadenas, E. Mitochondrial Nitric Oxide Synthase. *Trends Pharmacol. Sci.* **2005**, *26*, 190–195. [CrossRef]
70. Bergendi, L.; Beneš, L.; Ďuracková, Z.; Ferenčik, M. Chemistry, Physiology and Pathology of Free Radicals. *Life Sci.* **1999**, *65*, 1865–1874. [CrossRef]
71. Ignarro, L.J. Biosynthesis and Metabolism of Endothelium-Derived Nitric Oxide. *Annu. Rev. Pharmacol. Toxicol.* **1990**, *30*, 535–560. [CrossRef]
72. Di Meo, S.; Reed, T.T.; Venditti, P.; Victor, V.M. Role of ROS and RNS Sources in Physiological and Pathological Conditions. *Oxid. Med. Cell. Longev.* **2016**, *2016*, 1245049. [CrossRef]
73. Nathan, C.; Xie, Q.W. Nitric Oxide Synthases: Roles, Tolls, and Controls. *Cell* **1994**, *78*, 915–918. [CrossRef]
74. Cury, Y.; Picolo, G.; Gutierrez, V.P.; Ferreira, S.H. Pain and Analgesia: The Dual Effect of Nitric Oxide in the Nociceptive System. *Nitric Oxide* **2011**, *25*, 243–254. [CrossRef] [PubMed]
75. Radi, R. Nitric Oxide, Oxidants, and Protein Tyrosine Nitration. *Proc. Natl. Acad. Sci. USA* **2004**, *101*, 4003–4008. [CrossRef]
76. Bartesaghi, S.; Radi, R. Fundamentals on the Biochemistry of Peroxynitrite and Protein Tyrosine Nitration. *Redox Biol.* **2018**, *14*, 618–625. [CrossRef] [PubMed]
77. Carr, A.C.; McCall, M.R.; Frei, B. Oxidation of LDL by Myeloperoxidase and Reactive Nitrogen Species: Reaction Pathways and Antioxidant Protection. *Arter. Thromb. Vasc. Biol.* **2000**, *20*, 1716–1723. [CrossRef] [PubMed]
78. Ferrer-Sueta, G.; Radi, R. Chemical Biology of Peroxynitrite: Kinetics, Diffusion, and Radicals. *ACS Chem. Biol.* **2009**, *4*, 161–177. [CrossRef] [PubMed]
79. Radi, R.; Peluffo, G.; Alvarez, M.N.; Naviliat, M.; Cayota, A. Unraveling Peroxynitrite Formation in Biological Systems. *Free Radic. Biol. Med.* **2001**, *30*, 463–488. [CrossRef]
80. Szabó, C.; Ischiropoulos, H.; Radi, R. Peroxynitrite: Biochemistry, Pathophysiology and Development of Therapeutics. *Nat. Rev. Drug Discov.* **2007**, *6*, 662–680. [CrossRef]

81. Muscoli, C.; Cuzzocrea, S.; Riley, D.P.; Zweier, J.L.; Thiemermann, C.; Wang, Z.Q.; Salvemini, D. On the Selectivity of Superoxide Dismutase Mimetics and Its Importance in Pharmacological Studies. *Br. J. Pharmacol.* **2003**, *140*, 445–460. [CrossRef]
82. Julius, D. TRP Channels and Pain. *Annu. Rev. Cell Dev. Biol.* **2013**, *29*, 355–384. [CrossRef]
83. Alberts, B.; Johnson, A.; Lewis, J.; Raff, M.; Roberts, K.; Walter, P. The Endoplasmic Reticulum. In *Molecular Biology of the Cell*; Garland Science: New York, NY, USA, 2002.
84. Helenius, A.; Marquardt, T.; Braakman, I. The Endoplasmic Reticulum as a Protein-Folding Compartment. *Trends Cell Biol.* **1992**, *2*, 227–231. [CrossRef]
85. De Sousa, M.; Oliveira, S.J.; Pinto, J.P. ER Stress and Iron Homeostasis: A New Frontier for the UPR. *Biochem. Res. Int.* **2011**, *2011*, 896474. [CrossRef]
86. Napier, J.A.; Michaelson, L.V.; Sayanova, O. The Role of Cytochrome b 5 Fusion Desaturases in the Synthesis of Polyunsaturated Fatty Acids. *Prostaglandins Leukot. Essent. Fat. Acids* **2003**, *68*, 135–143. [CrossRef] [PubMed]
87. Wiercinska, P.; Lou, Y.; Squires, E.J. The Roles of Different Porcine Cytochrome P450 Enzymes and Cytochrome B5A in Skatole Metabolism. *Animal* **2012**, *6*, 834–845. [CrossRef] [PubMed]
88. Forrester, S.J.; Kikuchi, D.S.; Hernandes, M.S.; Xu, Q.; Griendling, K.K. Reactive Oxygen Species in Metabolic and Inflammatory Signaling. *Circ. Res.* **2018**, *122*, 877. [CrossRef]
89. Zangar, R.C.; Davydov, D.R.; Verma, S. Mechanisms That Regulate Production of Reactive Oxygen Species by Cytochrome P450. *Toxicol. Appl. Pharmacol.* **2004**, *199*, 316–331. [CrossRef]
90. Veith, A.; Moorthy, B. Role of Cytochrome P450s in the Generation and Metabolism of Reactive Oxygen Species. *Curr. Opin. Toxicol.* **2018**, *7*, 44–51. [CrossRef]
91. Bhandary, B.; Marahatta, A.; Kim, H.R.; Chae, H.J. An Involvement of Oxidative Stress in Endoplasmic Reticulum Stress and Its Associated Diseases. *Int. J. Mol. Sci.* **2012**, *14*, 434–456. [CrossRef]
92. Santos, C.X.C.; Tanaka, L.Y.; Wosniak, J.; Laurindo, F.R.M. Mechanisms and Implications of Reactive Oxygen Species Generation during the Unfolded Protein Response: Roles of Endoplasmic Reticulum Oxidoreductases, Mitochondrial Electron Transport, and NADPH Oxidase. *Antioxid. Redox Signal.* **2009**, *11*, 2409–2427. [CrossRef]
93. Higa, A.; Chevet, E. Redox Signaling Loops in the Unfolded Protein Response. *Cell Signal.* **2012**, *24*, 1548–1555. [CrossRef]
94. Kim, H.S.; Lee, D.; Shen, S. Endoplasmic Reticular Stress as an Emerging Therapeutic Target for Chronic Pain: A Narrative Review. *Br. J. Anaesth.* **2024**, *132*, 707–724. [CrossRef]
95. Lei, J.; Paul, J.; Wang, Y.; Gupta, M.; Vang, D.; Thompson, S.; Jha, R.; Nguyen, J.; Valverde, Y.; Lamarre, Y.; et al. Heme Causes Pain in Sickle Mice via Toll-Like Receptor 4-Mediated Reactive Oxygen Species- and Endoplasmic Reticulum Stress-Induced Glial Activation. *Antiox. Redox Signal.* **2021**, *34*, 279–293. [CrossRef] [PubMed]
96. Kawanaka, R.; Jin, H.; Aoe, T. Unraveling the Connection: Pain and Endoplasmic Reticulum Stress. *Int. J. Mol. Sci.* **2024**, *25*, 4995. [CrossRef] [PubMed]
97. Brand, M.D. Mitochondrial Generation of Superoxide and Hydrogen Peroxide as the Source of Mitochondrial Redox Signaling. *Free Radic. Biol. Med.* **2016**, *100*, 14–31. [CrossRef] [PubMed]
98. Navarro, A.; Boveris, A. The Mitochondrial Energy Transduction System and the Aging Process. *Am. J. Physiol. Cell Physiol.* **2007**, *292*, 670–686. [CrossRef]
99. Valko, M.; Leibfritz, D.; Moncol, J.; Cronin, M.T.D.; Mazur, M.; Telser, J. Free Radicals and Antioxidants in Normal Physiological Functions and Human Disease. *Int. J. Biochem. Cell Biol.* **2007**, *39*, 44–84. [CrossRef]
100. Brand, M.D.; Nicholls, D.G. Assessing Mitochondrial Dysfunction in Cells. *Biochem. J.* **2011**, *435*, 297. [CrossRef]
101. Li, B.; Yu, K.; Zhou, X.; Sun, J.; Qi, L.; Li, W.; Yang, T.; Li, W.; Wang, N.; Gu, X.; et al. Increased TSPO Alleviates Neuropathic Pain by Preventing Pyroptosis via the AMPK-PGC-1α Pathway. *J. Headache Pain* **2025**, *26*, 16. [CrossRef]
102. Zhang, Q.; Liang, X.C. Effects of Mitochondrial Dysfunction via AMPK/PGC-1 α Signal Pathway on Pathogenic Mechanism of Diabetic Peripheral Neuropathy and the Protective Effects of Chinese Medicine. *Chin. J. Integr. Med.* **2019**, *25*, 386–394. [CrossRef]
103. Kim, H.Y.; Chung, J.M.; Chung, K. Increased Production of Mitochondrial Superoxide in the Spinal Cord Induces Pain Behaviors in Mice: The Effect of Mitochondrial Electron Transport Complex Inhibitors. *Neurosci. Lett.* **2008**, *447*, 87–91. [CrossRef]
104. Silva Santos Ribeiro, P.; Willemen, H.L.D.M.; Eijkelkamp, N. Mitochondria and Sensory Processing in Inflammatory and Neuropathic Pain. *Front. Pain Res.* **2022**, *3*, 1013577. [CrossRef]
105. Skulachev, V.P. Why Are Mitochondria Involved in Apoptosis? Permeability Transition Pores and Apoptosis as Selective Mechanisms to Eliminate Superoxide-Producing Mitochondria and Cell. *FEBS Lett.* **1996**, *397*, 7–10. [CrossRef] [PubMed]
106. Vignais, P.V. The Superoxide-Generating NADPH Oxidase: Structural Aspects and Activation Mechanism. *Cell. Mol. Life Sci.* **2002**, *59*, 1428–1459. [CrossRef] [PubMed]
107. Lambeth, J.D. NOX Enzymes and the Biology of Reactive Oxygen. *Nat. Rev. Immunol.* **2004**, *4*, 181–189. [CrossRef] [PubMed]
108. Zhang, W.; Jiao, B.; Yu, S.; Zhang, K.; Sun, J.; Liu, B.; Zhang, X. Spinal AT1R Contributes to Neuroinflammation and Neuropathic Pain via NOX2-Dependent Redox Signaling in Microglia. *Free Radic. Biol. Med.* **2025**, *227*, 143–156. [CrossRef]

109. Zhang, W.; Yu, S.; Jiao, B.; Zhang, C.; Zhang, K.; Liu, B.; Zhang, X. Vitamin D3 Attenuates Neuropathic Pain via Suppression of Mitochondria-Associated Ferroptosis by Inhibiting PKCα/NOX4 Signaling Pathway. *CNS Neurosci. Ther.* **2024**, 30. [CrossRef]
110. Sorce, S.; Krause, K.H. NOX Enzymes in the Central Nervous System: From Signaling to Disease. *Antioxid. Redox Signal.* **2009**, *11*, 2481–2504. [CrossRef]
111. De Logu, F.; Nassini, R.; Materazzi, S.; Carvalho Gonçalves, M.; Nosi, D.; Rossi Degl'Innocenti, D.; Marone, I.M.; Ferreira, J.; Li Puma, S.; Benemei, S.; et al. Schwann Cell TRPA1 Mediates Neuroinflammation That Sustains Macrophage-Dependent Neuropathic Pain in Mice. *Nat. Commun.* **2017**, *8*, 1887. [CrossRef]
112. Kallenborn-Gerhardt, W.; Hohmann, S.W.; Syhr, K.M.J.; Schröder, K.; Sisignano, M.; Weigert, A.; Lorenz, J.E.; Lu, R.; Brüne, B.; Brandes, R.P.; et al. Nox2-Dependent Signaling between Macrophages and Sensory Neurons Contributes to Neuropathic Pain Hypersensitivity. *Pain* **2014**, *155*, 2161–2170. [CrossRef]
113. Berger, J.V.; Deumens, R.; Goursaud, S.; Schäfer, S.; Lavand'homme, P.; Joosten, E.A.; Hermans, E. Enhanced Neuroinflammation and Pain Hypersensitivity after Peripheral Nerve Injury in Rats Expressing Mutated Superoxide Dismutase 1. *J. Neuroinflamm.* **2011**, *8*, 33. [CrossRef]
114. Geis, C.; Geuss, E.; Sommer, C.; Schmidt, H.H.H.W.; Kleinschnitz, C. NOX4 Is an Early Initiator of Neuropathic Pain. *Exp. Neurol.* **2017**, *288*, 94–103. [CrossRef]
115. Fu, C.N.; Wei, H.; Gao, W.S.; Song, S.S.; Yue, S.W.; Qu, Y.J. Obesity Increases Neuropathic Pain via the AMPK-ERK-NOX4 Pathway in Rats. *Aging* **2021**, *13*, 18606–18619. [CrossRef] [PubMed]
116. Kallenborn-Gerhardt, W.; Schröder, K.; del Turco, D.; Lu, R.; Kynast, K.; Kosowski, J.; Niederberger, E.; Shah, A.M.; Brandes, R.P.; Geisslinger, G.; et al. NADPH Oxidase-4 Maintains Neuropathic Pain after Peripheral Nerve Injury. *J. Neurosci.* **2012**, *32*, 10136–10145. [CrossRef] [PubMed]
117. Rius-Pérez, S.; Pérez, S.; Toledano, M.B.; Sastre, J. Mitochondrial Reactive Oxygen Species and Lytic Programmed Cell Death in Acute Inflammation. *Antioxid. Redox Signal.* **2023**, *39*, 708–727. [CrossRef] [PubMed]
118. Tousoulis, D.; Kampoli, A.-M.; Tentolouris Nikolaos Papageorgiou, C.; Stefanadis, C. The Role of Nitric Oxide on Endothelial Function. *Curr. Vasc. Pharmacol.* **2012**, *10*, 4–18. [CrossRef]
119. Kasai, S.; Shimizu, S.; Tatara, Y.; Mimura, J.; Itoh, K. Regulation of Nrf2 by Mitochondrial Reactive Oxygen Species in Physiology and Pathology. *Biomolecules* **2020**, *10*, 320. [CrossRef]
120. Marinho, H.S.; Real, C.; Cyrne, L.; Soares, H.; Antunes, F. Hydrogen Peroxide Sensing, Signaling and Regulation of Transcription Factors. *Redox Biol.* **2014**, *2*, 535–562. [CrossRef]
121. Shi, T.; Dansen, T.B. Reactive Oxygen Species Induced P53 Activation: DNA Damage, Redox Signaling, or Both? *Antioxid. Redox Signal.* **2020**, *33*, 839–859. [CrossRef]
122. Zhang, J.; Wang, X.; Vikash, V.; Ye, Q.; Wu, D.; Liu, Y.; Dong, W. ROS and ROS-Mediated Cellular Signaling. *Oxid. Med. Cell. Longev.* **2016**, *2016*, 4350965. [CrossRef]
123. Bak, D.W.; Bechtel, T.J.; Falco, J.A.; Weerapana, E. Cysteine Reactivity Across the Sub-Cellular Universe. *Curr. Opin. Chem. Biol.* **2019**, *48*, 96. [CrossRef]
124. Behring, J.B.; van der Post, S.; Mooradian, A.D.; Egan, M.J.; Zimmerman, M.I.; Clements, J.L.; Bowman, G.R.; Held, J.M. Spatial and Temporal Alterations in Protein Structure by EGF Regulate Cryptic Cysteine Oxidation. *Sci. Signal.* **2020**, *13*, eaay7315. [CrossRef]
125. Roos, G.; Messens, J. Protein Sulfenic Acid Formation: From Cellular Damage to Redox Regulation. *Free Radic. Biol. Med.* **2011**, *51*, 314–326. [CrossRef] [PubMed]
126. Sen, C.K. Cellular Thiols and Redox-Regulated Signal Transduction. *Curr. Top. Cell Regul.* **2001**, *36*, 1–30. [CrossRef]
127. Chen, H.; Xie, K.; Chen, Y.; Wang, Y.; Wang, Y.; Lian, N.; Zhang, K.; Yu, Y. Nrf2/HO-1 Signaling Pathway Participated in the Protection of Hydrogen Sulfide on Neuropathic Pain in Rats. *Int. Immunopharmacol.* **2019**, *75*, 105746. [CrossRef] [PubMed]
128. Yardim, A.; Kandemir, F.M.; Ozdemir, S.; Kucukler, S.; Comakli, S.; Gur, C.; Celik, H. Quercetin Provides Protection against the Peripheral Nerve Damage Caused by Vincristine in Rats by Suppressing Caspase 3, NF-KB, ATF-6 Pathways and Activating Nrf2, Akt Pathways. *Neurotoxicology* **2020**, *81*, 137–146. [CrossRef]
129. Guo, X.; Kang, J.; Wang, Z.; Wang, Y.; Liu, M.; Zhu, D.; Yang, F.; Kang, X. Nrf2 Signaling in the Oxidative Stress Response After Spinal Cord Injury. *Neuroscience* **2022**, *498*, 311–324. [CrossRef]
130. Zhou, Y.Q.; Mei, W.; Tian, X.B.; Tian, Y.K.; Liu, D.Q.; Ye, D.W. The Therapeutic Potential of Nrf2 Inducers in Chronic Pain: Evidence from Preclinical Studies. *Pharmacol. Ther.* **2021**, *225*, 107846. [CrossRef]
131. Parvez, S.; Long, M.J.C.; Poganik, J.R.; Aye, Y. Redox Signaling by Reactive Electrophiles and Oxidants. *Chem. Rev.* **2018**, *118*, 8798–8888. [CrossRef]
132. Singh, S.; Vrishni, S.; Singh, B.K.; Rahman, I.; Kakkar, P. Nrf2-ARE Stress Response Mechanism: A Control Point in Oxidative Stress-Mediated Dysfunctions and Chronic Inflammatory Diseases. *Free Radic. Res.* **2010**, *44*, 1267–1288. [CrossRef]
133. Fourquet, S.; Guerois, R.; Biard, D.; Toledano, M.B. Activation of NRF2 by Nitrosative Agents and H_2O_2 Involves KEAP1 Disulfide Formation. *J. Biol. Chem.* **2010**, *285*, 8463. [CrossRef]

134. Lee, J.H.; Khor, T.O.; Shu, L.; Su, Z.Y.; Fuentes, F.; Kong, A.N.T. Dietary Phytochemicals and Cancer Prevention: Nrf2 Signaling, Epigenetics, and Cell Death Mechanisms in Blocking Cancer Initiation and Progression. *Pharmacol. Ther.* **2013**, *137*, 153–171. [CrossRef]
135. Vasconcelos, A.R.; Dos Santos, N.B.; Scavone, C.; Munhoz, C.D. Nrf2/ARE Pathway Modulation by Dietary Energy Regulation in Neurological Disorders. *Front. Pharmacol.* **2019**, *10*, 33. [CrossRef] [PubMed]
136. Kobayashi, A.; Kang, M.-I.; Watai, Y.; Tong, K.I.; Shibata, T.; Uchida, K.; Yamamoto, M. Oxidative and Electrophilic Stresses Activate Nrf2 through Inhibition of Ubiquitination Activity of Keap1. *Mol. Cell Biol.* **2006**, *26*, 221–229. [CrossRef]
137. Motohashi, H.; Yamamoto, M. Nrf2-Keap1 Defines a Physiologically Important Stress Response Mechanism. *Trends Mol. Med.* **2004**, *10*, 549–557. [CrossRef] [PubMed]
138. Yamamoto, M.; Kensler, T.W.; Motohashi, H. The KEAP1-NRF2 System: A Thiol-Based Sensor-Effector Apparatus for Maintaining Redox Homeostasis. *Physiol. Rev.* **2018**, *98*, 1169. [CrossRef] [PubMed]
139. Kaidery, N.A.; Banerjee, R.; Yang, L.; Smirnova, N.A.; Hushpulian, D.M.; Liby, K.T.; Williams, C.R.; Yamamoto, M.; Kensler, T.W.; Ratan, R.R.; et al. Targeting Nrf2-Mediated Gene Transcription by Extremely Potent Synthetic Triterpenoids Attenuate Dopaminergic Neurotoxicity in the MPTP Mouse Model of Parkinson's Disease. *Antioxid. Redox Signal.* **2012**, *18*, 139–157. [CrossRef] [PubMed]
140. Liu, L.; Vollmer, M.K.; Kelly, M.G.; Fernandez, V.M.; Fernandez, T.G.; Kim, H.; Doré, S. Reactive Gliosis Contributes to Nrf2-Dependent Neuroprotection by Pretreatment with Dimethyl Fumarate or Korean Red Ginseng against Hypoxic-Ischemia: Focus on Hippocampal Injury. *Mol. Neurobiol.* **2020**, *57*, 105. [CrossRef]
141. Itoh, K.; Mochizuki, M.; Ishii, Y.; Ishii, T.; Shibata, T.; Kawamoto, Y.; Kelly, V.; Sekizawa, K.; Uchida, K.; Yamamoto, M. Transcription Factor Nrf2 Regulates Inflammation by Mediating the Effect of 15-Deoxy-Δ12,14-Prostaglandin J2. *Mol. Cell Biol.* **2004**, *24*, 36. [CrossRef]
142. Li, S.; Yang, C.; Fang, X.; Zhan, G.; Huang, N.; Gao, J.; Xu, H.; Hashimoto, K.; Luo, A. Role of Keap1-Nrf2 Signaling in Anhedonia Symptoms in a Rat Model of Chronic Neuropathic Pain: Improvement with Sulforaphane. *Front. Pharmacol.* **2018**, *9*, 887. [CrossRef]
143. Ghosh, S.; May, M.J.; Kopp, E.B. NF-Kappa B and Rel Proteins: Evolutionarily Conserved Mediators of Immune Responses. *Annu. Rev. Immunol.* **1998**, *16*, 225–260. [CrossRef]
144. Popiolek-Barczyk, K.; Mika, J. Targeting the Microglial Signaling Pathways: New Insights in the Modulation of Neuropathic Pain. *Curr. Med. Chem.* **2016**, *23*, 2908. [CrossRef]
145. Bakalkin, G.Y.; Yakovleva, T.; Terenius, L. NF-Kappa B-like Factors in the Murine Brain. Developmentally-Regulated and Tissue-Specific Expression. *Brain Res. Mol. Brain Res.* **1993**, *20*, 137–146. [CrossRef] [PubMed]
146. Massa, P.T.; Aleyasin, H.; Park, D.S.; Mao, X.; Barger, S.W. NFκB in Neurons? The Uncertainty Principle in Neurobiology. *J. Neurochem.* **2006**, *97*, 607–618. [CrossRef] [PubMed]
147. Shih, R.H.; Wang, C.Y.; Yang, C.M. NF-KappaB Signaling Pathways in Neurological Inflammation: A Mini Review. *Front. Mol. Neurosci.* **2015**, *8*, 77. [CrossRef] [PubMed]
148. Gloire, G.; Legrand-Poels, S.; Piette, J. NF-KB Activation by Reactive Oxygen Species: Fifteen Years Later. *Biochem. Pharmacol.* **2006**, *72*, 1493–1505. [CrossRef]
149. Brigelius-Flohé, R.; Flohé, L. Basic Principles and Emerging Concepts in the Redox Control of Transcription Factors. *Antioxid. Redox Signal.* **2011**, *15*, 2335–2381. [CrossRef] [PubMed]
150. de Oliveira-Marques, V.; Cyrne, L.; Marinho, H.S.; Antunes, F. A Quantitative Study of NF-KappaB Activation by H_2O_2: Relevance in Inflammation and Synergy with TNF-Alpha. *J. Immunol.* **2007**, *178*, 3893–3902. [CrossRef]
151. Basak, S.; Hoffmann, A. Crosstalk via the NF-KB Signaling System. *Cytokine Growth Factor Rev.* **2008**, *19*, 187–197. [CrossRef]
152. Liu, T.; Zhang, L. GDF11 Mitigates Neuropathic Pain via Regulation of Microglial Polarization and Neuroinflammation through TGF-BR1/SMAD2/NF-KB Pathway in Male Mice. *J. Neuroimmune Pharmacol.* **2025**, *20*, 20. [CrossRef]
153. Chen, G.; Zhang, Y.Q.; Qadri, Y.J.; Serhan, C.N.; Ji, R.R. Microglia in Pain: Detrimental and Protective Roles in Pathogenesis and Resolution of Pain. *Neuron* **2018**, *100*, 1292–1311. [CrossRef]
154. Wang, L.; Yin, C.; Liu, T.; Abdul, M.; Zhou, Y.; Cao, J.L.; Lu, C. Pellino1 Regulates Neuropathic Pain as Well as Microglial Activation through the Regulation of MAPK/NF-KB Signaling in the Spinal Cord. *J. Neuroinflamm.* **2020**, *17*, 83. [CrossRef]
155. Amoutzias, G.D.; Veron, A.S.; Weiner, J.; Robinson-Rechavi, M.; Bornberg-Bauer, E.; Oliver, S.G.; Robertson, D.L. One Billion Years of BZIP Transcription Factor Evolution: Conservation and Change in Dimerization and DNA-Binding Site Specificity. *Mol. Biol. Evol.* **2007**, *24*, 827–835. [CrossRef] [PubMed]
156. Hong, Y.; Boiti, A.; Vallone, D.; Foulkes, N.S. Reactive Oxygen Species Signaling and Oxidative Stress: Transcriptional Regulation and Evolution. *Antioxidants* **2024**, *13*, 312. [CrossRef] [PubMed]
157. Karin, M.; Liu, Z.G.; Zandi, E. AP-1 Function and Regulation. *Curr. Opin. Cell Biol.* **1997**, *9*, 240–246. [CrossRef] [PubMed]
158. Bhosale, P.B.; Kim, H.H.; Abusaliya, A.; Vetrivel, P.; Ha, S.E.; Park, M.Y.; Lee, H.J.; Kim, G.S. Structural and Functional Properties of Activator Protein-1 in Cancer and Inflammation. *Evid. Based Complement. Altern. Med.* **2022**, *2022*, 9797929. [CrossRef]

159. Shaulian, E.; Karin, M. AP-1 as a Regulator of Cell Life and Death. *Nat. Cell Biol.* **2002**, *4*, E131–E136. [CrossRef]
160. Wisdom, R. AP-1: One Switch for Many Signals. *Exp. Cell Res.* **1999**, *253*, 180–185. [CrossRef]
161. Gius, D.; Botero, A.; Shah, S.; Curry, H.A. Intracellular Oxidation/Reduction Status in the Regulation of Transcription Factors NF-KappaB and AP-1. *Toxicol. Lett.* **1999**, *106*, 93–106. [CrossRef]
162. Xanthoudakis', S.; Miao, G.; Wang, F.; Pan, Y.-C.E.; Curran, T. Redox Activation of Fos-Jun DNA Binding Activity Is Mediated by a DNA Repair Enzyme. *EMBO J.* **1992**, *11*, 3323. [CrossRef]
163. Tanos, T.; Marinissen, M.J.; Leskow, F.C.; Hochbaum, D.; Martinetto, H.; Gutkind, J.S.; Coso, O.A. Phosphorylation of C-Fos by Members of the P38 MAPK Family. Role in the AP-1 Response to UV Light. *J. Biol. Chem.* **2005**, *280*, 18842–18852. [CrossRef]
164. Shaulian, E.; Karin, M. AP-1 in Cell Proliferation and Survival. *Oncogene* **2001**, *20*, 2390–2400. [CrossRef]
165. Zhuang, Z.Y.; Gerner, P.; Woolf, C.J.; Ji, R.R. ERK Is Sequentially Activated in Neurons, Microglia, and Astrocytes by Spinal Nerve Ligation and Contributes to Mechanical Allodynia in This Neuropathic Pain Model. *Pain* **2005**, *114*, 149–159. [CrossRef] [PubMed]
166. Jin, S.X.; Zhuang, Z.Y.; Woolf, C.J.; Ji, R.R. P38 Mitogen-Activated Protein Kinase Is Activated after a Spinal Nerve Ligation in Spinal Cord Microglia and Dorsal Root Ganglion Neurons and Contributes to the Generation of Neuropathic Pain. *J. Neurosci.* **2003**, *23*, 4017–4022. [CrossRef] [PubMed]
167. Wei, C.J.; Li, Y.L.; Zhu, Z.L.; Jia, D.M.; Fan, M.L.; Li, T.; Wang, X.J.; Li, Z.G.; Ma, H.S. Inhibition of Activator Protein 1 Attenuates Neuroinflammation and Brain Injury after Experimental Intracerebral Hemorrhage. *CNS Neurosci. Ther.* **2019**, *25*, 1182. [CrossRef] [PubMed]
168. Chinenov, Y.; Kerppola, T.K. Close Encounters of Many Kinds: Fos-Jun Interactions That Mediate Transcription Regulatory Specificity. *Oncogene* **2001**, *20*, 2438–2452. [CrossRef]
169. Satoh, T.; Nakatsuka, D.; Watanabe, Y.; Nagata, I.; Kikuchi, H.; Namura, S. Neuroprotection by MAPK/ERK Kinase Inhibition with U0126 against Oxidative Stress in a Mouse Neuronal Cell Line and Rat Primary Cultured Cortical Neurons. *Neurosci. Lett.* **2000**, *288*, 163–166. [CrossRef]
170. Samanta, S.; Perkinton, M.S.; Morgan, M.; Williams, R.J. Hydrogen Peroxide Enhances Signal-Responsive Arachidonic Acid Release from Neurons: Role of Mitogen-Activated Protein Kinase. *J. Neurochem.* **1998**, *70*, 2082–2090. [CrossRef]
171. Kim, E.K.; Choi, E.J. Compromised MAPK Signaling in Human Diseases: An Update. *Arch. Toxicol.* **2015**, *89*, 867–882. [CrossRef]
172. Plotnikov, A.; Zehorai, E.; Procaccia, S.; Seger, R. The MAPK Cascades: Signaling Components, Nuclear Roles and Mechanisms of Nuclear Translocation. *Biochim. Biophys. Acta* **2011**, *1813*, 1619–1633. [CrossRef]
173. Whitmarsh, A.J. Regulation of Gene Transcription by Mitogen-Activated Protein Kinase Signaling Pathways. *Biochim. Biophys. Acta* **2007**, *1773*, 1285–1298. [CrossRef]
174. Niault, T.S.; Baccarini, M. Targets of Raf in Tumorigenesis. *Carcinogenesis* **2010**, *31*, 1165–1174. [CrossRef]
175. Kim, E.K.; Choi, E.J. Pathological Roles of MAPK Signaling Pathways in Human Diseases. *Biochim. Biophys. Acta* **2010**, *1802*, 396–405. [CrossRef] [PubMed]
176. Yoon, S.; Seger, R. The Extracellular Signal-Regulated Kinase: Multiple Substrates Regulate Diverse Cellular Functions. *Growth Factors* **2006**, *24*, 21–44. [CrossRef] [PubMed]
177. Cheng, X.P.; Wang, B.R.; Liu, H.L.; You, S.W.; Huang, W.J.; Jiao, X.Y.; Ju, G. Phosphorylation of Extracellular Signal-Regulated Kinases 1/2 Is Predominantly Enhanced in the Microglia of the Rat Spinal Cord Following Dorsal Root Transection. *Neuroscience* **2003**, *119*, 701–712. [CrossRef] [PubMed]
178. Ji, R.R.; Befort, K.; Brenner, G.J.; Woolf, C.J. ERK MAP Kinase Activation in Superficial Spinal Cord Neurons Induces Prodynorphin and NK-1 Upregulation and Contributes to Persistent Inflammatory Pain Hypersensitivity. *J. Neurosci.* **2002**, *22*, 478–485. [CrossRef]
179. Wang, H.; Dai, Y.; Fukuoka, T.; Yamanaka, H.; Obata, K.; Tokunaga, A.; Noguchi, K. Enhancement of Stimulation-Induced ERK Activation in the Spinal Dorsal Horn and Gracile Nucleus Neurons in Rats with Peripheral Nerve Injury. *Eur. J. Neurosci.* **2004**, *19*, 884–890. [CrossRef]
180. Wang, Z.; Jia, S.; Kang, X.; Chen, S.; Zhang, L.; Tian, Z.K.; Liang, X.; Meng, C. Isoliquiritigenin Alleviates Neuropathic Pain by Reducing Microglia Inflammation through Inhibition of the ERK Signaling Pathway and Decreasing CEBPB Transcription Expression. *Int. Immunopharmacol.* **2024**, *143*, 113536. [CrossRef]
181. Weston, C.R.; Davis, R.J. The JNK Signal Transduction Pathway. *Curr. Opin. Cell Biol.* **2007**, *19*, 142–149. [CrossRef]
182. Johnson, G.L.; Nakamura, K. The C-Jun Kinase/Stress-Activated Pathway: Regulation, Function and Role in Human Disease. *Biochim. Biophys. Acta* **2007**, *1773*, 1341–1348. [CrossRef]
183. Haeusgen, W.; Boehm, R.; Zhao, Y.; Herdegen, T.; Waetzig, V. Specific Activities of Individual C-Jun N-Terminal Kinases in the Brain. *Neuroscience* **2009**, *161*, 951–959. [CrossRef]
184. Zhu, Z.; Chen, G.; He, J.; Xu, Y. The Protective Effects of Orexin B in Neuropathic Pain by Suppressing Inflammatory Response. *Neuropeptides* **2024**, *108*, 102458. [CrossRef]
185. Zarubin, T.; Han, J. Activation and Signaling of the P38 MAP Kinase Pathway. *Cell Res.* **2005**, *15*, 11–18. [CrossRef] [PubMed]

186. Whitmarsh, A.J.; Davis, R.J. Role of Mitogen-Activated Protein Kinase Kinase 4 in Cancer. *Oncogene* **2007**, *26*, 3172–3184. [CrossRef] [PubMed]
187. Ji, R.R.; Suter, M.R. P38 MAPK, Microglial Signaling, and Neuropathic Pain. *Mol. Pain* **2007**, *3*, 1–9. [CrossRef] [PubMed]
188. Lee, C.T.; Yu, L.E.; Wang, J.Y. Nitroxide Antioxidant as a Potential Strategy to Attenuate the Oxidative/Nitrosative Stress Induced by Hydrogen Peroxide plus Nitric Oxide in Cultured Neurons. *Nitric Oxide* **2016**, *54*, 38–50. [CrossRef] [PubMed]
189. Vanhaesebroeck, B.; Whitehead, M.A.; Piñeiro, R. Molecules in Medicine Mini-Review: Isoforms of PI3K in Biology and Disease. *J. Mol. Med.* **2016**, *94*, 5–11. [CrossRef]
190. Hawkins, P.T.; Stephens, L.R. Emerging Evidence of Signalling Roles for PI(3,4)P2 in Class I and II PI3K-Regulated Pathways. *Biochem. Soc. Trans.* **2016**, *44*, 307–314. [CrossRef]
191. Deane, J.A.; Fruman, D.A. Phosphoinositide 3-Kinase: Diverse Roles in Immune Cell Activation. *Annu. Rev. Immunol.* **2004**, *22*, 563–598. [CrossRef]
192. Manning, B.D.; Toker, A. AKT/PKB Signaling: Navigating the Network. *Cell* **2017**, *169*, 381–405. [CrossRef]
193. Chen, S.-P.; Zhou, Y.-Q.; Liu, D.-Q.; Zhang, W.; Manyande, A.; Guan, X.-H.; Tian, Y.; Ye, D.-W.; Omar, D.M. PI3K/Akt Pathway: A Potential Therapeutic Target for Chronic Pain. *Curr. Pharm. Des.* **2017**, *23*, 1860–1868. [CrossRef]
194. Cianciulli, A.; Porro, C.; Calvello, R.; Trotta, T.; Lofrumento, D.D.; Panaro, M.A. Microglia Mediated Neuroinflammation: Focus on PI3K Modulation. *Biomolecules* **2020**, *10*, 137. [CrossRef]
195. Saponaro, C.; Cianciulli, A.; Calvello, R.; Dragone, T.; Iacobazzi, F.; Panaro, M.A. The PI3K/Akt Pathway Is Required for LPS Activation of Microglial Cells. *Immunopharmacol. Immunotoxicol.* **2012**, *34*, 858–865. [CrossRef] [PubMed]
196. Ryan, C.B.; Choi, J.S.; Kang, B.; Herr, S.; Pereira, C.; Moraes, C.T.; Al-Ali, H.; Lee, J.K. PI3K Signaling Promotes Formation of Lipid-Laden Foamy Macrophages at the Spinal Cord Injury Site. *Neurobiol. Dis.* **2024**, *190*, 106370. [CrossRef] [PubMed]
197. Zhu, J.; Wang, M.; Cao, B.; Hou, T.; Mao, X. Targeting the Phosphatidylinositol 3-Kinase/AKT Pathway for the Treatment of Multiple Myeloma. *Curr. Med. Chem.* **2014**, *21*, 3173–3187. [CrossRef] [PubMed]
198. Whitehead, M.A.; Bombardieri, M.; Pitzalis, C.; Vanhaesebroeck, B. Isoform-Selective Induction of Human P110δ PI3K Expression by TNFα: Identification of a New and Inducible PIK3CD Promoter. *Biochem. J.* **2012**, *443*, 857. [CrossRef]
199. Amzel, L.M.; Huang, C.H.; Mandelker, D.; Lengauer, C.; Gabelli, S.B.; Vogelstein, B. Structural Comparisons of Class I Phosphoinositide 3-Kinases. *Nat. Rev. Cancer* **2008**, *8*, 665–669. [CrossRef]
200. Katso, R.; Okkenhaug, K.; Ahmadi, K.; White, S.; Timms, J.; Waterfield, M.D. Cellular Function of Phosphoinositide 3-Kinases: Implications for Development, Homeostasis, and Cancer. *Annu. Rev. Cell Dev. Biol.* **2001**, *17*, 615–675. [CrossRef]
201. Vanhaesebroeck, B.; Guillermet-Guibert, J.; Graupera, M.; Bilanges, B. The Emerging Mechanisms of Isoform-Specific PI3K Signalling. *Nat. Rev. Mol. Cell Biol.* **2010**, *11*, 329–341. [CrossRef]
202. Hawkins, P.T.; Stephens, L.R. PI3K Signalling in Inflammation. *Biochim. Biophys. Acta* **2015**, *1851*, 882–897. [CrossRef]
203. Fritsch, R.; Downward, J. SnapShot: Class I PI3K Isoform Signaling. *Cell* **2013**, *154*, 940.e1. [CrossRef]
204. Hemmings, B.A.; Restuccia, D.F. PI3K-PKB/Akt Pathway. *Cold Spring Harb. Perspect. Biol.* **2012**, *4*, a011189. [CrossRef]
205. Rauf, A.; Khalil, A.A.; Awadallah, S.; Khan, S.A.; Abu-Izneid, T.; Kamran, M.; Hemeg, H.A.; Mubarak, M.S.; Khalid, A.; Wilairatana, P. Reactive Oxygen Species in Biological Systems: Pathways, Associated Diseases, and Potential Inhibitors—A Review. *Food Sci. Nutr.* **2023**, *12*, 675–693. [CrossRef] [PubMed]
206. Kma, L.; Baruah, T.J. The Interplay of ROS and the PI3K/Akt Pathway in Autophagy Regulation. *Biotechnol. Appl. Biochem.* **2022**, *69*, 248–264. [CrossRef] [PubMed]
207. Troutman, T.D.; Bazan, J.F.; Pasare, C. Toll-like Receptors, Signaling Adapters and Regulation of the pro-Inflammatory Response by PI3K. *Cell Cycle* **2012**, *11*, 3559. [CrossRef] [PubMed]
208. Nakanishi, A.; Wada, Y.; Kitagishi, Y.; Matsuda, S. Link between PI3K/AKT/PTEN Pathway and NOX Proteinin Diseases. *Aging Dis.* **2014**, *5*, 203. [CrossRef]
209. Leslie, N.R.; Downes, C.P. PTEN: The down Side of PI 3-Kinase Signalling. *Cell Signal.* **2002**, *14*, 285–295. [CrossRef]
210. Innocenti, M.; Frittoli, E.; Ponzanelli, I.; Falck, J.R.; Brachmann, S.M.; Di Fiore, P.P.; Scita, G. Phosphoinositide 3-Kinase Activates Rac by Entering in a Complex with Eps8, Abi1, and Sos-1. *J. Cell Biol.* **2003**, *160*, 17. [CrossRef]
211. Lee, S.R.; Yang, K.S.; Kwon, J.; Lee, C.; Jeong, W.; Rhee, S.G. Reversible Inactivation of the Tumor Suppressor PTEN by H_2O_2. *J. Biol. Chem.* **2002**, *277*, 20336–20342. [CrossRef]
212. Murata, H.; Ihara, Y.; Nakamura, H.; Yodoi, J.; Sumikawa, K.; Kondo, T. Glutaredoxin Exerts an Antiapoptotic Effect by Regulating the Redox State of Akt. *J. Biol. Chem.* **2003**, *278*, 50226–50233. [CrossRef]
213. Arcaro, A.; Khanzada, U.K.; Vanhaesebroeck, B.; Tetley, T.D.; Waterfield, M.D.; Seckl, M.J. Two Distinct Phosphoinositide 3-Kinases Mediate Polypeptide Growth Factor-Stimulated PKB Activation. *EMBO J.* **2002**, *21*, 5097. [CrossRef]
214. Seitz, C.; Hugle, M.; Cristofanon, S.; Tchoghandjian, A.; Fulda, S. The Dual PI3K/MTOR Inhibitor NVP-BEZ235 and Chloroquine Synergize to Trigger Apoptosis via Mitochondrial-Lysosomal Cross-Talk. *Int. J. Cancer* **2013**, *132*, 2682–2693. [CrossRef]
215. Wang, L.; Chen, Y.; Sternberg, P.; Cai, J. Essential Roles of the PI3 Kinase/Akt Pathway in Regulating Nrf2-Dependent Antioxidant Functions in the RPE. *Investig. Ophthalmol. Vis. Sci.* **2008**, *49*, 1671–1678. [CrossRef] [PubMed]

216. Xu, B.; Guan, X.H.; Yu, J.X.; Lv, J.; Zhang, H.X.; Fu, Q.C.; Xiang, H.B.; Bu, H.L.; Shi, D.; Shu, B.; et al. Activation of Spinal Phosphatidylinositol 3-Kinase/Protein Kinase B Mediates Pain Behavior Induced by Plantar Incision in Mice. *Exp. Neurol.* **2014**, *255*, 71–82. [CrossRef]
217. Xu, J.T.; Tu, H.Y.; Xin, W.J.; Liu, X.G.; Zhang, G.H.; Zhai, C.H. Activation of Phosphatidylinositol 3-Kinase and Protein Kinase B/Akt in Dorsal Root Ganglia and Spinal Cord Contributes to the Neuropathic Pain Induced by Spinal Nerve Ligation in Rats. *Exp. Neurol.* **2007**, *206*, 269–279. [CrossRef] [PubMed]
218. Carvalho, T.T.; Flauzino, T.; Otaguiri, E.S.; Batistela, A.P.; Zarpelon, A.C.; Cunha, T.M.; Ferreira, S.H.; Cunha, F.Q.; Verri, W.A. Granulocyte-Colony Stimulating Factor (G-CSF) Induces Mechanical Hyperalgesia via Spinal Activation of MAP Kinases and PI3K in Mice. *Pharmacol. Biochem. Behav.* **2011**, *98*, 188–195. [CrossRef] [PubMed]
219. Guan, X.H.; Lu, X.F.; Zhang, H.X.; Wu, J.R.; Yuan, Y.; Bao, Q.; Ling, D.Y.; Cao, J.L. Phosphatidylinositol 3-Kinase Mediates Pain Behaviors Induced by Activation of Peripheral EphrinBs/EphBs Signaling in Mice. *Pharmacol. Biochem. Behav.* **2010**, *95*, 315–324. [CrossRef]
220. Song, X.J.; Zheng, J.H.; Cao, J.L.; Liu, W.T.; Song, X.S.; Huang, Z.J. EphrinB-EphB Receptor Signaling Contributes to Neuropathic Pain by Regulating Neural Excitability and Spinal Synaptic Plasticity in Rats. *Pain* **2008**, *139*, 168–180. [CrossRef]
221. Jiang, W.; Yu, P.; Yang, Y.; Cai, M.-T.; Gan, L.; Qu, K.; Cheng, Y.-Y.; Dong, M. PI3K-Mediated Kif1a DNA Methylation Contributes to Neuropathic Pain: An in Vivo Study. *Pain* **2025**, 1–15. [CrossRef]
222. Gleichmann, M.; Mattson, M.P. Neuronal Calcium Homeostasis and Dysregulation. *Antioxid. Redox Signal.* **2011**, *14*, 1261. [CrossRef]
223. Berridge, M.J.; Lipp, P.; Bootman, M.D. The Versatility and Universality of Calcium Signalling. *Nat. Rev. Mol. Cell Biol.* **2000**, *1*, 11–21. [CrossRef]
224. Brini, M.; Calì, T.; Ottolini, D.; Carafoli, E. Neuronal Calcium Signaling: Function and Dysfunction. *Cell. Mol. Life Sci.* **2014**, *71*, 2787. [CrossRef]
225. Baughman, J.M.; Perocchi, F.; Girgis, H.S.; Plovanich, M.; Belcher-Timme, C.A.; Sancak, Y.; Bao, X.R.; Strittmatter, L.; Goldberger, O.; Bogorad, R.L.; et al. Integrative Genomics Identifies MCU as an Essential Component of the Mitochondrial Calcium Uniporter. *Nature* **2011**, *476*, 341–345. [CrossRef] [PubMed]
226. Ott, M.; Gogvadze, V.; Orrenius, S.; Zhivotovsky, B. Mitochondria, Oxidative Stress and Cell Death. *Apoptosis* **2007**, *12*, 913–922. [CrossRef] [PubMed]
227. Angelova, P.R. Sources and Triggers of Oxidative Damage in Neurodegeneration. *Free Radic. Biol. Med.* **2021**, *173*, 52–63. [CrossRef] [PubMed]
228. Jiménez, A.; Zu, W.; Rawe, V.Y.; Pelto-Huikko, M.; Flickinger, C.J.; Sutovsky, P.; Gustafsson, J.Å.; Oko, R.; Miranda-Vizuete, A. Spermatocyte/Spermatid-Specific Thioredoxin-3, a Novel Golgi Apparatus-Associated Thioredoxin, Is a Specific Marker of Aberrant Spermatogenesis. *J. Biol. Chem.* **2004**, *279*, 34971–34982. [CrossRef]
229. Cosentino-Gomes, D.; Rocco-Machado, N.; Meyer-Fernandes, J.R. Cell Signaling through Protein Kinase C Oxidation and Activation. *Int. J. Mol. Sci.* **2012**, *13*, 10697. [CrossRef]
230. Angelova, P.R.; Abramov, A.Y. Functional Role of Mitochondrial Reactive Oxygen Species in Physiology. *Free Radic. Biol. Med.* **2016**, *100*, 81–85. [CrossRef]
231. Kania, E.; Roest, G.; Vervliet, T.; Parys, J.B.; Bultynck, G. IP3 Receptor-Mediated Calcium Signaling and Its Role in Autophagy in Cancer. *Front. Oncol.* **2017**, *7*, 140. [CrossRef]
232. Novikova, I.N.; Manole, A.; Zherebtsov, E.A.; Stavtsev, D.D.; Vukolova, M.N.; Dunaev, A.V.; Angelova, P.R.; Abramov, A.Y. Adrenaline Induces Calcium Signal in Astrocytes and Vasoconstriction via Activation of Monoamine Oxidase. *Free Radic. Biol. Med.* **2020**, *159*, 15–22. [CrossRef]
233. Görlach, A.; Bertram, K.; Hudecova, S.; Krizanova, O. Calcium and ROS: A Mutual Interplay. *Redox Biol.* **2015**, *6*, 260. [CrossRef]
234. Todorovic, S.M.; Jevtovic-Todorovic, V. Redox Regulation of Neuronal Voltage-Gated Calcium Channels. *Antioxid. Redox Signal.* **2014**, *21*, 880. [CrossRef]
235. Bogeski, I.; Kummerow, C.; Al-Ansary, D.; Schwarz, E.C.; Koehler, R.; Kozai, D.; Takahashi, N.; Peinelt, C.; Griesemer, D.; Bozem, M.; et al. Differential Redox Regulation of ORAI Ion Channels: A Mechanism to Tune Cellular Calcium Signaling. *Sci. Signal.* **2010**, *3*, 2000672. [CrossRef] [PubMed]
236. Perraud, A.L.; Takanishi, C.L.; Shen, B.; Kang, S.; Smith, M.K.; Schmitz, C.; Knowles, H.M.; Ferraris, D.; Li, W.; Zhang, J.; et al. Accumulation of Free ADP-Ribose from Mitochondria Mediates Oxidative Stress-Induced Gating of TRPM2 Cation Channels. *J. Biol. Chem.* **2005**, *280*, 6138–6148. [CrossRef] [PubMed]
237. Susankova, K.; Tousova, K.; Vyklicky, L.; Teisinger, J.; Vlachova, V. Reducing and Oxidizing Agents Sensitize Heat-Activated Vanilloid Receptor (TRPV1) Current. *Mol. Pharmacol.* **2006**, *70*, 383–394. [CrossRef] [PubMed]
238. Bogeski, I.; Kilch, T.; Niemeyer, B.A. ROS and SOCE: Recent Advances and Controversies in the Regulation of STIM and Orai. *J. Physiol.* **2012**, *590*, 4193–4200. [CrossRef]

239. Li, W.W.; Zhao, Y.; Liu, H.C.; Liu, J.; Chan, S.O.; Zhong, Y.F.; Zhang, T.Y.; Liu, Y.; Zhang, W.; Xia, Y.Q.; et al. Roles of Thermosensitive Transient Receptor Channels TRPV1 and TRPM8 in Paclitaxel-Induced Peripheral Neuropathic Pain. *Int. J. Mol. Sci.* **2024**, *25*, 5813. [CrossRef]
240. Xu, S.; Wang, Y. Transient Receptor Potential Channels: Multiple Modulators of Peripheral Neuropathic Pain in Several Rodent Models. *Neurochem. Res.* **2024**, *49*, 872–886. [CrossRef]
241. Chu, W.G.; Wang, F.D.; Sun, Z.C.; Ma, S.B.; Wang, X.; Han, W.J.; Wang, F.; Bai, Z.T.; Wu, S.X.; Freichel, M.; et al. TRPC1/4/5 Channels Contribute to Morphine-Induced Analgesic Tolerance and Hyperalgesia by Enhancing Spinal Synaptic Potentiation and Structural Plasticity. *FASEB J.* **2020**, *34*, 8526–8543. [CrossRef]
242. Alessandri-Haber, N.; Dina, O.A.; Chen, X.; Levine, J.D. TRPC1 and TRPC6 Channels Cooperate with TRPV4 to Mediate Mechanical Hyperalgesia and Nociceptor Sensitization. *J. Neurosci.* **2009**, *29*, 6217. [CrossRef]
243. Godoy, J.A.; Rios, J.A.; Picón-Pagès, P.; Herrera-Fernández, V.; Swaby, B.; Crepin, G.; Vicente, R.; Fernández-Fernández, J.M.; Muñoz, F.J. Mitostasis, Calcium and Free Radicals in Health, Aging and Neurodegeneration. *Biomolecules* **2021**, *11*, 1012. [CrossRef]
244. Gong, J.; Wang, X.Z.; Wang, T.; Chen, J.J.; Xie, X.Y.; Hu, H.; Yu, F.; Liu, H.L.; Jiang, X.Y.; Fan, H.D. Molecular Signal Networks and Regulating Mechanisms of the Unfolded Protein Response. *J. Zhejiang Univ. Sci. B* **2017**, *18*, 1–14. [CrossRef]
245. Malhotra, J.D.; Kaufman, R.J. Endoplasmic Reticulum Stress and Oxidative Stress: A Vicious Cycle or a Double-Edged Sword? *Antioxid. Redox Signal.* **2007**, *9*, 2277–2293. [CrossRef] [PubMed]
246. Chaudhari, N.; Talwar, P.; Parimisetty, A.; d'Hellencourt, C.L.; Ravanan, P. A Molecular Web: Endoplasmic Reticulum Stress, Inflammation, and Oxidative Stress. *Front. Cell. Neurosci.* **2014**, *8*, 213. [CrossRef] [PubMed]
247. Sokka, A.L.; Putkonen, N.; Mudo, G.; Pryazhnikov, E.; Reijonen, S.; Khiroug, L.; Belluardo, N.; Lindholm, D.; Korhonen, L. Endoplasmic Reticulum Stress Inhibition Protects against Excitotoxic Neuronal Injury in the Rat Brain. *J. Neurosci.* **2007**, *27*, 901–908. [CrossRef]
248. Niella, R.V.; Corrêa, J.M.X.; dos Santos, J.F.R.; Lima, L.F.; da Costa Marques, C.S.; Santos, L.C.; Santana, L.R.; Silva, Á.J.C.; Farias, K.S.; Pirovani, C.P.; et al. Post-Treatment with Maropitant Reduces Oxidative Stress, Endoplasmic Reticulum Stress and Neuroinflammation on Peripheral Nerve Injury in Rats. *PLoS ONE* **2024**, *19*, e287390. [CrossRef]
249. Zhou, F.; Zhang, W.; Zhou, J.; Li, M.; Zhong, F.; Zhang, Y.; Liu, Y.; Wang, Y. Involvement of Endoplasmic Reticulum Stress in Formalin-Induced Pain Is Attenuated by 4-Phenylbutyric Acid. *J. Pain Res.* **2017**, *10*, 653–662. [CrossRef]
250. Patel, S.; Pangarkar, A.; Mahajan, S.; Majumdar, A. Therapeutic Potential of Endoplasmic Reticulum Stress Inhibitors in the Treatment of Diabetic Peripheral Neuropathy. *Metab. Brain Dis.* **2023**, *38*, 1841–1856. [CrossRef]
251. Read, A.; Schröder, M. The Unfolded Protein Response: An Overview. *Biology* **2021**, *10*, 384. [CrossRef]
252. Borges, C.R.; Lake, D.F. Oxidative Protein Folding: Nature's Knotty Challenge. *Antioxid. Redox Signal.* **2014**, *21*, 392–395. [CrossRef]
253. Kozlov, G.; Määttänen, P.; Thomas, D.Y.; Gehring, K. A Structural Overview of the PDI Family of Proteins. *FEBS J.* **2010**, *277*, 3924–3936. [CrossRef]
254. Zhang, Z.; Zhang, L.; Zhou, L.; Lei, Y.; Zhang, Y.; Huang, C. Redox Signaling and Unfolded Protein Response Coordinate Cell Fate Decisions under ER Stress. *Redox Biol.* **2019**, *25*, 101047. [CrossRef]
255. Klappa, P.; Ruddock, L.W.; Darby, N.J.; Freedman, R.B. The b' Domain Provides the Principal Peptide-Binding Site of Protein Disulfide Isomerase but All Domains Contribute to Binding of Misfolded Proteins. *EMBO J.* **1998**, *17*, 927–935. [CrossRef] [PubMed]
256. Araki, K.; Iemura, S.I.; Kamiya, Y.; Ron, D.; Kato, K.; Natsume, T.; Nagata, K. Ero1-α and PDIs Constitute a Hierarchical Electron Transfer Network of Endoplasmic Reticulum Oxidoreductases. *J. Cell Biol.* **2013**, *202*, 861. [CrossRef] [PubMed]
257. Bhattarai, K.R.; Riaz, T.A.; Kim, H.R.; Chae, H.J. The Aftermath of the Interplay between the Endoplasmic Reticulum Stress Response and Redox Signaling. *Exp. Mol. Med.* **2021**, *53*, 151. [CrossRef]
258. Sevier, C.S. Erv2 and Quiescin Sulfhydryl Oxidases: Erv-Domain Enzymes Associated with the Secretory Pathway. *Antioxid. Redox Signal.* **2012**, *16*, 800–808. [CrossRef]
259. Zeeshan, H.M.A.; Lee, G.H.; Kim, H.R.; Chae, H.J. Endoplasmic Reticulum Stress and Associated ROS. *Int. J. Mol. Sci.* **2016**, *17*, 327. [CrossRef]
260. Eletto, D.; Chevet, E.; Argon, Y.; Appenzeller-Herzog, C. Redox Controls UPR to Control Redox. *J. Cell Sci.* **2014**, *127*, 3649–3658. [CrossRef]
261. Uehara, T.; Nakamura, T.; Yao, D.; Shi, Z.Q.; Gu, Z.; Ma, Y.; Masliah, E.; Nomura, Y.; Lipton, S.A. S-Nitrosylated Protein-Disulphide Isomerase Links Protein Misfolding to Neurodegeneration. *Nature* **2006**, *441*, 513–517. [CrossRef]
262. Vinokurov, A.Y.; Stelmashuk, O.A.; Ukolova, P.A.; Zherebtsov, E.A.; Abramov, A.Y. Brain Region Specificity in Reactive Oxygen Species Production and Maintenance of Redox Balance. *Free Radic. Biol. Med.* **2021**, *174*, 195–201. [CrossRef]
263. Lee, K.H.; Cha, M.; Lee, B.H. Neuroprotective Effect of Antioxidants in the Brain. *Int. J. Mol. Sci.* **2020**, *21*, 7152. [CrossRef]

264. Niedzielska, E.; Smaga, I.; Gawlik, M.; Moniczewski, A.; Stankowicz, P.; Pera, J.; Filip, M. Oxidative Stress in Neurodegenerative Diseases. *Mol. Neurobiol.* **2016**, *53*, 4094. [CrossRef]
265. Halliwell, B. How to Characterize a Biological Antioxidant. *Free Radic. Res. Commun.* **1990**, *9*, 1–32. [CrossRef] [PubMed]
266. Pisoschi, A.M.; Pop, A. The Role of Antioxidants in the Chemistry of Oxidative Stress: A Review. *Eur. J. Med. Chem.* **2015**, *97*, 55–74. [CrossRef]
267. Cadet, J.; Davies, K.J.A.; Medeiros, M.H.; Di Mascio, P.; Wagner, J.R. Formation and Repair of Oxidatively Generated Damage in Cellular DNA. *Free Radic. Biol. Med.* **2017**, *107*, 13–34. [CrossRef] [PubMed]
268. Jomova, K.; Alomar, S.Y.; Alwasel, S.H.; Nepovimova, E.; Kuca, K.; Valko, M. Several Lines of Antioxidant Defense against Oxidative Stress: Antioxidant Enzymes, Nanomaterials with Multiple Enzyme-Mimicking Activities, and Low-Molecular-Weight Antioxidants. *Arch. Toxicol.* **2024**, *98*, 1323–1367. [CrossRef] [PubMed]
269. Ighodaro, O.M.; Akinloye, O.A. First Line Defence Antioxidants-Superoxide Dismutase (SOD), Catalase (CAT) and Glutathione Peroxidase (GPX): Their Fundamental Role in the Entire Antioxidant Defence Grid. *Alex. J. Med.* **2018**, *54*, 287–293. [CrossRef]
270. Wang, Y.; Branicky, R.; Noë, A.; Hekimi, S. Superoxide Dismutases: Dual Roles in Controlling ROS Damage and Regulating ROS Signaling. *J. Cell Biol.* **2018**, *217*, 1915–1928. [CrossRef]
271. Davis, S.M.; Pennypacker, K.R. Targeting Antioxidant Enzyme Expression as a Therapeutic Strategy for Ischemic Stroke. *Neurochem. Int.* **2017**, *107*, 23. [CrossRef]
272. Adachi, T.; Wang, X.L. Association of Extracellular-Superoxide Dismutase Phenotype with the Endothelial Constitutive Nitric Oxide Synthase Polymorphism. *FEBS Lett.* **1998**, *433*, 166–168. [CrossRef]
273. Zelko, I.N.; Mariani, T.J.; Folz, R.J. Superoxide Dismutase Multigene Family: A Comparison of the CuZn-SOD (SOD1), Mn-SOD (SOD2), and EC-SOD (SOD3) Gene Structures, Evolution, and Expression. *Free Radic. Biol. Med.* **2002**, *33*, 337–349. [CrossRef]
274. Agrahari, G.; Sah, S.K.; Bang, C.H.; Kim, Y.H.; Kim, T.Y. Superoxide Dismutase 3 Controls the Activation and Differentiation of CD4+T Cells. *Front. Immunol.* **2021**, *12*, 628117. [CrossRef]
275. Mccords, J.M.; Fridovich, I. Superoxide Dismutase: An Enzymic Function for Erythrocuprein (Hemocuprein). *J. Biol. Chem.* **1969**, *244*, 6049–6065. [CrossRef]
276. Quist, D.A.; Diaz, D.E.; Liu, J.J.; Karlin, K.D. Activation of Dioxygen by Copper Metalloproteins and Insights from Model Complexes. *J. Biol. Inorg. Chem.* **2017**, *22*, 253. [CrossRef] [PubMed]
277. Fan, H.; Bai, Q.; Yang, Y.; Shi, X.; Du, G.; Yan, J.; Shi, J.; Wang, D. The Key Roles of Reactive Oxygen Species in Microglial Inflammatory Activation: Regulation by Endogenous Antioxidant System and Exogenous Sulfur-Containing Compounds. *Eur. J. Pharmacol.* **2023**, *956*, 175966. [CrossRef] [PubMed]
278. Hall, E.D.; Wang, J.A.; Miller, D.M.; Cebak, J.E.; Hill, R.L. Newer Pharmacological Approaches for Antioxidant Neuroprotection in Traumatic Brain Injury. *Neuropharmacology* **2019**, *145*, 247–258. [CrossRef]
279. Murakami, K.; Murata, N.; Noda, Y.; Tahara, S.; Kaneko, T.; Kinoshita, N.; Hatsuta, H.; Murayama, S.; Barnham, K.J.; Irie, K.; et al. SOD1 (Copper/Zinc Superoxide Dismutase) Deficiency Drives Amyloid β Protein Oligomerization and Memory Loss in Mouse Model of Alzheimer Disease. *J. Biol. Chem.* **2011**, *286*, 44557. [CrossRef]
280. Kirkman, H.N.; Gaetani, G.F. Catalase: A Tetrameric Enzyme with Four Tightly Bound Molecules of NADPH. *Proc. Natl. Acad. Sci. USA* **1984**, *81*, 4343–4347. [CrossRef]
281. Amir Aslani, B.; Ghobadi, S. Studies on Oxidants and Antioxidants with a Brief Glance at Their Relevance to the Immune System. *Life Sci.* **2016**, *146*, 163–173. [CrossRef]
282. Usui, S.; Komeima, K.; Lee, S.Y.; Jo, Y.J.; Ueno, S.; Rogers, B.S.; Wu, Z.; Shen, J.; Lu, L.; Oveson, B.C.; et al. Increased Expression of Catalase and Superoxide Dismutase 2 Reduces Cone Cell Death in Retinitis Pigmentosa. *Mol. Ther.* **2009**, *17*, 778–786. [CrossRef]
283. Uttara, B.; Singh, A.V.; Zamboni, P.; Mahajan, R.T. Oxidative Stress and Neurodegenerative Diseases: A Review of Upstream and Downstream Antioxidant Therapeutic Options. *Curr. Neuropharmacol.* **2009**, *7*, 65–74. [CrossRef]
284. Baker, A.; Lin, C.C.; Lett, C.; Karpinska, B.; Wright, M.H.; Foyer, C.H. Catalase: A Critical Node in the Regulation of Cell Fate. *Free Radic. Biol. Med.* **2023**, *199*, 56–66. [CrossRef]
285. Nandi, A.; Yan, L.J.; Jana, C.K.; Das, N. Role of Catalase in Oxidative Stress- and Age-Associated Degenerative Diseases. *Oxid. Med. Cell. Longev.* **2019**, 2019. [CrossRef]
286. Dutta, R.K.; Lee, J.N.; Maharjan, Y.; Park, C.; Choe, S.K.; Ho, Y.S.; Park, R. Catalase Deficiency Facilitates the Shuttling of Free Fatty Acid to Brown Adipose Tissue through Lipolysis Mediated by ROS during Sustained Fasting. *Cell Biosci.* **2021**, *11*, 201. [CrossRef] [PubMed]
287. Kumar Kalvala, A.; Bagde, A.; Arthur, P.; Kumar Surapaneni, S.; Ramesh, N.; Nathani, A.; Singh, M. Role of Cannabidiol and Tetrahydrocannabivarin on Paclitaxel-Induced Neuropathic Pain in Rodents. *Int. Immunopharmacol.* **2022**, *107*, 108693. [CrossRef] [PubMed]
288. Luangwattananun, P.; Yainoy, S.; Eiamphungporn, W.; Songtawee, N.; Bülow, L.; Ayudhya, C.I.N.; Prachayasittikul, V. Engineering of a Novel Tri-Functional Enzyme with MnSOD, Catalase and Cell-Permeable Activities. *Int. J. Biol. Macromol.* **2016**, *85*, 451–459. [CrossRef] [PubMed]

289. Pei, J.; Pan, X.; Wei, G.; Hua, Y. Research Progress of Glutathione Peroxidase Family (GPX) in Redoxidation. *Front. Pharmacol.* **2023**, *14*, 1147414. [CrossRef] [PubMed]
290. Handy, D.E.; Loscalzo, J. The Role of Glutathione Peroxidase-1 in Health and Disease. *Free Radic. Biol. Med.* **2022**, *188*, 146–161. [CrossRef]
291. Weaver, K.; Skouta, R. The Selenoprotein Glutathione Peroxidase 4: From Molecular Mechanisms to Novel Therapeutic Opportunities. *Biomedicines* **2022**, *10*, 891. [CrossRef]
292. Brigelius-Flohé, R.; Flohé, L. Regulatory Phenomena in the Glutathione Peroxidase Superfamily. *Antioxid. Redox Signal.* **2020**, *33*, 498–516. [CrossRef]
293. Li, J.; Cao, F.; Yin, H.L.; Huang, Z.J.; Lin, Z.T.; Mao, N.; Sun, B.; Wang, G. Ferroptosis: Past, Present and Future. *Cell Death Dis.* **2020**, *11*, 88. [CrossRef]
294. Hazzaa, S.M.; Abdou, A.G.; Ibraheim, E.O.; Salem, E.A.; Hassan, M.H.A.; Abdel-Razek, H.A.D. Effect of L-Carnitine and Atorvastatin on a Rat Model of Ischemia-Reperfusion Injury of Spinal Cord. *J. Immunoass. Immunochem.* **2021**, *42*, 596–619. [CrossRef]
295. Drummond, I.S.A.; de Oliveira, J.N.S.; Niella, R.V.; Silva, Á.J.C.; de Oliveira, I.S.; de Souza, S.S.; da Costa Marques, C.S.; Corrêa, J.M.X.; Silva, J.F.; de Lavor, M.S.L. Evaluation of the Therapeutic Potential of Amantadine in a Vincristine-Induced Peripheral Neuropathy Model in Rats. *Animals* **2024**, *14*, 1941. [CrossRef] [PubMed]
296. Liu, T.; Wang, R.; Qi, W.; Jia, L.; Ma, K.; Si, J.; Yin, J.; Zhao, Y.; Dai, Z.; Yin, J. Methyl Ferulic Acid Alleviates Neuropathic Pain by Inhibiting Nox4-Induced Ferroptosis in Dorsal Root Ganglia Neurons in Rats. *Mol. Neurobiol.* **2023**, *60*, 3175–3189. [CrossRef]
297. Zhang, X.; Song, T.; Zhao, M.; Tao, X.; Zhang, B.; Sun, C.; Wang, P.; Wang, K.; Zhao, L. Sirtuin 2 Alleviates Chronic Neuropathic Pain by Suppressing Ferroptosis in Rats. *Front. Pharmacol.* **2022**, *13*, 827016. [CrossRef] [PubMed]
298. Wang, H.; Huo, X.; Han, C.; Ning, J.; Chen, H.; Li, B.; Liu, J.; Ma, W.; Li, Q.; Yu, Y.; et al. Ferroptosis Is Involved in the Development of Neuropathic Pain and Allodynia. *Mol. Cell. Biochem.* **2021**, *476*, 3149–3161. [CrossRef]
299. Guo, Y.; Du, J.; Xiao, C.; Xiang, P.; Deng, Y.; Hei, Z.; Li, X. Inhibition of Ferroptosis-like Cell Death Attenuates Neuropathic Pain Reactions Induced by Peripheral Nerve Injury in Rats. *Eur. J. Pain* **2021**, *25*, 1227–1240. [CrossRef] [PubMed]
300. Fan, S.; Wang, K.; Zhang, T.; Deng, D.; Shen, J.; Zhao, B.; Fu, D.; Chen, X. Mechanisms and Therapeutic Potential of GPX4 in Pain Modulation. *Pain. Ther.* **2025**, *14*, 21–45. [CrossRef]
301. Bjørklund, G.; Zou, L.; Peana, M.; Chasapis, C.T.; Hangan, T.; Lu, J.; Maes, M. The Role of the Thioredoxin System in Brain Diseases. *Antioxidants* **2022**, *11*, 2161. [CrossRef]
302. Burke-Gaffney, A.; Callister, M.E.J.; Nakamura, H. Thioredoxin: Friend or Foe in Human Disease? *Trends Pharmacol. Sci.* **2005**, *26*, 398–404. [CrossRef]
303. Arnér, E.S.J.; Holmgren, A. Physiological Functions of Thioredoxin and Thioredoxin Reductase. *Eur. J. Biochem.* **2000**, *267*, 6102–6109. [CrossRef]
304. Hasan, A.A.; Kalinina, E.; Tatarskiy, V.; Shtil, A. The Thioredoxin System of Mammalian Cells and Its Modulators. *Biomedicines* **2022**, *10*, 1757. [CrossRef]
305. Monteiro, H.P.; Ogata, F.T.; Stern, A. Thioredoxin Promotes Survival Signaling Events under Nitrosative/Oxidative Stress Associated with Cancer Development. *Biomed. J.* **2017**, *40*, 189–199. [CrossRef] [PubMed]
306. Silva-Adaya, D.; Gonsebatt, M.E.; Guevara, J. Thioredoxin System Regulation in the Central Nervous System: Experimental Models and Clinical Evidence. *Oxid. Med. Cell. Longev.* **2014**, *2014*, 590808. [CrossRef] [PubMed]
307. Ren, X.; Zou, L.; Zhang, X.; Branco, V.; Wang, J.; Carvalho, C.; Holmgren, A.; Lu, J. Redox Signaling Mediated by Thioredoxin and Glutathione Systems in the Central Nervous System. *Antioxid. Redox Signal.* **2017**, *27*, 989–1010. [CrossRef] [PubMed]
308. Vilhardt, F.; Haslund-Vinding, J.; Jaquet, V.; McBean, G. Microglia Antioxidant Systems and Redox Signalling. *Br. J. Pharmacol.* **2017**, *174*, 1719–1732. [CrossRef] [PubMed]
309. Valek, L.; Kanngießer, M.; Häussler, A.; Agarwal, N.; Lillig, C.H.; Tegeder, I. Redoxins in Peripheral Neurons after Sciatic Nerve Injury. *Free Radic. Biol. Med.* **2015**, *89*, 581–592. [CrossRef]
310. Ji, R.R.; Chamessian, A.; Zhang, Y.Q. Pain Regulation by Non-Neuronal Cells and Inflammation. *Science 1979* **2016**, *354*, 572–577. [CrossRef]
311. Fernandes, V.; Sharma, D.; Vaidya, S.; Shantanu, P.A.; Guan, Y.; Kalia, K.; Tiwari, V. Cellular and Molecular Mechanisms Driving Neuropathic Pain: Recent Advancements and Challenges. *Expert. Opin. Ther. Targets* **2018**, *22*, 131–142. [CrossRef]
312. Jensen, T.S.; Baron, R.; Haanpää, M.; Kalso, E.; Loeser, J.D.; Rice, A.S.C.; Treede, R.D. A New Definition of Neuropathic Pain. *Pain* **2011**, *152*, 2204–2205. [CrossRef]
313. Bouhassira, D. Neuropathic Pain: Definition, Assessment and Epidemiology. *Rev. Neurol.* **2019**, *175*, 16–25. [CrossRef]
314. Bouhassira, D.; Lantéri-Minet, M.; Attal, N.; Laurent, B.; Touboul, C. Prevalence of Chronic Pain with Neuropathic Characteristics in the General Population. *Pain* **2008**, *136*, 380–387. [CrossRef]
315. Ji, R.R.; Berta, T.; Nedergaard, M. Glia and Pain: Is Chronic Pain a Gliopathy? *Pain* **2013**, *154*, S10–S28. [CrossRef] [PubMed]

316. Ji, R.R.; Xu, Z.Z.; Gao, Y.J. Emerging Targets in Neuroinflammation-Driven Chronic Pain. *Nat. Rev. Drug Discov.* **2014**, *13*, 533–548. [CrossRef] [PubMed]
317. Zhou, Y.Q.; Liu, Z.; Liu, Z.H.; Chen, S.P.; Li, M.; Shahveranov, A.; Ye, D.W.; Tian, Y.K. Interleukin-6: An Emerging Regulator of Pathological Pain. *J. Neuroinflamm.* **2016**, *13*, 141. [CrossRef] [PubMed]
318. Gwak, Y.S.; Kang, J.; Unabia, G.C.; Hulsebosch, C.E. Spatial and Temporal Activation of Spinal Glial Cells: Role of Gliopathy in Central Neuropathic Pain Following Spinal Cord Injury in Rats. *Exp. Neurol.* **2012**, *234*, 362–372. [CrossRef]
319. Raghavendra, V.; Tanga, F.; Deleo, J.A. Inhibition of Microglial Activation Attenuates the Development but Not Existing Hypersensitivity in a Rat Model of Neuropathy. *J. Pharmacol. Exp. Ther.* **2003**, *306*, 624–630. [CrossRef]
320. Peng, J.; Gu, N.; Zhou, L.; Eyo, U.B.; Murugan, M.; Gan, W.B.; Wu, L.J. Microglia and Monocytes Synergistically Promote the Transition from Acute to Chronic Pain after Nerve Injury. *Nat. Commun.* **2016**, *7*, 12029. [CrossRef]
321. Ellis, A.; Bennett, D.L.H. Neuroinflammation and the Generation of Neuropathic Pain. *Br. J. Anaesth.* **2013**, *111*, 26–37. [CrossRef]
322. Town, T.; Nikolic, V.; Tan, J. The Microglial "Activation" Continuum: From Innate to Adaptive Responses. *J. Neuroinflamm.* **2005**, *2*, 24. [CrossRef]
323. Hu, X.; Leak, R.K.; Shi, Y.; Suenaga, J.; Gao, Y.; Zheng, P.; Chen, J. Microglial and Macrophage Polarization—New Prospects for Brain Repair. *Nat. Rev. Neurol.* **2015**, *11*, 56–64. [CrossRef]
324. Schreiner, B.; Romanelli, E.; Liberski, P.; Ingold-Heppner, B.; Sobottka-Brillout, B.; Hartwig, T.; Chandrasekar, V.; Johannssen, H.; Zeilhofer, H.U.; Aguzzi, A.; et al. Astrocyte Depletion Impairs Redox Homeostasis and Triggers Neuronal Loss in the Adult CNS. *Cell Rep.* **2015**, *12*, 1377–1384. [CrossRef]
325. Gan, L.; Vargas, M.R.; Johnson, D.A.; Johnson, J.A. Astrocyte-Specific Overexpression of Nrf2 Delays Motor Pathology and Synuclein Aggregation throughout the CNS in the Alpha-Synuclein Mutant (A53T) Mouse Model. *J. Neurosci.* **2012**, *32*, 17775. [CrossRef]
326. Wang, D.; Couture, R.; Hong, Y. Activated Microglia in the Spinal Cord Underlies Diabetic Neuropathic Pain. *Eur. J. Pharmacol.* **2014**, *728*, 59–66. [CrossRef] [PubMed]
327. Liu, B.; Gao, H.M.; Wang, J.Y.; Jeohn, G.H.; Cooper, C.L.; Hong, J.S. Role of Nitric Oxide in Inflammation-Mediated Neurodegeneration. *Ann. N. Y. Acad. Sci.* **2002**, *962*, 318–331. [CrossRef] [PubMed]
328. Lee, K.H.; Cha, M.; Lee, B.H. Crosstalk between Neuron and Glial Cells in Oxidative Injury and Neuroprotection. *Int. J. Mol. Sci.* **2021**, *22*, 13315. [CrossRef] [PubMed]
329. Choi, S.H.; Aid, S.; Kim, H.W.; Jackson, S.H.; Bosetti, F. Inhibition of NADPH Oxidase Promotes Alternative and Anti-Inflammatory Microglial Activation during Neuroinflammation. *J. Neurochem.* **2012**, *120*, 292–301. [CrossRef] [PubMed]
330. Lijia, Z.; Zhao, S.; Wang, X.; Wu, C.; Yang, J. A Self-Propelling Cycle Mediated by Reactive Oxide Species and Nitric Oxide Exists in LPS-Activated Microglia. *Neurochem. Int.* **2012**, *61*, 1220–1230. [CrossRef]
331. Abramov, A.Y.; Canevari, L.; Duchen, M.R. Beta-Amyloid Peptides Induce Mitochondrial Dysfunction and Oxidative Stress in Astrocytes and Death of Neurons through Activation of NADPH Oxidase. *J. Neurosci.* **2004**, *24*, 565–575. [CrossRef]
332. Teismann, P.; Schulz, J.B. Cellular Pathology of Parkinson's Disease: Astrocytes, Microglia and Inflammation. *Cell Tissue Res.* **2004**, *318*, 149–161. [CrossRef]
333. Markiewicz, I.; Lukomska, B. The Role of Astrocytes in the Physiology and Pathology of the Central Nervous System. *Acta Neurobiol. Exp.* **2006**, *66*, 343–358. [CrossRef]
334. Wang, A.; Xu, C. The Role of Connexin43 in Neuropathic Pain Induced by Spinal Cord Injury. *Acta Biochim. Biophys. Sin.* **2019**, *51*, 555–561. [CrossRef]
335. Shih, E.K.; Robinson, M.B. Role of Astrocytic Mitochondria in Limiting Ischemic Brain Injury? *Physiology* **2018**, *33*, 99–112. [CrossRef] [PubMed]
336. Chen, X.; Guan, T.; Li, C.; Shang, H.; Cui, L.; Li, X.M.; Kong, J. SOD1 Aggregation in Astrocytes Following Ischemia/Reperfusion Injury: A Role of NO-Mediated S-Nitrosylation of Protein Disulfide Isomerase (PDI). *J. Neuroinflamm.* **2012**, *9*, 237. [CrossRef] [PubMed]
337. Biagioni, F.; Vivacqua, G.; Lazzeri, G.; Ferese, R.; Iannacone, S.; Onori, P.; Morini, S.; D'Este, L.; Fornai, F. Chronic MPTP in Mice Damage-Specific Neuronal Phenotypes within Dorsal Laminae of the Spinal Cord. *Neurotoxic. Res.* **2021**, *39*, 156–169. [CrossRef] [PubMed]
338. Qin, L.; Liu, Y.; Wang, T.; Wei, S.J.; Block, M.L.; Wilson, B.; Liu, B.; Hong, J.S. NADPH Oxidase Mediates Lipopolysaccharide-Induced Neurotoxicity and Proinflammatory Gene Expression in Activated Microglia. *J. Biol. Chem.* **2004**, *279*, 1415–1421. [CrossRef]
339. Andersson, D.A.; Filipovic, M.R.; Gentry, C.; Eberhardt, M.; Vastani, N.; Leffler, A.; Reeh, P.; Bevan, S. Streptozotocin Stimulates the Ion Channel TRPA1 Directly: Involvement of Peroxynitrite. *J. Biol. Chem.* **2015**, *290*, 15185–15196. [CrossRef]
340. Chen, S.J.; Zhang, W.; Tong, Q.; Conrad, K.; Hirschler-Laszkiewicz, I.; Bayerl, M.; Kim, J.K.; Cheung, J.Y.; Miller, B.A. Role of TRPM2 in Cell Proliferation and Susceptibility to Oxidative Stress. *Am. J. Physiol. Cell Physiol.* **2013**, *304*, 548–560. [CrossRef]

341. Özdemir, Ü.S.; Nazıroğlu, M.; Şenol, N.; Ghazizadeh, V. Hypericum Perforatum Attenuates Spinal Cord Injury-Induced Oxidative Stress and Apoptosis in the Dorsal Root Ganglion of Rats: Involvement of TRPM2 and TRPV1 Channels. *Mol. Neurobiol.* **2016**, *53*, 3540–3551. [CrossRef]
342. Haraguchi, K.; Kawamoto, A.; Isami, K.; Maeda, S.; Kusano, A.; Asakura, K.; Shirakawa, H.; Mori, Y.; Nakagawa, T.; Kaneko, S. TRPM2 Contributes to Inflammatory and Neuropathic Pain through the Aggravation of Pronociceptive Inflammatory Responses in Mice. *J. Neurosci.* **2012**, *32*, 3931–3941. [CrossRef]
343. Dangi, A.; Sharma, S.S. Pharmacological Agents Targeting Transient Receptor Potential (TRP) Channels in Neuropathic Pain: Preclinical and Clinical Status. *Eur. J. Pharmacol.* **2024**, *980*, 176845. [CrossRef]
344. Rodrigues, P.; Ruviaro, N.A.; Trevisan, G. TRPV4 Role in Neuropathic Pain Mechanisms in Rodents. *Antioxidants* **2022**, *12*, 24. [CrossRef]
345. Antoniazzi, C.T.D.D.; Nassini, R.; Rigo, F.K.; Milioli, A.M.; Bellinaso, F.; Camponogara, C.; Silva, C.R.; de Almeida, A.S.; Rossato, M.F.; De Logu, F.; et al. Transient Receptor Potential Ankyrin 1 (TRPA1) Plays a Critical Role in a Mouse Model of Cancer Pain. *Int. J. Cancer* **2019**, *144*, 355–365. [CrossRef] [PubMed]
346. Mobasheri, A.; Simon, F.; Bello, A.; Soriano, S.; Nazıro, M.; Braidy, N. Thermo-Sensitive TRP Channels: Novel Targets for Treating Chemotherapy-Induced Peripheral Pain. *Front. Physiol.* **2017**, *8*, 1040. [CrossRef]
347. Chondrogianni, N.; Petropoulos, I.; Grimm, S.; Georgila, K.; Catalgol, B.; Friguet, B.; Grune, T.; Gonos, E.S. Protein Damage, Repair and Proteolysis. *Mol. Asp. Med.* **2014**, *35*, 1–71. [CrossRef]
348. Winterbourn, C.C.; Hampton, M.B. Thiol Chemistry and Specificity in Redox Signaling. *Free Radic. Biol. Med.* **2008**, *45*, 549–561. [CrossRef]
349. Go, Y.M.; Chandler, J.D.; Jones, D.P. The Cysteine Proteome. *Free Radic. Biol. Med.* **2015**, *84*, 227–245. [CrossRef]
350. Chandran, S.; Binninger, D. Role of Oxidative Stress, Methionine Oxidation and Methionine Sulfoxide Reductases (MSR) in Alzheimer's Disease. *Antioxidants* **2023**, *13*, 21. [CrossRef]
351. Rayner, B.S.; Love, D.T.; Hawkins, C.L. Comparative Reactivity of Myeloperoxidase-Derived Oxidants with Mammalian Cells. *Free Radic. Biol. Med.* **2014**, *71*, 240–255. [CrossRef]
352. Fujiwara, N.; Nakano, M.; Kato, S.; Yoshihara, D.; Ookawara, T.; Eguchi, H.; Taniguchi, N.; Suzuki, K. Oxidative Modification to Cysteine Sulfonic Acid of Cys111 in Human Copper-Zinc Superoxide Dismutase. *J. Biol. Chem.* **2007**, *282*, 35933–35944. [CrossRef]
353. Rubio, M.A.; Herrando-Grabulosa, M.; Vilches, J.J.; Navarro, X. Involvement of Sensory Innervation in the Skin of SOD1G93A ALS Mice. *J. Peripher. Nerv. Syst.* **2016**, *21*, 88–95. [CrossRef]
354. Sharp, P.S.; Dick, J.R.T.; Greensmith, L. The Effect of Peripheral Nerve Injury on Disease Progression in the SOD1 (G93A) Mouse Model of Amyotrophic Lateral Sclerosis. *Neuroscience* **2005**, *130*, 897–910. [CrossRef]
355. Bombaci, A.; Lupica, A.; Pozzi, F.E.; Remoli, G.; Manera, U.; Di Stefano, V. Sensory Neuropathy in Amyotrophic Lateral Sclerosis: A Systematic Review. *J. Neurol.* **2023**, *270*, 5677–5691. [CrossRef] [PubMed]
356. Fang, T.; Jozsa, F.; Al-Chalabi, A. Nonmotor Symptoms in Amyotrophic Lateral Sclerosis: A Systematic Review. *Int. Rev. Neurobiol.* **2017**, *134*, 1409–1441. [CrossRef] [PubMed]
357. Xiong, Y.; Uys, J.D.; Tew, K.D.; Townsend, D.M. S-Glutathionylation: From Molecular Mechanisms to Health Outcomes. *Antioxid. Redox Signal.* **2011**, *15*, 233. [CrossRef]
358. Tu, C.; Wang, S.-C.; Dai, M.-X.; Lai, S.-Q.; Huang, Z.-W.; Yu, Y.-P.; Chen, Y.-B.; Zeng, J.-H.; Wang, L.; Zhong, Z.-M. Accumulation of Advanced Oxidative Protein Products Exacerbate Satellite Glial Cells Activation and Neuropathic Pain. *Mol. Med.* **2025**, *31*, 25. [CrossRef] [PubMed]
359. Nuriel, T.; Hansler, A.; Gross, S.S. Protein Nitrotryptophan: Formation, Significance and Identification. *J. Proteom.* **2011**, *74*, 2300. [CrossRef] [PubMed]
360. Salvemini, D.; Doyle, T.; Chen, Z.; Muscoli, C.; Bryant, L.; Esposito, E.; Cuzzocrea, S.; Dagostino, C.; Ryerse, J.; Rausaria, S.; et al. Targeting the Overproduction of Peroxynitrite for the Prevention and Reversal of Paclitaxel-Induced Neuropathic Pain. *J. Neurosci.* **2012**, *32*, 6149–6160. [CrossRef]
361. Salvemini, D.; Neumann, W. Targeting Peroxynitrite Driven Nitroxidative Stress with Synzymes: A Novel Therapeutic Approach in Chronic Pain Management. *Life Sci.* **2010**, *86*, 604–614. [CrossRef]
362. Salvemini, D.; Kim, S.F.; Mollace, V. Reciprocal Regulation of the Nitric Oxide and Cyclooxygenase Pathway in Pathophysiology: Relevance and Clinical Implications. *Am. J. Physiol. Regul. Integr. Comp. Physiol.* **2013**, *304*, 473–487. [CrossRef]
363. Ilari, S.; Giancotti, L.A.; Lauro, F.; Dagostino, C.; Gliozzi, M.; Malafoglia, V.; Sansone, L.; Palma, E.; Tafani, M.; Russo, M.A.; et al. Antioxidant Modulation of Sirtuin 3 during Acute Inflammatory Pain: The ROS Control. *Pharmacol. Res.* **2020**, *157*, 104851. [CrossRef]
364. Liao, M.F.; Lu, K.T.; Hsu, J.L.; Lee, C.H.; Cheng, M.Y.; Ro, L.S. The Role of Autophagy and Apoptosis in Neuropathic Pain Formation. *Int. J. Mol. Sci.* **2022**, *23*, 2685. [CrossRef]
365. Valluru, L.; Diao, Y.; Hachmeister, J.E.; Liu, D. Mn (III) Tetrakis (4-Benzoic Acid) Porphyrin Protects against Neuronal and Glial Oxidative Stress and Death after Spinal Cord Injury. *CNS Neurol. Disord. Drug Targets* **2012**, *11*, 774–790. [CrossRef] [PubMed]

366. Ntalouka, F.; Tsirivakou, A. Luteolin: A Promising Natural Agent in Management of Pain in Chronic Conditions. *Front. Pain Res.* **2023**, *4*, 1114428. [CrossRef] [PubMed]
367. Frediani, J.K.; Lal, A.A.; Kim, E.; Leslie, S.L.; Boorman, D.W.; Singh, V. The Role of Diet and Non-Pharmacologic Supplements in the Treatment of Chronic Neuropathic Pain: A Systematic Review. *Pain Pract.* **2024**, *24*, 186–210. [CrossRef] [PubMed]
368. Batinic-Haberle, I.; Tovmasyan, A.; Roberts, E.R.H.; Vujaskovic, Z.; Leong, K.W.; Spasojevic, I. SOD Therapeutics: Latest Insights into Their Structure-Activity Relationships and Impact on the Cellular Redox-Based Signaling Pathways. *Antioxid. Redox Signal.* **2014**, *20*, 2372–2415. [CrossRef]
369. Bonetta, R. Potential Therapeutic Applications of MnSODs and SOD-Mimetics. *Chem.—A Eur. J.* **2018**, *24*, 5032–5041. [CrossRef]
370. Buettner, G.R.; Wagner, B.A.; Rodgers, V.G.J. Quantitative Redox Biology: An Approach to Understand the Role of Reactive Species in Defining the Cellular Redox Environment. *Cell Biochem. Biophys.* **2013**, *67*, 477–483. [CrossRef]
371. Policar, C.; Bouvet, J.; Bertrand, H.C.; Delsuc, N. SOD Mimics: From the Tool Box of the Chemists to Cellular Studies. *Curr. Opin. Chem. Biol.* **2022**, *67*, 102109. [CrossRef]
372. Eckshtain, M.; Zilbermann, I.; Mahammed, A.; Saltsman, I.; Okun, Z.; Maimon, E.; Cohen, H.; Meyerstein, D.; Gross, Z. Superoxide Dismutase Activity of Corrole Metal Complexes. *Dalton Trans.* **2009**, *38*, 7879–7882. [CrossRef]
373. Franke, A.; Scheitler, A.; Kenkel, I.; Lippert, R.; Zahl, A.; Balbinot, D.; Jux, N.; Ivanović-Burmazović, I. Positive Charge on Porphyrin Ligand and Nature of Metal Center Define Basic Physicochemical Properties of Cationic Manganese and Iron Porphyrins in Aqueous Solution. *Inorg. Chem.* **2019**, *58*, 9618–9630. [CrossRef]
374. Batinić-Haberle, I.; Rebouças, J.S.; Spasojević, I. Superoxide Dismutase Mimics: Chemistry, Pharmacology, and Therapeutic Potential. *Antioxid. Redox Signal.* **2010**, *13*, 877–918. [CrossRef]
375. Batinic-Haberle, I.; Tome, M.E. Thiol Regulation by Mn Porphyrins, Commonly Known as SOD Mimics. *Redox Biol.* **2019**, *25*, 101139. [CrossRef] [PubMed]
376. Day, B.J. Catalase and Glutathione Peroxidase Mimics. *Biochem. Pharmacol.* **2009**, *77*, 285–296. [CrossRef] [PubMed]
377. Rebouças, J.S.; Spasojević, I.; Batinić-Haberle, I. Quality of Potent Mn Porphyrin-Based SOD Mimics and Peroxynitrite Scavengers for Pre-Clinical Mechanistic/Therapeutic Purposes. *J. Pharm. Biomed. Anal.* **2008**, *48*, 1046. [CrossRef] [PubMed]
378. Pollard, J.M.; Reboucas, J.S.; Durazo, A.; Kos, I.; Fike, F.; Panni, M.; Gralla, E.B.; Valentine, J.S.; Batinic-Haberle, I.; Gatti, R.A. Radio-Protective Effects of Manganese-Containing Superoxide Dismutase Mimics on Ataxia Telangiectasia Cells. *Free Radic. Biol. Med.* **2009**, *47*, 250. [CrossRef]
379. Ferrer-Sueta, G.; Hannibal, L.; Batinic-Haberle, I.; Radi, R. Reduction of Manganese Porphyrins by Flavoenzymes and Submitochondrial Particles: A Catalytic Cycle for the Reduction of Peroxynitrite. *Free Radic. Biol. Med.* **2006**, *41*, 503–512. [CrossRef]
380. Wang, C.; Li, S.; Shang, D.J.; Wang, X.L.; You, Z.L.; Li, H.B. Antihyperglycemic and Neuroprotective Effects of One Novel Cu-Zn SOD Mimetic. *Bioorg Med. Chem. Lett.* **2011**, *21*, 4320–4324. [CrossRef]
381. Celic, T.; Španjol, J.; Bobinac, M.; Tovmasyan, A.; Vukelic, I.; Reboucas, J.S.; Batinic-Haberle, I.; Bobinac, D. Mn Porphyrin-Based SOD Mimic, MnTnHex-2-PyP(5+), and Non-SOD Mimic, MnTBAP(3−), Suppressed Rat Spinal Cord Ischemia/Reperfusion Injury via NF-KB Pathways. *Free Radic. Res.* **2014**, *48*, 1426–1442. [CrossRef]
382. Tse, H.M.; Milton, M.J.; Piganelli, J.D. Mechanistic Analysis of the Immunomodulatory Effects of a Catalytic Antioxidant on Antigen-Presenting Cells: Implication for Their Use in Targeting Oxidation–Reduction Reactions in Innate Immunity. *Free Radic. Biol. Med.* **2004**, *36*, 233–247. [CrossRef]
383. Ferrer-Sueta, G.; Vitturi, D.; Batinić-Haberle, I.; Fridovich, I.; Goldstein, S.; Czapski, G.; Radi, R. Reactions of Manganese Porphyrins with Peroxynitrite and Carbonate Radical Anion. *J. Biol. Chem.* **2003**, *278*, 27432–27438. [CrossRef]
384. Ferrer-Sueta, G.; Batinić-Haberle, I.; Spasojević, I.; Fridovich, I.; Radi, R. Catalytic Scavenging of Peroxynitrite by Isomeric Mn(III) N-Methylpyridylporphyrins in the Presence of Reductants. *Chem. Res. Toxicol.* **1999**, *12*, 442–449. [CrossRef]
385. Campo, G.M.; Avenoso, A.; D'Ascola, A.; Scuruchi, M.; Nastasi, G.; Micali, A.; Puzzolo, D.; Pisani, A.; Calatroni, A.; Campo, S. The SOD Mimic MnTM-2-PyP(5+) Reduces Hyaluronan Degradation-Induced Inflammation in Mouse Articular Chondrocytes Stimulated with Fe (II) plus Ascorbate. *Int. J. Biochem. Cell Biol.* **2013**, *45*, 1610–1619. [CrossRef] [PubMed]
386. Janes, K.; Doyle, T.; Bryant, L.; Esposito, E.; Cuzzocrea, S.; Ryerse, J.; Bennett, G.J.; Salvemini, D. Bioenergetic Deficits in Peripheral Nerve Sensory Axons during Chemotherapy-Induced Neuropathic Pain Resulting from Peroxynitrite-Mediated Post-Translational Nitration of Mitochondrial Superoxide Dismutase. *Pain* **2013**, *154*, 2432–2440. [CrossRef] [PubMed]
387. Negi, G.; Kumar, A.; Sharma, S.S. Concurrent Targeting of Nitrosative Stress-PARP Pathway Corrects Functional, Behavioral and Biochemical Deficits in Experimental Diabetic Neuropathy. *Biochem. Biophys. Res. Commun.* **2010**, *391*, 102–106. [CrossRef] [PubMed]
388. Arora, M.; Kumar, A.; Kaundal, R.K.; Sharma, S.S. Amelioration of Neurological and Biochemical Deficits by Peroxynitrite Decomposition Catalysts in Experimental Diabetic Neuropathy. *Eur. J. Pharmacol.* **2008**, *596*, 77–83. [CrossRef]
389. Drel, V.R.; Pacher, P.; Vareniuk, I.; Pavlov, I.; Ilnytska, O.; Lyzogubov, V.V.; Tibrewala, J.; Groves, J.T.; Obrosova, I.G. A Peroxynitrite Decomposition Catalyst Counteracts Sensory Neuropathy in Streptozotocin-Diabetic Mice. *Eur. J. Pharmacol.* **2007**, *569*, 48–58. [CrossRef]

390. Kwak, K.H.; Jung, H.; Park, J.M.; Yeo, J.S.; Kim, H.; Lee, H.C.; Byun, S.H.; Kim, J.C.; Park, S.S.; Lim, D.G. A Peroxynitrite Decomposition Catalyst Prevents Mechanical Allodynia and NMDA Receptor Activation in the Hind-Paw Ischemia Reperfusion Injury Rats. *Exp. Ther. Med.* **2014**, *7*, 508–512. [CrossRef]
391. Komirishetty, P.; Areti, A.; Arruri, V.K.; Sistla, R.; Gogoi, R.; Kumar, A. FeTMPyP a Peroxynitrite Decomposition Catalyst Ameliorated Functional and Behavioral Deficits in Chronic Constriction Injury Induced Neuropathic Pain in Rats. *Free Radic. Res.* **2021**, *55*, 1005–1017. [CrossRef]

Disclaimer/Publisher's Note: The statements, opinions and data contained in all publications are solely those of the individual author(s) and contributor(s) and not of MDPI and/or the editor(s). MDPI and/or the editor(s) disclaim responsibility for any injury to people or property resulting from any ideas, methods, instructions or products referred to in the content.

International Journal of *Molecular Sciences*

MDPI

Review

Pathological Interplay between Inflammation and Mitochondria Aggravates Glutamate Toxicity

Annette Vaglio-Garro [1,2], Andrey V. Kozlov [1,2,*], Yuliya D. Smirnova [1,3] and Adelheid Weidinger [1,2]

1 Ludwig Boltzmann Institute for Traumatology, The Research Center in Cooperation with AUVA, 1200 Vienna, Austria; annette.vaglio@trauma.lbg.ac.at (A.V.-G.); yuliya.dvoretskaya@trauma.lbg.ac.at (Y.D.S.); adelheid.weidinger@trauma.lbg.ac.at (A.W.)

2 Austrian Cluster for Tissue Regeneration, 1200 Vienna, Austria

3 Laboratory of Metagenomics and Food Biotechnology, Voronezh State University of Engineering Technologies, 394036 Voronezh, Russia

* Correspondence: andrey.kozlov@trauma.lbg.ac.at; Tel.: +43-5-9393-41980

Abstract: Mitochondrial dysfunction and glutamate toxicity are associated with neural disorders, including brain trauma. A review of the literature suggests that toxic and transmission actions of neuronal glutamate are spatially and functionally separated. The transmission pathway utilizes synaptic GluN2A receptors, rapidly released pool of glutamate, evoked release of glutamate mediated by Synaptotagmin 1 and the amount of extracellular glutamate regulated by astrocytes. The toxic pathway utilizes extrasynaptic GluN2B receptors and a cytoplasmic pool of glutamate, which results from the spontaneous release of glutamate mediated by Synaptotagmin 7 and the neuronal 2-oxoglutarate dehydrogenase complex (OGDHC), a tricarboxylic acid (TCA) cycle enzyme. Additionally, the inhibition of OGDHC observed upon neuro-inflammation is due to an excessive release of reactive oxygen/nitrogen species by immune cells. The loss of OGDHC inhibits uptake of glutamate by mitochondria, thus facilitating its extracellular accumulation and stimulating toxic glutamate pathway without affecting transmission. High levels of extracellular glutamate lead to dysregulation of intracellular redox homeostasis and cause ferroptosis, excitotoxicity, and mitochondrial dysfunction. The latter affects the transmission pathway demanding high-energy supply and leading to cell death. Mitochondria aggravate glutamate toxicity due to impairments in the TCA cycle and become a victim of glutamate toxicity, which disrupts oxidative phosphorylation. Thus, therapies targeting the TCA cycle in neurological disorders may be more efficient than attempting to preserve mitochondrial oxidative phosphorylation.

Keywords: ferroptosis; glutamate; neuronal death; mitochondrial dysfunction; TCA cycle

1. Metabolism of Glutamate in Neuronal Tissue

Glutamate in the nervous system interfaces with many metabolic pathways, such as oxidation of energy, incorporation into proteins, and the formation of glutamine, gamma-aminobutyric acid (GABA), and glutathione. However, signal transmission via the glutamate-glutamine cycle is the major physiological function of glutamate in the nervous system [1] (Figure 1). Nevertheless, glutamate can also be toxic if occurring at high concentrations in extracellular fluid (Figure 1). Toxic extracellular concentrations of glutamate have been documented in a number of central nervous system (CNS) diseases [2] and brain injury [3]. There are two toxic pathways killing neurons upon elevated extracellular glutamate levels. The first one is excitotoxicity, which kills neurons via sustained activation of glutamate receptors and an excessive influx of Ca^{2+} into neurons [4]. The second, more recently described mechanism, ferroptosis, is a kind of programmed cell death mediated by iron [5]. A high concentration of glutamate in the extracellular fluid blocks glutamate/cysteine antiporters and reduces the intracellular levels of cysteine and, subsequently, the synthesis of glutathione. The latter inhibits glutathione peroxidase activity and induces neuronal death via iron-mediated oxidation of lipids [6].

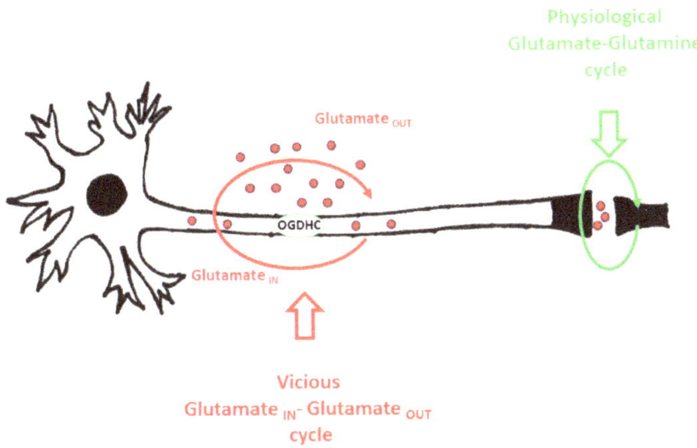

Figure 1. Graphical abstract. The glutamateglutamate vicious toxic cycle (shown in red) occurs at the extrasynaptic site. Physiological glutamate-glutamine cycle (shown in green) is located in the synaptic cleft.

In intact neurons, the major part of glutamate is packed into intracellular glutamate vesicles known to serve in signal transmission. Glutamate is released into the synaptic cleft upon activation of neurons in a process called evoked release. Images obtained by electron microscopy showed that astrocytes almost completely close to the synaptic cleft prevent the release of glutamate into the extracellular fluid [7], suggesting that only astrocytes located in close vicinity to synapses predominantly take up intrasynaptic glutamate. Astrocytes convert glutamate to glutamine, which is released into the extracellular fluid. Glutamine does not activate neuronal receptors, but neurons take up glutamine, convert it back to glutamate, and refill glutamate vesicles [8,9]. However, upon brain trauma and neurodegenerative diseases, the glutamate concentration in the extracellular fluid drastically increases due to the release of glutamate from damaged cells and from activated microglia and reactive astrocytes [10,11]. The distribution of glutamate in different compartments of the neuron is shown in Figure 2. The activation of glial cells always accompanies neuroinflammation, which occurs in most central nervous system (CNS) disorders [12]. In contrast to synapses, distant parts of the neuronal bodies have no tided contact with astrocytes. Subsequently, both neuronal and astrocytic bodies, which are not part of the synapsis structure, are exposed to extracellular glutamate [13–15]. Excessive activation of glutamate receptors in neurons causes their dysfunction and death [4]. There are several types of glutamate receptors described below but not all of them mediate neuronal death.

N-methyl-D-aspartate (NMDA) glutamate receptors can be formed by different types of isoforms: GluN2A, GluN2B, and GluN2C/2D. NMDA glutamate receptors with GluN2B and GluN2C/2D subunits are mainly located in the extrasynaptic site, whereas NMDA glutamate receptors rich in GluN2A are normally working in the signal transmission process located in the synaptic cleft.

Under physiological conditions, glutamate will be packed in the so-called readily releasable pool. Under pathological conditions, for example after brain trauma, intra- and extracellular concentrations will change, leading to the diffusion of intracellular glutamate into the extracellular fluid and induction of excitotoxicity.

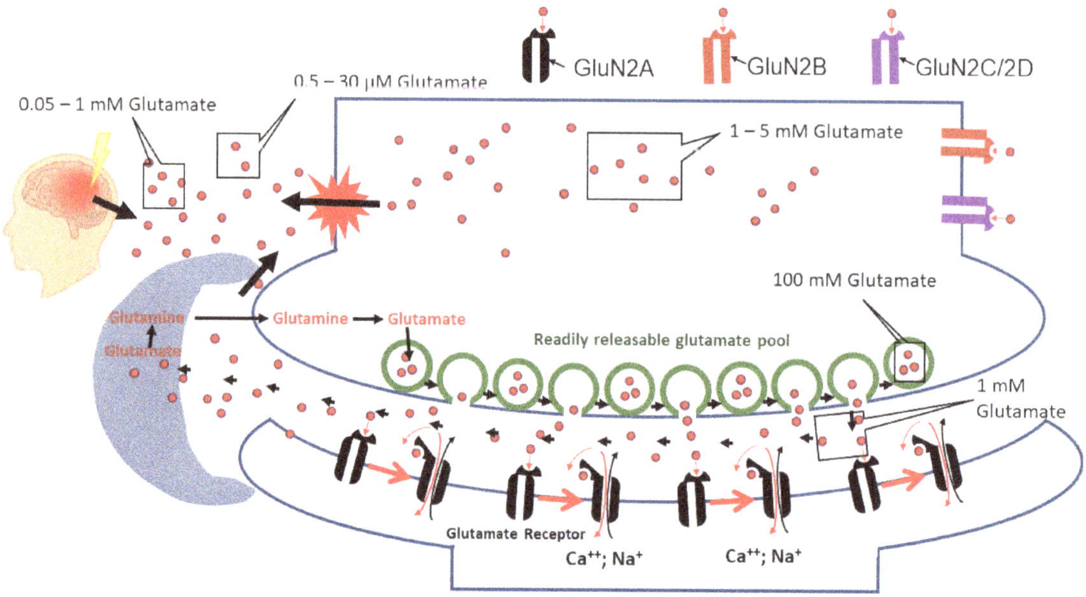

Figure 2. Distribution of glutamate in the nervous tissue, and the glutamate-glutamine cycle and its changes upon brain injury.

2. Heterogeneity of Glutamate Receptors

Several types of ionotropic glutamate receptors have been identified. They include ligand-gated ion channels called N-methyl-D-aspartate (NMDA) receptors, α-amino-3-hydroxy-5-methyl-4-isoxazolepropionic acid (AMPA) receptors, and Kainate receptors (KARs) [16–18]. All of them allow Na^+ and K^+ to flow through the plasma membrane of neurons upon activation [19]. Within these three types, only NMDA receptors, in addition to Na^+ and K^+, allow the entrance of Ca^{2+} into neurons. Inside neurons, Ca^{2+} activates numerous intracellular signaling cascades [20]; overwhelmed Ca^{2+} influx is the key factor causing excitotoxicity [21]. There are several types of NMDA receptors. All of them are tetramers with two glycine-binding subunits GluN1; two glutamate-binding subunits, such as GluN2A, GluN2B, or GluN2c/2D; as well as a combination of GluN2 and glycine-binding GluN3 subunits, such as GluN3A or GluN3B [22]. The major function of the NMDA receptors is to transmit the neuronal signal through the synaptic cleft. However, NMDA receptors are not only located at the synaptic cleft but also at the extrasynaptic site of neurons [23–25]. Synaptic and extrasynaptic NMDA receptors have different chemical structures. Synaptic receptors are enriched with the GluN2A subunit, while extrasynaptic receptors contain predominantly the GluN2B or GluN2C/2D subunits [25,26] (Figure 2). There is a body of evidence suggesting that extrasynaptic NMDA receptors containing GluN2B are primarily linked to excitotoxic effects of glutamate upon stroke and in a number of neurodegenerative diseases, as described in Table 1 [25,27,28], while activation of GluN2A subunits has a neuroprotective effect [29]. In general, the lowest agonist and co-agonist potency was found at GluN1/GluN2A NMDARs and the highest at GluN1/GluN2D NMDARs [30]. However, deactivation rates are fastest at GluN1/GluN2A NMDARs and slowest for GluN1/GluN2D NMDARs [30]. Similar deactivation rates were observed for NMDARs containing the GluN2B and GluN2C subunits [30]. These data suggest that transmission and the toxic effects of glutamate undergo different molecular pathways, resulting in the activation of either GluN2A or GluN2B, respectively.

Table 1. Examples of neurological disorders related to glutamate toxicity.

	Pathology	Mechanism	Reference
1.	Amyotrophic lateral sclerosis (ALS)	Increased glutamatergic neurotransmission is present in ALS and might contribute to neurodegeneration. Downregulation of EAAT2 and upregulation of system Xc cause increased activation of glutamate receptors.	[31–33]
2.	Alzheimer's disease (AD)	Increased glutamate concentrations, changes in calcium homeostasis, and increased sensitivity of NMDA receptors make neurons more sensitive to excitotoxicity in AD.	[34–36]
3.	Parkinson's disease	Impaired mitochondrial function can make the substantia nigra sensitive to glutamate.	[37]
4.	Huntington's disease	In HD, there is a redistribution of NMDA receptors to the extrasynaptic compartment, which might activate signaling pathways that foster excitotoxicity and neurodegeneration.	[38]
5.	Multiple sclerosis	Glutamate excitotoxicity mediated by the AMPA/Kainate type of glutamate receptors is an important mechanism in autoimmune demyelination.	[39]

Reasonably, in an attempt to prevent sustained activation of glutamate receptors, astrocytes and neurons should take up glutamate from the extracellular fluid. Glutamate transporter 1 (GLT-1) is the major transport system for glutamate in the CNS. Although GLT-1 was found predominantly in astrocytes, it is also expressed in neuronal bodies and axons [40]. The mechanisms mediating and regulating extracellular glutamate are not very clear. Astrocytes have a higher capacity to take up glutamate than neurons, but since only local glutamate levels close to neurons induce excitotoxicity, the uptake of glutamate by neurons may be critical for glutamate toxicity. The further fate of glutamate in neurons and astrocytes is also different.

3. Intracellular Metabolism of Glutamate

In the majority of cells, including neurons, intracellular glutamate is predominantly used for ATP synthesis and converted into other organic molecules. One specific function of glutamate in neurons is the activation of postsynaptic receptors, playing a central role in neuronal signal transduction. In contrast, the major fate of glutamate in astrocytes is its conversion into glutamine [41–45] (Figure 2). It has been unequivocally accepted that astrocytes do not require glutamate and aerobic metabolism through mitochondria to meet their bioenergetics demands [46]. Glutamate is converted to glutamine (GLN) by glutamine synthetase (GS). GS is expressed exclusively in astrocytes, which is an important segment of the glutamate-glutamine cycle [47]. Thus, the glutamate consumed by astrocytes will not be extracted from the total glutamate pool, and it will, again and again, reappear in neurons as glutamine, which will be converted back to glutamate [1]; the latter can theoretically contribute to both signal transduction and toxicity. Being a part of blood–brain barrier, astrocytes take up glucose from the blood, convert it to pyruvate and then to lactate by glycolysis. Lactate will be transported to neurons (lactate shuttle) and transformed in neurons back to pyruvate, which will be used as a mitochondrial substrate to provide ATP [48].

In contrast to astrocytes, neurons strictlyy require mitochondria with active oxidative phosphorylation (OxPhos) and the tricarboxylic acid cycle (TCA-cycle). They have higher energetic demands to support extensive ion flows accompanying the activation of neurons and signal transmission and, consequently, mitochondria play a key role in the energy metabolism in neurons [49,50]. This suggests that neurons, rather than astrocytes, can burn glutamate in mitochondria, subsequently extracting it from the glutamate-glutamine pool [51,52], thus preventing glutamate toxicity. Divakaruni et al. reported the first strong experimental evidence supporting this assumption. The authors inhibited the pyruvate transporter in neurons and reported three important observations: the inhibition of the mitochondrial pyruvate transporter in primary cortical cultures (1) shifted neuronal metabolism in favor of glutamate consumption, (2) increased the consumption of glutamate in neurons, which caused a drop in the extracellular glutamate levels, and (3) reduced cell death rate [53]. This study shed new light on how neurons can cope with glutamate excitotoxicity, namely increasing the rate of glutamate oxidation in mitochondria. Logically, one can therefore expect that the inhibition of glutamate uptake by mitochondria causes the opposite effect, i.e., an increase in extracellular glutamate levels and aggravation of glutamate toxicity.

Indeed, a recent publication by Weidinger et al. [54] investigated the segment of the TCA cycle responsible for glutamate uptake in mitochondria and showed that inhibition of this TCA cycle segment in neurons resulted in an increase of extracellular glutamate levels and elevated neuron death rate, which is in line with Divakaruni et al. [53]. In fact, indirect evidence of the involvement of the glutamate segment of the TCA cycle in glutamate toxicity was already previously reported under different pathological situations. A remarkable inhibition of mitochondrial respiration by glutamate was observed in the cortex of rats subjected to traumatic brain injury (TBI), while pyruvate respiration was affected much less [55], suggesting a defect at either the glutamate segment of the TCA cycle or complex I of the respiratory chain of mitochondria. A follow-up study showed that the mechanical injury of the cortex, occurring upon TBI, resulted in the inhibition of the oxoglutarate dehydrogenase complex enzyme (OGDHC), one out of two enzymes responsible for the uptake of glutamate in mitochondria [56]. The oxidation of glutamate in the mitochondria requires glutamate dehydrogenase (GDH) and OGDHC. However, in contrast to OGDHC, no changes in GDH activity were found upon TBI, suggesting that OGDHC is more susceptible to inhibition upon brain injury [54], although a decrease in GDH activity was reported in a penetrating ballistic-like brain injury model [57]. Other data shed light on the possible mechanism underlying the inhibition of OGDHC upon brain injury. It has been shown that the inhibition of glutamate uptake by mitochondria coincides with increased cortical nitric oxide (NO) levels [55]. A similar relationship between brain NO and glutamate levels was observed in patients with subarachnoid hemorrhages (SAHs) [58]. The clinical status of these patients was monitored by examination of microdialysate obtained from brain tissue close to the injured area. It was shown that, in the acute phase of SAH, these patients manifested elevated levels of glutamate, which were accompanied by a substantial increase in the levels of NO in microdialysate [58]. The assumption that NO has a critical role is in line with the fact that there are many sources of NO in the brain, which can be activated upon CNS disorders, such as brain injury [59], Parkinson's disease, and Alzheimer's disease [60]. A body of literature suggests that NO plays important physiologic [61,62] as well as pathologic [63] roles in CNS disorders. It has been assumed that the pathologic role of NO is due to S-nitrosylation of proteins, which lose their structure and function, contributing to the development of neurodegenerative diseases [64–66]. The majority of the existing reports on the pathologic impact of NO in CNS disorders consider reversible and/or irreversible inhibition of the mitochondrial respiratory chain by NO and its derivatives, impairing the bioenergetics status of neurons [67,68]. However, recently, Weidinger et al., have shown that the most sensitive target of NO in mitochondria is not the electron transport chain (ETC), but OGDHC [54]. Weidinger et al. showed that NO, at concentrations previously determined in vivo [55], inhibits OGDHC rather than ETC. Of note, the inhibition of OGDHC does not necessarily cause inhibition of the ETC because mitochondrial respiration and ATP synthesis can be recovered by switching to other substrates or bridging the electron transport via extra-mitochondrial pathways [69–71]. In contrast, glutamate uptake cannot be bridged by other pathways and fully depends on the activity of OGDHC. This leads us to the fundamental assumption that mitochondria play a boomerang-like role in the CNS. They induce glutamate toxicity, slowing down the oxidation of glutamate, but later they suffer from glutamate toxicity, reducing respiratory activity and ATP synthesis. The fact that the drop in bioenergetics capacity of mitochondria is secondary to glutamate toxicity has already been supported by a body of literature [72–74]. The fact that the primary mitochondrial defect is inhibition of mitochondrial OGDHC, but not a defect at the ETC level, substantially changes the view on therapeutic strategies.

In contrast to the defects occurring in the ETC, which can be hardly recovered (perhaps only genetically), the activity of OGDHC is easier to compensate by using thiamine (vitamin B1), a cofactor of this enzyme. Indeed, it has been shown that treatment with thiamine restored OGDHC activity and mitochondrial respiration in rats impaired by TBI [56]. A similar relationship between NO levels, OGDHC activity, and glutamate toxicity was already observed in patients with Wernicke–Korsakow syndrome. This CNS disorder is characterized by thiamine deficiency and can be efficiently reversed by thiamine substitution [75,76]. This suggests thiamine is a valid therapeutic option in humans. Of note, the activities of pyruvate dehydrogenase (PDH), another TCA cycle enzyme, which has thiamine as a cofactor, and GDH (the second enzyme contributing to glutamate uptake) did not change in the experimental models described above [54,56], suggesting the high vulnerability of OGDHC. These data suggest that the OGDHC of neurons attenuates extracellular glutamate levels and glutamate toxicity. Below, we consider how changes in OGDHC activity can regulate the levels of extracellular glutamate.

In humans, glutamate toxicity can be addressed by the inhibition of ionotropic glutamate receptors or by the decrease of extracellular glutamate. Magnesium is one of the most common inhibitors of Ca^{2+} flux through NMDA receptors, meaning that in its presence, glutamate can only act on the AMPA receptor [77,78]. Some other blockers of glutamate receptors are used in preclinical models [38,79]. The data analyzed in this review address the reduction of extracellular glutamate by re-activation of OGDHC by its cofactor, thiamine, which is used in clinical practice for the treatment of Wernicke–Korsakov syndrome. Our data suggest that thiamine is a valid therapeutic option for all neurodegenerative diseases and brain injury in humans [54,56].

4. Mechanisms Regulating Glutamate Absorption in Neurons

GLT-1 is the major glutamate transport system, which accounts for approx. 95% of the total glutamate uptake activity in the brain tissue [80–82]. GLT-1 co-transports one glutamate with 3 Na^+ and 1 H^+ and with a counter transport of 1 K^+ into the extracellular fluid per cycle (Figure 3). GLT-1 is highly expressed in astrocytes but also in neurons. However, the function of neuronal GLT-1 is relatively unexplored [40,83]. There is evidence suggesting that, in neurons, GLT-1-mediated glutamate uptake provides glutamate predominantly for energy metabolism [83] (in contrast to the glutamine-glutamate cycle operating in close vicinity to synapses). Perhaps, extracellular glutamate can be located on the body of neurons far from synapses and be absorbed by GLT-1, which facilitates its uptake by mitochondria before it diffuses to synapses through the axon. GLT-1 transport can be regulated, but the regulation mechanisms have been investigated predominantly in astrocytes because they carry most of this transporter in the body; similar mechanisms can also operate in neurons [40,84]. It has been shown that the rate of glutamate transport in astrocytes can be regulated by GLT-1 expression/activity and by their lateral diffusion to specific locations with high glutamate levels [85]. The change in the rate of glutamate absorption by GLT-1 can regulate not only its toxic effects but also the magnitude and duration of the activation of postsynaptic neurons [85,86]. Also, approx. 5% of non-GLT-1 transporters regulate extracellular glutamate persistence and the duration of NMDA receptor activation [87]. The rate of glutamate uptake in astrocytes is substantially higher than in neurons (approx. 4 nmol Glu/min/mg protein) [88] vs. approx. 0.2 nmol Glu/min/mg protein as recalculated based on data in Rink et al. [89]. However, since neuron glutamate receptors are responsible for the induction of toxic effects, the regulation of glutamate concentration close to the neuronal plasma membrane by neuronal transporters can be important.

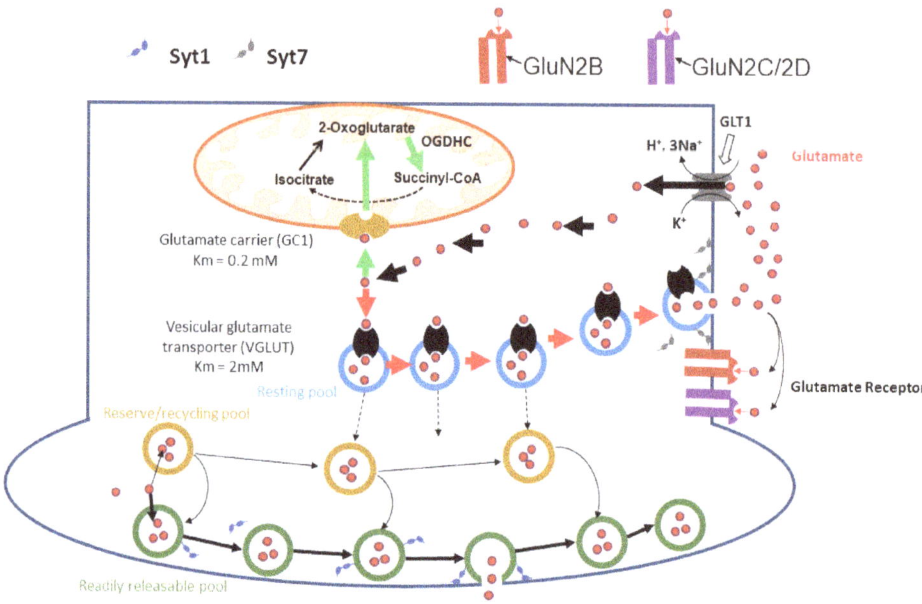

Figure 3. Evoked and spontaneous glutamate release, resulting in a balance between extracellular and intracellular glutamate pools. Spontaneous release and uptake of glutamate by GLT1 create a vicious $\text{Glutamate}_{\text{in}}$-$\text{Glutamate}_{\text{out}}$ cycle.

Theoretically, since an increased concentration of extracellular glutamate is often due to tissue damage accompanied by the release of K^+, high levels of extracellular K^+, and high concentration of intracellular glutamate, if this compound is not taken up by mitochondria and vesicles, this process can slow down the absorption of glutamate and increase glutamate toxicity. This, however, can be compensated by an increased number of GTL-1 by moving them to the spot with high glutamate concentration [86]. However, there is no clear data on how the increase in intracellular glutamate levels can influence the flow of glutamate through GLT-1. This mechanism can be linked to the intracellular distribution of glutamate between different intracellular pools. Rimmele et al. reported an interesting observation. They showed that the conditional knockout (KO) of GTL-1 in neuronal axons (predominantly located in presynaptic terminals) induces excitotoxic death of neurons, while it does not affect the stimulus of evoked glutamate release [40]. This suggests that two different pools of glutamate are responsible for the transmission and toxic effects of glutamate and that neurons contribute to the clearance of the toxic glutamate pool. Moreover, electron microscopy combined with immunolabeling showed that GTL-1 extends beyond the terminal portion of axons [40]. This suggests that neurons, via their GTL-1 transporters, are able to clean up the extrasynaptic glutamate pool responsible for toxic effects. This, however, still does not bring a link involving OGDHC in the regulation of glutamate toxicity. The key to understanding this link comes by considering the intracellular destinations of glutamate.

Glutamate has several different purposes in the body: (1) In mitochondria, it can be used as a source of energy when it is fed into the tricarboxylic acid cycle. (2) In the cytoplasm, it can be packed in vesicles (resting pool) and reach different fates. For example, in the synaptic cleft, the membrane of the vesicles of the readily releasable pool fuses with the cell membrane by interacting with Synaptotagmin 1 (Syt1) to form the synapsis structure and release the glutamate (2a). Additionally, at the extrasynaptic site, the membrane of the vesicles of the resting pool interacts with Synaptotagmin 7 (Syt7) of the cell membrane to release glutamate in the extracellular fluid outside of the synaptic cleft (2b). The latter can interact with some glutamate transporters, like GLUT1, as a returning pathway into the intracellular fluid or activate extrasynaptic NMDA glutamate receptors. Here, Km stands for the half-saturation transport constant, which is also known as the apparent affinity constant (Ka).

5. Mechanisms Regulating the Balance between Intracellular and Extracellular Glutamate Pools

After entering the neuron, glutamate can undergo two major pathways, i.e., energy conservation in mitochondria and signal transmission via synaptic vesicles (Figure 3). The distribution of glutamate between these two pathways is regulated based on their relative affinities to the mitochondrial and the vesicular glutamate transporters (VGLUTs). The mitochondrial transporter has a much higher affinity to glutamate [K_m = 0.2 mM [90]] than VGLUT [K_m = 2 mM [91]]. Following this fact, glutamate entering neurons through GTL-1 will be predominantly taken up by mitochondria and used for ATP synthesis. However, if mitochondria do not efficiently consume glutamate, glutamate will enter glutamate vesicles. However, the vesicle pool in neurons is not homogeneous and the fate of glutamate depends on the vesicle pool it enters. There are three pools of glutamate vesicles in neurons: (1) the readily releasable pool (RRP), (2) the recycling pool, and (3) the reserve pool [92]. There are two types of glutamate release from vesicles, evoked (stimulated, including stimulation with sucrose) [93] and spontaneous [94–96]. The evoked release is well known and serves for neuronal signal transmission, while spontaneous release is less explored. It is known that the spontaneous release plays a physiological role in supporting synaptic maturation and homeostasis (for review, see [97]) but it is also associated with glutamate excitotoxicity [98]. It has been shown that evoked glutamate release originates from RRP, while spontaneous release originates from the resting pool [97,99]. In contrast to evoked release in the synapse, the spontaneous glutamate release occurs in the axon membrane located out of the active zone of synapses [99]. The spontaneous release of glutamate, which is mediated by extrasynaptic glutamate receptors, is linked to the toxic effects of glutamate. For example, the elevated spontaneous release is associated with upregulation of VGLUT, thus causing an increase in intravesicular glutamate levels and suggesting that neurons attempt to release an excess of glutamate in extracellular fluid, which may multiply the toxic effects of glutamate. Genetic upregulation of VGLUT facilitates glutamate transport into vesicles but simultaneously increases spontaneous release of glutamate from these vesicles before they enter RRP and induces excitotoxic effects on neurons [98]. Evoked and spontaneous glutamate release are regulated by different mechanisms (Figure 3). The release of glutamate from vesicles is regulated by synaptotagmin, a family of Ca^{2+} sensor proteins regulating the release of glutamate from the vesicles and hormone secretion. It has been shown that the evoked release of glutamate is regulated by Syt1, which simultaneously inhibits the spontaneous glutamate release [100], while Syt7 triggers spontaneous glutamate release and efficiently activates GluN2B glutamate receptors [101] (Figure 3). Action potential-evoked neurotransmitter release is associated with the transduction of neuronal signals, while spontaneous release has been associated with neurological disorders. Also, under certain circumstances, glutamate that is released in the synapses can diffuse out of the synaptic cleft and activate nearby extrasynaptic glutamate receptors [102]. Collectively, these results support the premise that spontaneous and evoked neurotransmission activate distinct sets of NMDA receptors, which are responsible for neurotransmission and toxic effects, respectively [103]. Below, we show the list of evidence suggesting that transmission and toxic effects of glutamate undergo different pathways, i.e., intrasynaptic and extrasynaptic, respectively (Table 2). We assume that both pathways either do not communicate or have only limited communication between each other. There is a logical biological explanation for this observation. By separating these two pathways, a neuron preserves its transmission function; when the extracellular concentration of glutamate overpasses the threshold and the neuron cannot offset anymore the process, the $glutamate_{IN}$-$glutamate_{OUT}$

cycle will be activated to induce neuronal death. In this way, single neurons will die before the transmission function is seriously impaired. OGDHC plays a critical role in turning on the vicious glutamate$_{IN}$-glutamate$_{OUT}$ cycle, reducing the consumption of extracellular glutamate and re-directing an excess of intracellular glutamate back to the extracellular pool via spontaneous release (Figure 3).

Table 2. Evidence suggesting that transmission and toxic effects of glutamate undergo spatially and functionally separated pathways.

	Evidence	Reference
1.	Transmission and toxic pathways engage NMDA receptors characterized by two different chemical structures, GluN2A and GluN2B, respectively.	[22]
2.	Transmission and toxic pathways engage NMDA receptors characterized by two different locations, synaptic and extra-synaptic, respectively.	[23–25]
3.	Glutamate absorbed by neurons enters the cytoplasmic pool and is used by mitochondria rather than packed in RRP vesicles.	[97,99]
4.	Transmission and toxic pathways engage two different pools of vesicles, RRP and reserve pool, respectively.	[40]
5.	Transmission and toxic pathways are mediated by two different pathways of glutamate release evoked and spontaneous, respectively.	[93–96]
6.	Evoked and spontaneous glutamate release are regulated by two different mechanisms, mediated by Syt1 and Syt7, respectively.	[100,101]
7.	Evoked and spontaneous glutamate release activate two different pools of glutamate receptors, intra- and extra-synaptic NMDA receptors, respectively.	[98,99,103]
8.	The activity of OGDHC balances intra- and extra-cellular glutamate levels and glutamate toxicity.	[54,104]
9.	Low OGDHC activity causes an increase in extracellular glutamate and cell death, while high activity reduces the levels of extracellular glutamate and cell death rate.	[105–108]

6. Glutamate-Mediated Death of Neurons

Glutamate can induce different death pathways. Primary pathways include excitotoxicity [18] and ferroptosis [109,110]. The excitotoxic pathway is characterized by excessive excitation of neurons and influx of Ca^{2+} into the cell. Ca^{2+} activates different enzymatic systems, such as proteases, phospholipases, NOS, and other Ca^{2+}-dependent enzymes. An excess of Ca^{2+} also stimulates the formation of mitochondrial permeability transition pores (mPTPs) and subsequent mitochondrial dysfunction accompanied by reduced ATP levels and damage to the mitochondrial membrane [74] (Figure 4). The damage of mitochondrial membranes facilitates the release of cytochrome c and the induction of apoptosis (Figure 4). More recently, it has also been shown that ferroptotic cell death can be activated. High levels of extracellular glutamate block the glutamate cysteine antiporter, causing the depletion of intracellular cysteine and then glutathione. The latter inhibits glutathione peroxidases and induces oxidation of lipids and ferroptosis [110], resulting in iron-dependent non-apoptotic cell death [110]. The Xc-antiporter system consists of a light chain subunit SLC7A11 and heavy chain subunits SLC3A2. Typically, stress-induced SLC7A11 expression is mediated by two major transcription factors: activating transcription factor 4 (ATF4) and/or nuclear factor erythroid 2-related factor 2 (NRF2). However, recently, a study by Zhang et al. showed that erastin (a ferroptosis-inducing agent) induces SLC7A11 expression through an NRF2- or ATF4-independent mechanism [111]. It has also been shown that p53 is able to inhibit cystine uptake by repressing SLC7A11 transcription, which limits the production of intracellular glutathione and increases the sensitivity of cells to ferroptosis [112]. All this indicates a critical role for SLC7A11-mediated cystine uptake in the suppression of ferroptosis, including in neuronal cells.

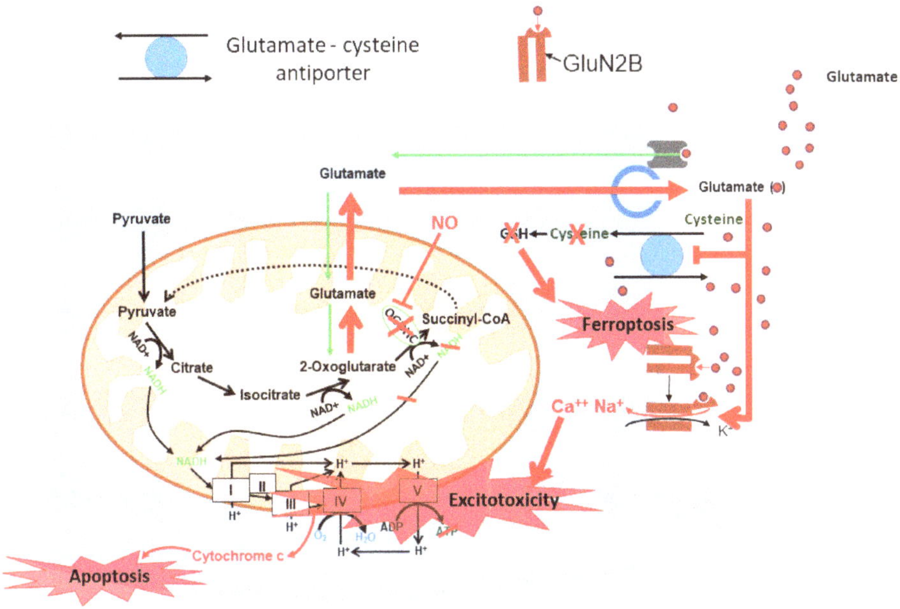

Figure 4. The impact of mitochondria in mechanisms of glutamate-mediated neuronal death. When the oxoglutarate dehydrogenase enzyme is blocked in mitochondria, the glutamate accumulates. Inside the mitochondria, it can disrupt the normal function of the electron transport chain, where the remaining glutamate will leak into the cytoplasm and secreted into the extracellular fluid. The high concentration of glutamate in the extracellular fluid will affect other transporters like (1) the glutamate-cysteine antiporter, leading to the reduction of intracellular cysteine, and thus reducing the activity of enzymes of the family of the glutathione peroxidase, causing the accumulation of lipid peroxides that, together with iron, can cause ferroptosis. Additionally, (2) glutamate can cause excitotoxicity via activated extrasynaptically NMDA receptors.

Fe^{2+} concentrations and expression of genes associated with ferroptosis are altered during intracerebral hemorrhages. The potential antioxidant agent crocin has been shown to inhibit neuronal cell ferroptosis through the expression of Nrf2 and SLC7A11, thereby alleviating intracerebral hemorrhages [113]. In an oligodendrocyte model of RSL3-induced ferroptosis, it was shown that RSL3 inhibits the main anti-ferroptosis pathway, i.e., the SLC7A11/glutathione/glutathione peroxidase 4 pathway [114]. A study by Zhao et al. on mouse hippocampal HT-22 cells treated with ferric ammonium citrate showed that Isorhynchophylline (alkaloid isolated from *Uncaria rhynchophylla*) is able to protect neurocytes from ferroptosis through the miR-122-5p/TP53/SLC7A11 pathway, which could become a potential therapeutic target for intracerebral hemorrhages [115]. Another study showed that oxygen-glucose deprivation/reperfusion (OGD/R)-induced neuronal injury induces ferroptosis by reducing the levels of SLC7A11 and glutathione peroxidase 4 (GPX4) and also reduces the levels of endogenous antioxidants, including NADPH, glutathione, and superoxide dismutase in neurons [116].

All these data point to the involvement of SLC7A11 and ferroptosis in the pathogenesis of many neurological diseases and suggest ferroptosis as an attractive therapeutic target.

7. Interplay between Glutamate Mediated Excitation and Intracellular Glutamate Signaling

Above, we discussed the toxic action of glutamate mediated by glutamate receptors and transporters. However, additionally, glutamate is also involved in a number of intracellular signaling cascades, which can interfere with the action of glutamate described above. Previously mentioned mechanisms include the action of ionotropic glutamate receptors, which are ligand-gated transmembrane channels that selectively transport cations. This not only induces action potential but also impacts several intracellular signaling cascades. In neurons, ionotropic glutamate receptors-mediated calcium influx activates Src family kinases, PI3K and ERK [117,118]. In addition to ionotropic glutamate receptors, neurons and a number of glial cells express three groups of metabotropic glutamate receptors (mGluRs). mGluRs are members of the G-protein-coupled receptor (GPCR) superfamily and they are coupled to inositol trisphosphate (group I) IP3 or to cAMP (group II and III), mediating intracellular signaling pathways. Group I mGluRs activate a range of molecular pathways, including phospholipase D and several protein kinases. Group II and III inhibit adenylyl cyclase and permeability of Ca^{2+} channels. Additionally, group III stimulates cGMP-dependent pathways [119]. Group I mGluRs are often localized postsynaptically and, in general, increase neuronal excitability, while group II and group III are localized on presynaptic axons and inhibit neurotransmitter release and excitation. In addition, the neurotransmitter release causes a relatively short but strong acidification of the synaptic cleft [120,121]. In addition, glutamine, which is synthesized during the glutamate–glutamine cycle, is a very important amino acid for the regulation of the acid–base cellular equilibrium [122,123]. These multiple intracellular pathways of glutamate can modulate the excitation, neurosignal transmission, as well as death cell pathways operating in neurons upon high levels of extracellular glutamate. We expect that the interaction between these three major functions of glutamate, neuronal signal transmission, toxicity, and intracellular signaling pathways will be an important area of research in the following years.

8. Conclusions

Glutamate toxicity plays a pivotal role in brain injury and neurodegenerative diseases. An excess of glutamate kills neurons via excitotoxic and ferroptotic pathways. The recent literature shed light on the mechanisms controlling glutamate-mediated neuronal death. The toxic effects of glutamate, at least partially, are controlled by OGDHC, a TCA cycle enzyme in neurons. Interestingly, insufficiency of this enzyme predominantly influences the turnover of extracellular glutamate but not OxPhos. Impaired OxPhos, a common feature of neuronal diseases, appears to be secondary to glutamate toxicity. This toxic effect is spatially and functionally separated from the transmission function of glutamate. This suggests that glutamate toxicity can kill neurons without affecting the transmission function. In contrast to the respiratory function of mitochondria, which cannot be easily improved, the activity of OGDHC can be recovered by treatment with its cofactor thiamine. We assume that there is only a very weak communication between the transmission and the toxic pathways of glutamate, aiming to preserve transmission function in toxic environments.

Author Contributions: Conceptualization, A.V.K., A.V.-G., Y.D.S. and A.W.; software and illustrations, A.V.-G. and A.V.K.; validation, A.V.K., A.V.-G., Y.D.S. and A.W.; writing—original draft preparation, A.V.-G. and A.V.K.; writing—review and editing, A.V.K., A.V.-G., Y.D.S. and A.W.; visualization, A.V.-G. and Y.D.S.; supervision, A.V.K. and A.W.; funding acquisition, A.V.K. All authors have read and agreed to the published version of the manuscript.

Funding: This research was funded by the Austrian Science Fund, grant number FWF-33799, and the article processing charges (APC) were funded by the Austrian Science Fund.

Institutional Review Board Statement: Not applicable.

Informed Consent Statement: Not applicable.

Data Availability Statement: No new data were created or analyzed in this study. Data sharing is not applicable to this article.

Conflicts of Interest: The authors declare no conflicts of interest. The funders had no role in the design of the study; in the collection, analyses, or interpretation of data; in the writing of the manuscript; or in the decision to publish the results.

References

1. McKenna, M.C. The Glutamate-Glutamine Cycle Is Not Stoichiometric: Fates of Glutamate in Brain. *J. Neurosci. Res.* **2007**, *85*, 3347–3358. [CrossRef] [PubMed]
2. Andrew, R.D.; Farkas, E.; Hartings, J.A.; Brennan, K.C.; Herreras, O.; Müller, M.; Kirov, S.A.; Ayata, C.; Ollen-Bittle, N.; Reiffurth, C.; et al. Questioning Glutamate Excitotoxicity in Acute Brain Damage: The Importance of Spreading Depolarization. *Neurocrit. Care* **2022**, *37*, 11–30. [CrossRef] [PubMed]
3. Cantu, D.; Walker, K.; Andresen, L.; Taylor-Weiner, A.; Hampton, D.; Tesco, G.; Dulla, C.G. Traumatic Brain Injury Increases Cortical Glutamate Network Activity by Compromising GABAergic Control. *Cereb. Cortex* **2015**, *25*, 2306–2320. [CrossRef] [PubMed]
4. Mattson, M.P. Excitotoxicity. In *Stress: Physiology, Biochemistry, and Pathology Handbook of Stress Series*; Elsevier: Amsterdam, The Netherlands, 2019; Volume 3, pp. 125–134. ISBN 9780128131466.
5. Dixon, S.J.; Lemberg, K.M.; Lamprecht, M.R.; Skouta, R.; Zaitsev, E.M.; Gleason, C.E.; Patel, D.N.; Bauer, A.J.; Cantley, A.M.; Yang, W.S.; et al. Ferroptosis: An Iron-Dependent Form of Nonapoptotic Cell Death. *Cell* **2012**, *149*, 1060–1072. [CrossRef] [PubMed]
6. Li, J.; Cao, F.; Yin, H.L.; Huang, Z.J.; Lin, Z.T.; Mao, N.; Sun, B.; Wang, G. Ferroptosis: Past, Present and Future. *Cell Death Dis.* **2020**, *11*, 88. [CrossRef] [PubMed]
7. Araque, A.; Parpura, V.; Sanzgiri, R.; Haydon, P. Tripartite Synapses: Glia, the Unacknowledged Partner. *Trends Neurosci.* **1999**, *22*, 208–215. [CrossRef]
8. Bröer, S.; Brookes, N. Transfer of Glutamine between Astrocytes and Neurons. *J. Neurochem.* **2001**, *77*, 705–719. [CrossRef]
9. Rothstein, J.D.; Tabakoff, B. Alteration of Striatal Glutamate Release After Glutamine Synthetase Inhibition. *J. Neurochem.* **1984**, *43*, 1438–1446. [CrossRef]
10. Takeuchi, H.; Jin, S.; Wang, J.; Zhang, G.; Kawanokuchi, J.; Kuno, R.; Sonobe, Y.; Mizuno, T.; Suzumura, A. Tumor Necrosis Factor-α Induces Neurotoxicity via Glutamate Release from Hemichannels of Activated Microglia in an Autocrine Manner. *J. Biol. Chem.* **2006**, *281*, 21362–21368. [CrossRef]
11. Wetherington, J.; Serrano, G.; Dingledine, R. Astrocytes in the Epileptic Brain. *Neuron* **2008**, *58*, 168–178. [CrossRef]
12. Kwon, H.S.; Koh, S.H. Neuroinflammation in Neurodegenerative Disorders: The Roles of Microglia and Astrocytes. *Transl Neurodegener* **2020**, *9*, 42. [CrossRef]
13. Anderson, C.M.; Swanson, R.A. Astrocyte Glutamate Transport: Review of Properties, Regulation, and Physiological Functions. *Glia* **2000**, *32*, 1–14. [CrossRef] [PubMed]
14. Lehre, K.P.; Danbolt, N.C. The Number of Glutamate Transporter Subtype Molecules at Glutamatergic Synapses: Chemical and Stereological Quantification in Young Adult Rat Brain. *J. Neurosci.* **1998**, *18*, 8751–8757. [CrossRef]
15. Mahmoud, S.; Gharagozloo, M.; Simard, C.; Gris, D. Astrocytes Maintain Glutamate Homeostasis in the Cns by Controlling the Balance between Glutamate Uptake and Release. *Cells* **2019**, *8*, 184. [CrossRef] [PubMed]
16. Monaghan, D.T.; Bridges, R.J.; Cotman, C.W. THE EXCITATORY AMINO ACID RECEPTORS: Their Classes, Pharmacology, and Distinct Properties in the Function of the Central Nervous System. *Annu. Rev. Pharmacol. Toxicol* **1989**, *29*, 365402. [CrossRef]
17. Young, A.; Fagg, G. Excitatory Amino Acid Receptors in the Brain: Membrane Binding and Receptor Autoradiographic Approaches. *Biochem. Biophys. Res. Commun* **1990**, *56*, 126–133. [CrossRef]
18. Lau, A.; Tymianski, M. Glutamate Receptors, Neurotoxicity and Neurodegeneration. *Pflugers Arch* **2010**, *460*, 525–542. [CrossRef] [PubMed]
19. Lüscher, C.; Malenka, R.C. NMDA Receptor-Dependent Long-Term Potentiation and Long-Term Depression (LTP/LTD). *Cold Spring Harb. Perspect. Biol.* **2012**, *4*, a005710. [CrossRef]
20. Wyllie, D.J.A.; Bowie, D. Ionotropic Glutamate Receptors: Structure, Function and Dysfunction. *J. Physiol.* **2022**, *600*, 175–179. [CrossRef]
21. Mattson, M.P. Excitotoxic and Excitoprotective Mechanisms Abundant Targets for the Prevention and Treatment of Neurodegenerative Disorders. *Neuromolecular Med.* **2003**, *65*, 65–94. [CrossRef]
22. Yamamoto, H.; Hagino, Y.; Kasai, S.; Ikeda, K. Specific Roles of NMDA Receptor Subunits in Mental Disorders. *Curr. Mol. Med.* **2015**, *15*, 193–205. [CrossRef]
23. Hardingham, G.E.; Fukunaga, Y.; Bading, H. Extrasynaptic NMDARs Oppose Synaptic NMDARs by Triggering CREB Shut-off and Cell Death Pathways. *Nat. Neurosci.* **2002**, *5*, 405–414. [CrossRef]
24. Papouin, T.; Oliet, S.H.R. Organization, Control and Function of Extrasynaptic NMDA Receptors. *Philos. Trans. R. Soc. B Biol. Sci.* **2014**, *369*, 20130601. [CrossRef]
25. Yu, S.P.; Jiang, M.Q.; Shim, S.S.; Pourkhodadad, S.; Wei, L. Extrasynaptic NMDA Receptors in Acute and Chronic Excitotoxicity: Implications for Preventive Treatments of Ischemic Stroke and Late-Onset Alzheimer's Disease. *Mol. Neurodegener.* **2023**, *18*, 43. [CrossRef]

26. Brickley, S.; Revilla, V.; Cull-Candy, S.; Wisden, W.; Farrant, M. Adaptive Regulation of Neuronal Excitability by a Voltage-Independent Potassium Conductance. *Nature* **2001**, *409*, 88–92. [CrossRef]
27. Bading, H. Therapeutic Targeting of the Pathological Triad of Extrasynaptic NMDA Receptor Signaling in Neurodegenerations. *J. Exp. Med.* **2017**, *214*, 569–578. [CrossRef]
28. Léveillé, F.; El Gaamouch, F.; Gouix, E.; Lecocq, M.; Lobner, D.; Nicole, O.; Buisson, A. Neuronal Viability Is Controlled by a Functional Relation between Synaptic and Extrasynaptic NMDA Receptors. *FASEB J.* **2008**, *22*, 4258–4271. [CrossRef]
29. Brassai, A.; Suvanjeiev, R.G.; Bán, E.G.; Lakatos, M. Role of Synaptic and Nonsynaptic Glutamate Receptors in Ischaemia Induced Neurotoxicity. *Brain Res. Bull.* **2015**, *112*, 1–6. [CrossRef]
30. Wyllie, D.J.A.; Livesey, M.R.; Hardingham, G.E. Influence of GluN2 Subunit Identity on NMDA Receptor Function. *Neuropharmacology* **2013**, *74*, 4–17. [CrossRef]
31. Howland, D.S.; Liu, J.; She, Y.; Goad, B.; Maragakis, N.J.; Kim, B.; Erickson, J.; Kulik, J.; Devito, L.; Psaltis, G.; et al. Focal Loss of the Glutamate Transporter EAAT2 in a Transgenic Rat Model of SOD1 Mutant-Mediated Amyotrophic Lateral Sclerosis (ALS). *Proc. Natl. Acad. Sci. USA* **2002**, *99*, 1604–1609. [CrossRef]
32. Sasaki, S.; Komori, T.; Iwata, M. Excitatory Amino Acid Transporter 1 and 2 Immunoreactivity. *Acta Neuropathol.* **2000**, *100*, 138–144. [CrossRef]
33. Albano, R.; Liu, X.Q.; Lobner, D. Regulation of System Xc- in the SOD1-G93A Mouse Model of ALS. *Exp. Neurol.* **2013**, *250*, 69–73. [CrossRef]
34. Mattson, M.P.; Cheng, B.; Davis, D.; Bryant, K.; Lieberburg, I.; Rydel, R.E. P-Amyloid Peptides Destabilize Calcium Homeostasis and Render Human Cortical Neurons Vulnerable to Excitotoxicity. *J. Neurosci.* **1992**, *12*, 376–389. [CrossRef]
35. Deng, Y.; Xiong, Z.; Chen, P.; Wei, J.; Chen, S.; Yan, Z. β-Amyloid Impairs the Regulation of N-Methyl-D-Aspartate Receptors by Glycogen Synthase Kinase 3. *Neurobiol. Aging* **2014**, *35*, 449–459. [CrossRef]
36. Scimemi, A.; Meabon, J.S.; Woltjer, R.L.; Sullivan, J.M.; Diamond, J.S.; Cook, D.G. Amyloid-B1-42 Slows Clearance of Synaptically Released Glutamate by Mislocalizing Astrocytic GLT-1. *J. Neurosci.* **2013**, *158*, 5312–5318. [CrossRef]
37. Blandini, F.; Porter, R.H.P.; Greenamyre, J.T. Glutamate and Parkinson's Disease. *Mol. Neurobiol.* **1996**, *12*, 73–94. [CrossRef]
38. Lewerenz, J.; Maher, P. Chronic Glutamate Toxicity in Neurodegenerative Diseases-What Is the Evidence? *Front. Neurosci.* **2015**, *9*, 469. [CrossRef] [PubMed]
39. Pitt, D.; Werner, P.; Raine, C.S. Glutamate Excitotoxicity in a Model of Multiple Sclerosis. *Nat. Med.* **2000**, *6*, 67–70. [CrossRef] [PubMed]
40. Rimmele, T.S.; Rosenberg, P.A. GLT-1: The Elusive Presynaptic Glutamate Transporter. *Neurochem. Int.* **2016**, *98*, 19–28. [CrossRef] [PubMed]
41. Anlauf, E.; Derouiche, A. Glutamine Synthetase as an Astrocytic Marker: Its Cell Type and Vesicle Localization. *Front. Endocrinol.* **2013**, *4*, 144. [CrossRef] [PubMed]
42. Bergles, D.E.; Jahr, C.E. Synaptic Activation of Glutamate Transporters in Hippocampal Astrocytes. *Neuron* **1997**, *19*, 1297–1308. [CrossRef] [PubMed]
43. Huyghe, D.; Nakamura, Y.; Terunuma, M.; Faideau, M.; Haydon, P.; Pangalos, M.N.; Moss, S.J. Glutamine Synthetase Stability and Ubcellular Distribution in Astrocytes Are Regulated by γ-Aminobutyric Type B Receptors. *J. Biol. Chem.* **2014**, *289*, 28808–28815. [CrossRef] [PubMed]
44. Lehre, K.P.; Levy, L.M.; Ottersen, O.P.; Storm-Mathisen, J.; Danbolt, N.C. Differential Expression of Two Glial Glutamate Transporters in the Rat Brain: Quantitative and Lmmunocytochemical Observations. *J. Neurosci.* **1995**, *15*, 1835–1853. [CrossRef] [PubMed]
45. Yamada, D.; Kawabe, K.; Tosa, I.; Tsukamoto, S.; Nakazato, R.; Kou, M.; Fujikawa, K.; Nakamura, S.; Ono, M.; Oohashi, T.; et al. Inhibition of the Glutamine Transporter SNAT1 Confers Neuroprotection in Mice by Modulating the MTOR-Autophagy System. *Commun. Biol.* **2019**, *2*, 346. [CrossRef] [PubMed]
46. Rose, J.; Brian, C.; Pappa, A.; Panayiotidis, M.I.; Franco, R. Mitochondrial Metabolism in Astrocytes Regulates Brain Bioenergetics, Neurotransmission and Redox Balance. *Front. Neurosci.* **2020**, *14*, 536682. [CrossRef] [PubMed]
47. Schousboe, A.; Scafidi, S.; Bak, L.K.; Waagepetersen, H.S.; McKenna, M.C. Glutamate Metabolism in the Brain Focusing on Astrocytes. *Adv. Neurobiol.* **2014**, *11*, 13–30.
48. Mason, S. Lactate Shuttles in Neuroenergetics-Homeostasis, Allostasis and Beyond. *Front. Neurosci.* **2017**, *11*, 43. [CrossRef]
49. Walker, B.R.; Moraes, C.T. Nuclear-Mitochondrial Interactions. *Biomolecules* **2022**, *12*, 427. [CrossRef]
50. Chang, D.T.W.; Honick, A.S.; Reynolds, I.J. Mitochondrial Trafficking to Synapses in Cultured Primary Cortical Neurons. *J. Neurosci.* **2006**, *26*, 7035–7045. [CrossRef]
51. Verstreken, P.; Ly, C.V.; Venken, K.J.T.; Koh, T.W.; Zhou, Y.; Bellen, H.J. Synaptic Mitochondria Are Critical for Mobilization of Reserve Pool Vesicles at Drosophila Neuromuscular Junctions. *Neuron* **2005**, *47*, 365–378. [CrossRef]
52. Tang, Y.-G.; Zucker, R.S. Mitochondrial Involvement in Post-Tetanic Potentiation of Synaptic Transmission. *Neuron* **1997**, *18*, 483–491. [CrossRef] [PubMed]
53. Divakaruni, A.S.; Wallace, M.; Buren, C.; Martyniuk, K.; Andreyev, A.Y.; Li, E.; Fields, J.A.; Cordes, T.; Reynolds, I.J.; Bloodgood, B.L.; et al. Inhibition of the Mitochondrial Pyruvate Carrier Protects from Excitotoxic Neuronal Death. *J. Cell Biol.* **2017**, *216*, 1091–1105. [CrossRef] [PubMed]

54. Weidinger, A.; Milivojev, N.; Hosmann, A.; Duvigneau, J.C.; Szabo, C.; Törö, G.; Rauter, L.; Vaglio-Garro, A.; Mkrtchyan, G.V.; Trofimova, L.; et al. Oxoglutarate Dehydrogenase Complex Controls Glutamate-Mediated Neuronal Death. *Redox Biol.* **2023**, *62*, 102669. [CrossRef]
55. Üçal, M.; Kraitsy, K.; Weidinger, A.; Paier-Pourani, J.; Patz, S.; Fink, B.; Molcanyi, M.; Schäfer, U. Comprehensive Profiling of Modulation of Nitric Oxide Levels and Mitochondrial Activity in the Injured Brain: An Experimental Study Based on the Fluid Percussion Injury Model in Rats. *J. Neurotrauma* **2017**, *34*, 475–486. [CrossRef]
56. Mkrtchyan, G.V.; Üçal, M.; Müllebner, A.; Dumitrescu, S.; Kames, M.; Moldzio, R.; Molcanyi, M.; Schaefer, S.; Weidinger, A.; Schaefer, U.; et al. Thiamine Preserves Mitochondrial Function in a Rat Model of Traumatic Brain Injury, Preventing Inactivation of the 2-Oxoglutarate Dehydrogenase Complex. *Biochim. Biophys. Acta Bioenerg.* **2018**, *1859*, 925–931. [CrossRef]
57. Pandya, J.D.; Leung, L.Y.; Yang, X.; Flerlage, W.J.; Gilsdorf, J.S.; Deng-Bryant, Y.; Shear, D.A. Comprehensive Profile of Acute Mitochondrial Dysfunction in a Preclinical Model of Severe Penetrating TBI. *Front. Neurol.* **2019**, *10*, 605. [CrossRef] [PubMed]
58. Hosmann, A.; Milivojev, N.; Dumitrescu, S.; Reinprecht, A.; Weidinger, A.; Kozlov, A. V Cerebral Nitric Oxide and Mitochondrial Function in Patients Suffering Aneurysmal Subarachnoid Hemorrhage-a Translational Approach. *Acta Neurochirurgica* **2021**, *163*, 139–149. [CrossRef]
59. Kozlov, A.V.; Bahrami, S.; Redl, H.; Szabo, C. Alterations in Nitric Oxide Homeostasis during Traumatic Brain Injury. *Biochim. Biophys. Acta Mol. Basis Dis.* **2017**, *1863*, 2627–2632. [CrossRef]
60. Iova, O.M.; Marin, G.E.; Lazar, I.; Stanescu, I.; Dogaru, G.; Nicula, C.A.; Bulboacă, A.E. Nitric Oxide/Nitric Oxide Synthase System in the Pathogenesis of Neurodegenerative Disorders—An Overview. *Antioxidants* **2023**, *12*, 753. [CrossRef]
61. Archer, S. Measurement of Nitric Oxide in. Biological Models. *FASEB J.* **1993**, *7*, 349–360. [CrossRef]
62. Rosselli, M.; Keller, P.J.; Dubey, R.K. Role of Nitric Oxide in the Biology, Physiology and Pathophysiology of Reproduction. *Hum. Reprod. Update* **1998**, *4*, 3–24. [CrossRef] [PubMed]
63. Marín, J.; Rodríguez-Martínez, M.A. Role of vascular nitric oxide in physiological and pathological conditions. *Pharmacol. Ther.* **1997**, *75*, 111–134. [CrossRef] [PubMed]
64. Cho, D.H.; Nakamura, T.; Fang, J.; Cieplak, P.; Godzik, A.; Gu, Z.; Lipton, S.A. S-Nitrosylation of Drp1 Mediates Beta-Amyloid-Related Mitochondrial Fission and Neuronal Injury. *Science* **2009**, *324*, 102–105. [CrossRef] [PubMed]
65. Nakamura, T.; Wang, L.; Wong, C.C.L.; Scott, F.L.; Eckelman, B.P.; Han, X.; Tzitzilonis, C.; Meng, F.; Gu, Z.; Holland, E.A.; et al. Transnitrosylation of XIAP Regulates Caspase-Dependent Neuronal Cell Death. *Mol. Cell* **2010**, *39*, 184–195. [CrossRef] [PubMed]
66. Uehara, T.; Nakamura, T.; Yao, D.; Shi, Z.Q.; Gu, Z.; Ma, Y.; Masliah, E.; Nomura, Y.; Lipton, S.A. S-Nitrosylated Protein-Disulphide Isomerase Links Protein Misfolding to Neurodegeneration. *Nature* **2006**, *441*, 513–517. [CrossRef] [PubMed]
67. Duvigneau, C.; Kozlov, A. Pathological Impact of the Interaction of NO and CO with Mitochondria in Critical Care Diseases. *Front. Med.* **2017**, *4*, 223. [CrossRef] [PubMed]
68. Stewart, V.C.; Heales, S.J.R. Nitric Oxide-Induced Mitochondrial Dysfunction: Implications for Neurodegeneration. *Free Radic. Biol. Med.* **2003**, *34*, 287–303. [CrossRef]
69. Kang, W.; Suzuki, M.; Saito, T.; Miyado, K. Emerging Role of Tca Cycle-Related Enzymes in Human Diseases. *Int. J. Mol. Sci.* **2021**, *22*, 13057. [CrossRef]
70. Kiss, G.; Konrad, C.; Doczi, J.; Starkov, A.A.; Kawamata, H.; Manfredi, G.; Zhang, S.F.; Gibson, G.E.; Beal, M.F.; Adam-Vizi, V.; et al. The Negative Impact of α-Ketoglutarate Dehydrogenase Complex Deficiency on Matrix Substrate-Level Phosphorylation. *FASEB J.* **2013**, *27*, 2393–2406. [CrossRef]
71. Scialò, F.; Fernández-Ayala, D.J.; Sanz, A. Role of Mitochondrial Reverse Electron Transport in ROS Signaling: Potential Roles in Health and Disease. *Front. Physiol.* **2017**, *8*, 428. [CrossRef]
72. Stout, A.K.; Raphael, H.M.; Kanterewicz, B.I.; Klann, E.; Reynolds, I.J. Glutamate-Induced Neuron Death Requires Mitochondrial Calcium Uptake. *Nat. Neurosci.* **1998**, *1*, 366–373. [CrossRef] [PubMed]
73. Angelova, P.R.; Vinogradova, D.; Neganova, M.E.; Serkova, T.P.; Sokolov, V.V.; Bachurin, S.O.; Shevtsova, E.F.; Abramov, A.Y. Pharmacological Sequestration of Mitochondrial Calcium Uptake Protects Neurons Against Glutamate Excitotoxicity. *Mol. Neurobiol.* **2019**, *56*, 2244–2255. [CrossRef] [PubMed]
74. Verma, M.; Lizama, B.N.; Chu, C.T. Excitotoxicity, Calcium and Mitochondria: A Triad in Synaptic Neurodegeneration. *Transl. Neurodegener.* **2022**, *11*, 3. [CrossRef] [PubMed]
75. Desjardins, P.; Butterworth, R. Desjardins2005. *Mol. Neurobiol.* **2005**, *31*, 17–25. [CrossRef] [PubMed]
76. Hazell, A.S.; Faim, S.; Wertheimer, G.; Silva, V.R.; Marques, C.S. The Impact of Oxidative Stress in Thiamine Deficiency: A Multifactorial Targeting Issue. *Neurochem. Int.* **2013**, *62*, 796–802. [CrossRef]
77. Mathew, A.A.; Panonnummal, R. 'Magnesium'-the Master Cation-as a Drug—Possibilities and Evidences. *BioMetals* **2021**, *34*, 955–986. [CrossRef]
78. Chollat, C.; Sentilhes, L.; Marret, S. Fetal Neuroprotection by Magnesium Sulfate: From Translational Research to Clinical Application. *Front. Neurol.* **2018**, *9*, 247. [CrossRef]
79. Molinuevo, J.L.; Lladó, A.; Rami, L. Memantine Targeting Glutamate Excitotoxicity in Alzheimer's Disease and Other Dementias. *Am. J. Alzheimer's Dis. Other Dement.* **2005**, *20*, 77–85. [CrossRef]
80. Danbolt, N.C.; Storm-Mathisen, J.; Kanner, B.I. An [Na^+ + K^+]Coupled L-Glutamate Transporter Purified from Rat Brain Is Located in Glial Cell Processes. *Neuroscience* **1992**, *51*, 295–310. [CrossRef] [PubMed]

81. Haugeto, Ø.; Ullensvang, K.; Levy, L.M.; Chaudhry, F.A.; Honoré, T.; Nielsen, M.; Lehre, K.P.; Danbolt, N.C. Brain Glutamate Transporter Proteins Form Homomultimers. *J. Biol. Chem.* **1996**, *271*, 27715–27722. [CrossRef] [PubMed]
82. Tanaka, K.; Watase, K.; Manabe, T.; Yamada, K.; Watanabe, M.; Takahashi, K.; Iwama, H.; Nishikawa, T.; Ichihara, N.; Kikuchi, T.; et al. Epilepsy and Exacerbation of Brain Injury in Mice Lacking the Glutamate Transporter. *Science* **1997**, *276*, 1699–1702. [CrossRef] [PubMed]
83. McNair, L.F.; Andersen, J.V.; Aldana, B.I.; Hohnholt, M.C.; Nissen, J.D.; Sun, Y.; Fischer, K.D.; Sonnewald, U.; Nyberg, N.; Webster, S.C.; et al. Deletion of Neuronal GLT-1 in Mice Reveals Its Role in Synaptic Glutamate Homeostasis and Mitochondrial Function. *J. Neurosci.* **2019**, *39*, 4847–4863. [CrossRef] [PubMed]
84. Mennerick, S.; Dhond, R.P.; Benz, A.; Xu, W.; Rothstein, J.D.; Danbolt, N.C.; Isenberg, K.E.; Zorumski, C.F. Neuronal Expression of the Glutamate Transporter GLT-1 in Hippocampal Microcultures. *J. Neurosci.* **1998**, *18*, 4490–4499. [CrossRef]
85. Yao, L.; Grand, T.; Hanson, J.; Paoletti, P.; Zhou, Q. Higher Ambient Synaptic Glutamate at Inhibitory versus Excitatory Neurons Differentially Impacts NMDA Receptor Activity. *Nat. Commun.* **2018**, *9*, 4000. [CrossRef]
86. Murphy-Royal, C.; Dupuis, J.P.; Varela, J.A.; Panatier, A.; Pinson, B.; Baufreton, J.; Groc, L.; Oliet, S.H.R. Surface Diffusion of Astrocytic Glutamate Transporters Shapes Synaptic Transmission. *Nat. Neurosci.* **2015**, *18*, 219–226. [CrossRef] [PubMed]
87. Romanos, J.; Benke, D.; Saab, A.S.; Zeilhofer, H.U.; Santello, M. Differences in Glutamate Uptake between Cortical Regions Impact Neuronal NMDA Receptor Activation. *Commun. Biol.* **2019**, *2*, 127. [CrossRef] [PubMed]
88. Swanson, R.A.; Liu, J.; Miller, J.W.; Rothstein, J.D.; Farrell, K.; Stein, B.A.; Longuemare, M.C. Neuronal Regulation of Glutamate Transporter Subtype Expression in Astrocytes. *J. Neurosci.* **1997**, *17*, 932–940. [CrossRef]
89. Rink, C.; Gnyawali, S.; Stewart, R.; Teplitsky, S.; Harris, H.; Roy, S.; Sen, C.K.; Khanna, S. Glutamate Oxaloacetate Transaminase Enables Anaplerotic Refilling of TCA Cycle Intermediates in Stroke-Affected Brain. *FASEB J.* **2017**, *31*, 1709–1718. [CrossRef]
90. Monné, M.; Palmieri, F. Antiporters of the Mitochondrial Carrier Family. In *Current Topics in Membranes*; Academic Press Inc.: Cambridge, MA, USA, 2014; Volume 73, pp. 289–320.
91. Anne, C.; Gasnier, B. Vesicular Neurotransmitter Transporters: Mechanistic Aspects. In *Current Topics in Membranes*; Academic Press Inc.: Cambridge, MA, USA, 2014; Volume 73, pp. 149–174.
92. Denker, A.; Rizzoli, S.O. Synaptic Vesicle Pools: An Update. *Front. Synaptic Neurosci.* **2010**, *2*, 135. [CrossRef]
93. Moulder, K.L.; Mennerick, S. Reluctant Vesicles Contribute to the Total Readily Releasable Pool in Glutamatergic Hippocampal Neurons. *J. Neurosci.* **2005**, *25*, 3842–3850. [CrossRef]
94. Elmqvist, D.; Feldman, S. Spontaneous Activity at a Mammalian Neuromuscular Junction in Tetrodotoxin. *Acta Physiol. Scand.* **1965**, *64*, 475–476. [CrossRef]
95. Llinas, R.; Steinberg, I.Z.; Walton, K. Presynaptic Calcium Currents and Their Relation to Synaptic Transmission: Voltage Clamp Study in Squid Giant Synapse and Theoretical Model for the Calcium Gate (Mathematical Model/3-Aminopyridine). *Physiology* **1976**, *73*, 2918–2922. [CrossRef]
96. Vyleta, N.P.; Smith, S.M. Spontaneous Glutamate Release Is Independent of Calcium Influx and Tonically Activated by the Calcium-Sensing Receptor. *J. Neurosci.* **2011**, *31*, 4593–4606. [CrossRef]
97. Kavalali, E.T. The Mechanisms and Functions of Spontaneous Neurotransmitter Release. *Nat. Rev. Neurosci.* **2015**, *16*, 5–16. [CrossRef]
98. Daniels, R.W.; Miller, B.R.; DiAntonio, A. Increased Vesicular Glutamate Transporter Expression Causes Excitotoxic Neurodegeneration. *Neurobiol. Dis.* **2011**, *41*, 415–420. [CrossRef]
99. Ben Fredj, N.; Burrone, J. A Resting Pool of Vesicles Is Responsible for Spontaneous Vesicle Fusion at the Synapse. *Nat. Neurosci.* **2009**, *12*, 751–758. [CrossRef]
100. Tagliatti, E.; Bello, O.D.; Mendonça, P.R.F.; Kotzadimitriou, D.; Nicholson, E.; Coleman, J.; Timofeeva, Y.; Rothman, J.E.; Krishnakumar, S.S.; Volynski, K.E. Synaptotagmin 1 Oligomers Clamp and Regulate Different Modes of Neurotransmitter Release. *Proc. Natl. Acad. Sci. USA* **2020**, *117*, 3819–3827. [CrossRef] [PubMed]
101. Wang, Q.W.; Wang, Y.H.; Wang, B.; Chen, Y.; Lu, S.Y.; Yao, J. Synaptotagmin-7-Mediated Activation of Spontaneous NMDAR Currents Is Disrupted in Bipolar Disorder Susceptibility Variants. *PLoS Biol.* **2021**, *19*, e3001323. [CrossRef] [PubMed]
102. Salin, P.-A.; Lledo, P.-M.; Vincent, J.-D.; Charpak, S. Dendritic Glutamate Autoreceptors Modulate Signal Processing in Rat Mitral Cells. *J. Neurophysiol.* **2001**, *85*, 1275–1282. [CrossRef] [PubMed]
103. Atasoy, D.; Ertunc, M.; Moulder, K.L.; Blackwell, J.; Chung, C.; Su, J.; Kavalali, E.T. Spontaneous and Evoked Glutamate Release Activates Two Populations of NMDA Receptors with Limited Overlap. *J. Neurosci.* **2008**, *28*, 10151–10166. [CrossRef] [PubMed]
104. Graf, A.; Trofimova, L.; Loshinskaja, A.; Mkrtchyan, G.; Strokina, A.; Lovat, M.; Tylicky, A.; Strumilo, S.; Bettendorff, L.; Bunik, V.I. Up-Regulation of 2-Oxoglutarate Dehydrogenase as a Stress Response. *Int. J. Biochem. Cell Biol.* **2013**, *45*, 175–189. [CrossRef]
105. Artiukhov, A.V.; Graf, A.V.; Kazantsev, A.V.; Boyko, A.I.; Aleshin, V.A.; Ksenofontov, A.L.; Bunik, V.I. Increasing Inhibition of the Rat Brain 2-Oxoglutarate Dehydrogenase Decreases Glutathione Redox State, Elevating Anxiety and Perturbing Stress Adaptation. *Pharmaceuticals* **2022**, *15*, 182. [CrossRef] [PubMed]
106. Bunik, V.I.; Mkrtchyan, G.; Grabarska, A.; Oppermann, H.; Daloso, D.; Araujo, W.L.; Juszczak, M.; Rzeski, W.; Bettendorff, L.; Fernie, A.R.; et al. Inhibition of Mitochondrial 2-Oxoglutarate Dehydrogenase Impairs Viability of Cancer Cells in a Cell-Specific Metabolism-Dependent Manner. *Oncotarget* **2016**, *7*, 26400–26421. [CrossRef] [PubMed]

107. Liu, P.; Hu, Y.; Li, J.; Chen, X.; Guo, D.H.; Li, Q.S.; Rahman, K. Protection of SH-SY5Y Neuronal Cells from Glutamate-Induced Apoptosis by 3,6 ′-Disinapoyl Sucrose, a Bioactive Compound Isolated from Radix Polygala. *J. Biomed. Biotechnol.* **2012**, *2012*, 728342. [CrossRef]
108. Sun, Z.W.; Zhang, L.; Zhu, S.J.; Chen, W.C.; Mei, B. Excitotoxicity Effects of Glutamate on Human Neuroblastoma SH-SY5Y Cells via Oxidative Damage. *Neurosci. Bull.* **2010**, *26*, 8–16. [CrossRef] [PubMed]
109. Kang, Y.; Tiziani, S.; Park, G.; Kaul, M.; Paternostro, G. Cellular Protection Using Flt3 and PI3Ka Inhibitors Demonstrates Multiple Mechanisms of Oxidative Glutamate Toxicity. *Nat. Commun.* **2014**, *5*, 3672. [CrossRef] [PubMed]
110. Wang, Y.; Li, H.J.; He, Q.X.; Zou, R.; Cai, J.R.; Zhang, L. Ferroptosis: Underlying Mechanisms and Involvement in Neurodegenerative Diseases. *Apoptosis* **2023**, *29*, 3–21. [CrossRef] [PubMed]
111. Zhang, Y.; Koppula, P.; Gan, B. Regulation of H2A Ubiquitination and SLC7A11 Expression by BAP1 and PRC1. *Cell Cycle* **2019**, *18*, 773–783. [CrossRef]
112. Jiang, L.; Kon, N.; Li, T.; Wang, S.J.; Su, T.; Hibshoosh, H.; Baer, R.; Gu, W. Ferroptosis as a P53-Mediated Activity during Tumour Suppression. *Nature* **2015**, *520*, 57–62. [CrossRef]
113. Wang, F.; Li, W.-L.; Shen, L.J.; Jiang, T.T.; Xia, J.J.; You, D.L.; Hu, S.Y.; Wang, L.; Wu, X. Crocin Alleviates Intracerebral Hemorrhage–Induced Neuronal Ferroptosis by Facilitating Nrf2 Nuclear Translocation. *Neurotox. Res.* **2022**, *40*, 596–604. [CrossRef]
114. Fan, B.Y.; Pang, Y.L.; Li, W.X.; Zhao, C.X.; Zhang, Y.; Wang, X.; Ning, G.Z.; Kong, X.H.; Liu, C.; Yao, X.; et al. Liproxstatin-1 Is an Effective Inhibitor of Oligodendrocyte Ferroptosis Induced by Inhibition of Glutathione Peroxidase 4. *Neural Regen. Res.* **2021**, *16*, 561–566. [CrossRef]
115. Zhao, H.; Li, X.; Yang, L.; Zhang, L.; Jiang, X.; Gao, W.; Chen, P.; Cheng, Y.; Wang, F.; Liu, J. Isorhynchophylline Relieves Ferroptosis-Induced Nerve Damage after Intracerebral Hemorrhage Via MiR-122-5p/TP53/SLC7A11 Pathway. *Neurochem. Res.* **2021**, *46*, 1981–1994. [CrossRef] [PubMed]
116. Yuan, Y.; Zhai, Y.; Chen, J.; Xu, X.; Wang, H. Kaempferol Ameliorates Oxygen-Glucose Deprivation/Reoxygenation-Induced Neuronal Ferroptosis by Activating Nrf2/Slc7a11/Gpx4 Axis. *Biomolecules* **2021**, *11*, 923. [CrossRef] [PubMed]
117. Perkinton, M.S.; Ip, J.K.; Wood, G.L.; Crossthwaite, A.J.; Williams, R.J. Phosphatidylinositol 3-Kinase Is a Central Mediator of NMDA Receptor Signalling to MAP Kinase (Erk1/2), Akt/PKB and CREB in Striatal Neurones. *J. Neurochem.* **2002**, *80*, 239–254. [CrossRef] [PubMed]
118. Crossthwaite, A.J.; Valli, H.; Williams, R.J. Inhibiting Src Family Tyrosine Kinase Activity Blocks Glutamate Signalling to ERK1/2 and Akt/PKB but Not JNK in Cultured Striatal Neurones. *J. Neurochem.* **2004**, *88*, 1127–1139. [CrossRef] [PubMed]
119. Niswender, C.M.; Conn, P.J. Metabotropic Glutamate Receptors: Physiology, Pharmacology, and Disease. *Annu. Rev. Pharmacol. Toxicol.* **2010**, *50*, 295–322. [CrossRef] [PubMed]
120. Forgac, M. Vacuolar ATPases: Rotary Proton Pumps in Physiology and Pathophysiology. *Nat. Rev. Mol. Cell Biol.* **2007**, *8*, 917–929. [CrossRef] [PubMed]
121. Zhang, Z.; Nguyen, K.T.; Barrett, E.F.; David, G. Vesicular ATPase Inserted into the Plasma Membrane of Motor Terminals by Exocytosis Alkalinizes Cytosolic PH and Facilitates Endocytosis. *Neuron* **2010**, *68*, 1097–1108. [CrossRef]
122. Curi, R.; Newsholme, P.; Marzuca-Nassr, G.N.; Takahashi, H.K.; Hirabara, S.M.; Cruzat, V.; Krause, M.; De Bittencourt, P.I.H. Regulatory Principles in Metabolism -Then and Now. *Biochem. J.* **2016**, *473*, 1845–1857. [CrossRef]
123. Curi, R.; Lagranha, C.J.; Doi, S.Q.; Sellitti, D.F.; Procopio, J.; Pithon-Curi, T.C.; Corless, M.; Newsholme, P. Molecular Mechanisms of Glutamine Action. *J. Cell Physiol.* **2005**, *204*, 392–401. [CrossRef]

Disclaimer/Publisher's Note: The statements, opinions and data contained in all publications are solely those of the individual author(s) and contributor(s) and not of MDPI and/or the editor(s). MDPI and/or the editor(s) disclaim responsibility for any injury to people or property resulting from any ideas, methods, instructions or products referred to in the content.

MDPI AG
Grosspeteranlage 5
4052 Basel
Switzerland
Tel.: +41 61 683 77 34

International Journal of Molecular Sciences Editorial Office
E-mail: ijms@mdpi.com
www.mdpi.com/journal/ijms

Disclaimer/Publisher's Note: The title and front matter of this reprint are at the discretion of the Guest Editor. The publisher is not responsible for their content or any associated concerns. The statements, opinions and data contained in all individual articles are solely those of the individual Editor and contributors and not of MDPI. MDPI disclaims responsibility for any injury to people or property resulting from any ideas, methods, instructions or products referred to in the content.

www.ingramcontent.com/pod-product-compliance
Lightning Source LLC
Chambersburg PA
CBHW040745250926
51850CB00007BB/1416
* 9 7 8 3 7 2 5 8 6 5 3 0 7 *